SCIENCE IN CHINA
1600–1900

Essays By Benjamin A. Elman

SCIENCE IN CHINA 1600–1900

Essays By Benjamin A. Elman

Editor

Ho Yi Kai

Published by

World Century Publishing Corp.
27 Warren Street
Suite 401-402
Hackensack, NJ 07601
USA

and

World Scientific Publishing Co. Pte. Ltd.
5 Toh Tuck Link, Singapore 596224
USA office: 27 Warren Street, Suite 401-402, Hackensack, NJ 07601
UK office: 57 Shelton Street, Covent Garden, London WC2H 9HE

Library of Congress Cataloging-in-Publication Data
Ho, Yi Kai.
Science in China, 1600–1900 : essays by Benjamin A. Elman / Ho Yi Kai.
pages cm
Includes bibliographical references and index.
ISBN 978-9814651103 (alk. paper)
1. Elman, Benjamin A., 1946– On their own terms. 2. Science--China--History--17th century. 3. Science--China--History--18th century. 4. Science--China--History--19th century. I. Title.
Q127.C5H6325 2015
509.51'0903--dc23
2014038710

British Library Cataloguing-in-Publication Data
A catalogue record for this book is available from the British Library.

In-house Editor: Qi Xiao

Typeset by Stallion Press
Email: enquiries@stallionpress.com

Printed in Singapore by B & Jo Enterprise Pte Ltd

Editor's Note

Editing this book is an honor for young scholars like me, and could only have stayed a dream if Professor Elman had not kindly granted me an audience during his visit to Singapore. He graciously accepted my proposal to put together a selection of revised versions of his lesser known essays on a common theme, and so we have here a solid collection of studies on science in China.

The eight essays are structured in a chronological manner, with the first as an introduction of scientific studies in late imperial China, focusing on Phonology. It is followed by two essays covering the Ming–Qing era, discussing issues of natural studies and Jesuit learning. We then move on from the High Qing to the Late Qing in the following three essays, continuing on the Jesuit roles and Western learning, and venturing further on the promotion of modern science. The last two essays examine China from the late Qing to the 21st century. The penultimate essay looks at the "fall of China" in comparison to the "rise of Japan", with the last essay an apt closing, as the author reflects on traditional Chinese science and medicine in the 20th and 21st centuries.

Of notable value to the book might be the appendices, consisting of three interviews by the media in China, translated and introduced to English readers for the first time. The interesting flow of issues presented in the interviews captures Professor Elman's reflective thoughts and a gist of his extensive studies.

I would like to extend my heartfelt gratitude to Professor Elman, to have given me this invaluable opportunity and learning experience. I must also thank our in-house editor Mr. Qi Xiao, for his meticulous effort and extra miles he went to ensure the highest quality of the product. This notwithstanding, I should still bear the full responsibility for any errors and inadequacies in the book.

Ho Yi Kai
Assistant Editor-in-Chief
World Century Publishing Corp.
January 2015

Author's Acknowledgments

I wish to acknowledge the warm welcome I received from my Japanese and visiting colleagues at the Institute for the Advanced Study of Asia (IASA, 東洋文化研究所) at The University of Tokyo where I was appointed from July 1, 2014, until January 31, 2015, as the *Tōbunken* Visiting Chair Professor (東文研特聘講座教授). I especially want to thank the Director of the Institute at the time, Professor Ōki Yasushi 大木康, for inviting me to join his faculty as a Visiting Professor and to teach a graduate student research seminar in fall 2014, which was cross-listed with my Fall 2014 Princeton graduate seminar. I also very much appreciate the help and support I received from The University of Tokyo Vice-President for the Humanities, Professor Haneda Masashi 羽田正, previously also the Director of the IASA.

This volume was compiled and completed for the World Scientific Publishing Company and World Century Publishing Corporation during my stay at the IASA. Special thanks are due my editor, Dr. Ho Yi Kai, and Mr. Ng Chin Choon for the cover design.

For full citations of Chinese, English, and Japanese sources, see the Bibliography.

Benjamin A. Elman, Princeton University
Gordon Wu '58 Professor of Chinese Studies
Professor of East Asian Studies and History

Tokyo University, January 2015

Contents

Chapter 1

Introduction — From Value to Fact: The Emergence of Phonology as a Precise Discipline in Late Imperial China

In his pioneering article "The Abortiveness of Empiricism in Early Ch'ing Thought", which he included in his magnum opus on modern Chinese intellectual history, Joseph Levenson contended that the philological turn among Chinese literati scholars in the 17th and 18th century could not have developed a "scientific temper" on its own without the decisive intrusion of Western industrialism in the 19th century. Because the empirical attitudes of early and mid-Qing dynasty classicists were not scientific in and of themselves, their critique of idealism resembled Abelard's nominalism more than Francis Bacon's inductive and empirical science, according to Levenson. The late imperial textual focus celebrated the Chinese classics as repositories of knowledge and thus represented a dead-end from the standpoint of development of science.[1]

Levenson's influential conclusions were drawn, however, after he had refracted Qing philological scholarship through a "Western" prism that was calibrated according to an idealized development of modern science in Europe. He argued that the use of "scientific" to describe Qing philology was simply a metaphor drawn from the natural sciences. Europeans, he claimed, had proceeded from natural science to thinking scientifically about philological problems. Consequently, we cannot turn this natural development in Europe inside out and expect that the Chinese would have proceeded from sound philology to think philologically about natural science.

The "traditions of scholarship in an age of science" are not as clear cut as Levenson presented them, however. The polemical history of Western humanism favored by Bacon (1561–1626) and Descartes (1596–1650),

[1] Joseph Levenson, *Confucian China and Its Modern Fate: A Trilogy*, 1–14.

which Levenson uncritically accepted, has over-determined the allegedly antagonistic relations between Renaissance philology and early modern science. Levenson, unfortunately, never realized that men like Kepler (1571–1630), as well as Newton (1642–1727), were both humanists and scientists. They frequently interpreted references to natural phenomena in classical texts and used astronomy to date events in ancient history. Accordingly, humanist scholarship and science were a single pursuit, not the polar opposites that Levenson presented.[2]

The emergence of phonology as a key discipline during the Qing dynasty was closely tied to the triumph of precise empirical techniques of philological analysis championed by participants in the evidential research movement over Song–Ming, 1200–1600, moral philosophy. In phonology, such applications stressed the reconstruction of archaic finals through an examination of ancient rhyme schemes. In the late 18th century, significant steps were taken to investigate archaic initials as well. Such pioneering studies established the foundations of modern Chinese linguistics and at the same time provided Western linguists with much of the necessary data and tools needed to refine earlier reconstructions of ancient Chinese phonology.

During the Song and Ming dynasties, the scholarly goal of literati was the cultivation of moral perfection. Their ideal was a life of intense and unremitting effort, a life they felt would successfully emulate the ancient sages. After the fall of the Ming dynasty in 1644, however, this ideal was taken less and less literally. The Qing dynasty heirs of these fervent groups of Song-Ming literati were members of a secular academic community, which encouraged original and critical scholarship. For the Song–Ming scholars, the classical canon had been the repository of moral truth that transcended time and place. The reaction of Qing textual scholars against the unquestioned authority of the classics was most evident in their precise studies in linguistics, astronomy, mathematics, geography, and epigraphy. Scholars of the 17th and 18th century applied these fields of research to verify or controvert important elements of the classical legacy. They were dissatisfied with the unverifiable moral ideals that pervaded the Song–Ming vision of antiquity.

---

[2] See Anthony Grafton, *Defenders of the Text: The Traditions of Scholarship in an Age of Science, 1450–1800.*

Relying on systematic gathering of materials that they would then critically scrutinize and in certain cases even quantify, Qing scholars combined evidential research methods with data collection and organization. The modern scholar Liang Qichao 梁启超 (1873–1929), for example, has estimated that Qian Daxin 钱大昕 (1728–1804) recorded over one hundred items in his notation book before he attempted to shed new light on the phenomenon of labiodentals recorded in ancient Chinese texts. Qian presented his data within a systematic discussion of ancient pronunciation. Moreover, during the Qing period, there was noticeable progress in *kaozheng* 考证 fields of inquiry. Such progress was possible because evidential scholars, unlike their Song and Ming precursors, stressed research topics that lent themselves to cumulative results. Accompanying this sense of the continuity of academic progress was a quest for originality.

As in the West, the history of linguistics in China presents an interesting analogy to the evolution of empirical methods of verification in the natural sciences. The development of language study and the emergence of historical and comparative linguistics are not uniquely Western achievements. Qing evidential scholars in particular established the foundations of modern Chinese linguistic science. The papers in this volume confirm the limitations in such one-sided accounts of Qing philology and Renaissance humanism. Levenson not only isolated philology from natural studies, which is untenable for both late imperial China and early modern Europe, but he also underestimated how Qing classicists, like their Renaissance counterparts, had integrated mathematics and astronomy in their efforts to reconstruct antiquity and restore its traditions of natural studies.[3]

Narrative accounts of the history of science worldwide from 1500 to 1800, such as that of Joseph Levenson, have been portrayed mainly through European frames of reference, even when comparative themes are stressed. Hence, even though the emergence of "modern science" in industrializing portions of Western Europe is uncontested, the contested nature of the interaction since 1550 between late imperial Chinese and

[3] Benjamin A. Elman, "From Value to Fact: The Emergence of Phonology as a Precise Discipline in Late Imperial China," *Journal of the American Oriental Society*, 102, 3 (July–October 1982): 493–500.

early modern Europeans over the meaning and significance of natural studies remains a little-known story. Eurocentric portraits of the rise of modern science, while not monolithic or one-dimensional, usually represent variations of a single-minded historical teleology of Western European scientific "success", and, by comparison, non-Western "failure".[4]

Comparisons of early modern Europe and late imperial China suggest a number of ways that the comparative history of science can lead us away from such teleologies. First and foremost, historicizing the Western scientific revolution makes it possible to compare the ongoing role played by classical languages (Latin in Europe, ancient Chinese in China, and Sanskrit in India) as cultural mediums during the transition from natural philosophy to early modern science. Secondly, differential studies that wield appropriate concepts and categories for comparing precise historical situations are mandatory. In particular, the case studies provided in the papers below successfully integrate scientific contents and historical contexts as the key to moving from the local to the global and back again. A global account that is misinformed about local or regional realities will miss the mark.[5]

To paraphrase the views of Peter Winch, we must first acknowledge that as yet we do not have appropriate categories of learning that resemble the pre-modern Chinese frames for what we call "natural studies" or "natural history", according to which Chinese literati evaluated Jesuit *scientia* during the Ming and Qing dynasties. Moreover, as Donald Lach has pointed out, an analytical ordering of early modern European scholarship, such as the Jesuit *scientia* or Schweigger's crystal electrical theory of matter, within the framework of modern learning is equally problematic and anachronistic.[6]

To understand pre-modern Chinese frames of knowledge for the natural world, as for early modern Europe, we should first extend our own

---

[4] Benjamin A. Elman, *From Philosophy to Philology: Intellectual and Social Aspects of Change in Late Imperial China*.

[5] See Michel Paty, "Comparative History of Modern Science and the Context of Dependency", *Science, Technology, & Society* 4.2 (1999): 178, 184, 196.

[6] Peter Winch, "Understanding a Primitive Society," in Bryon Wilson, ed., *Rationality*, 93–102, and Donald F. Lach, *Asia in the Making of Europe. Volume II. A Century of Wonder*, *Book 3*: *The Scholarly Disciplines*, 395.

contemporary understanding and make room for them not as a "comedy of errors" but as a plausible set of ideas and beliefs. Placing natural studies in imperial China and pre-modern Europe within their own internal and external contexts allows us to reconstruct their historical communities of interpretation and how those communities — through interaction — constructed science on their own terms. By better understanding early modern European and Chinese interest in nature, technology, and medicine, we are more perceptive about ourselves and the cultural, economic, political, and social values that undergird our contemporary versions of modern science.

For over a century, Europeans have heralded the success of Western science and assumed the failure of science elsewhere. Such views until recently preempted positive narratives about early modern Chinese, Islamic, and Sanskrit exact studies. The rehabilitation of the exact sciences in the pre-modern non-Western world is a long-term precondition for balancing the historiographical playing field. In the decades since Needham, we have increasingly acknowledged that our focus on the "failure" of Chinese science to develop into modern science is heuristically interesting but historiographically misguided. We are now forced to reassess how the history of science globally should be rewritten.[7]

The emergence of phonology as a key discipline during the Qing dynasty was closely tied to the triumph of precise empirical techniques of philological analysis championed by participants in the evidential research movement over Song–Ming moral discourses.

---

[7] Sheldon Pollock's *The Language of the Gods in the World of Men: Sanskrit, Culture, and Power in Premodern India* successfully recaptures the traditions of Sanskrit exact learning that generations of British imperialism disavowed. See also Christopher Minkowski, "Competing Cosmologies in Early Modern Indian Astronomy," in Charles Burnett, Jan Hogendijk and Kim Plofker, eds., *Ketuprakasa: studies in the history of the exact sciences in honor of David Pingree*, 349–385.

Chapter 2

Native Traditions of Natural Studies during the Ming–Qing Transition, 1600–1800

When Europeans reached China during the "age of exploration", *scientia* did not yet connote "natural science" among humanists, Jesuits, or more secular early modern European scholars. A medieval French term, *science*, which was synonymous with "accurate and systematized knowledge", became, when Latinized, "*scientia*" and represented among early modern scholastics the specialized branches of Aristotelian moral and natural philosophy. Included were the seven sciences of medieval learning: grammar, logic, rhetoric, arithmetic, music, geometry, and astronomy. These seven liberal arts in Roman times had served educationally as preparation for more specialized training in philosophy, medicine, or law.

In medieval times, Boethius's (ca. 475–524) pioneering translations of Aristotle (384–322 B.C.) into Latin, for example, named the four mathematical disciplines (arithmetic, geometry, music, and astronomy) for elementary education as the *quadrivium* (four roads to wisdom), which balanced the three disciplines of logic (grammar, dialectics, and rhetoric) known later as the *trivium* (three roads). After the Muslim harvest of classical learning was rediscovered in Europe by the time of Thomas Acquinas (1225–1274) in Paris, the preferred order of Aristotelian learning was set for the Renaissance scholars and bookmen: (1) logic; (2) mathematics; (3) natural science; (4) moral philosophy; and (5) metaphysics.[1]

---

[1] See James Weisheipl, "Classification of the Sciences in Medieval Thought", *Medieval Studies* 27 (1965): 54–55, 58–68, 81–90, and David C. Lindberg, "The Transmission of Greek and Arabic Learning to the West", in David C. Lindberg, ed. *Science in the Middle Ages*, 52–90. Compare Sydney Ross, "*Scientist*: The Story of a Word", *Annals of Science* 18.2 (June 1962): 65–71, who notes that the term "scientist" was not commonly used in English until the 19th century.

Similarly in Ming China, when terms such as *scientia* were translated by the Jesuits and their Chinese colleagues from Latin into classical Chinese, the elite written language of Chinese literati, the translations reflected the views and frames for the natural studies of the 16th century in China and Europe — not the "science" of more modern times. *Xuewen* 学问 was the classical Chinese equivalent to correlate native categories of specialized learning with the *scientia* of the Jesuits.[2] Moreover, during the Jesuit–Chinese interaction in the 17th century, the "investigation of things" (*gewu* 格物), "exhaustively mastering principles" (*qiongli* 穷理), and "knowing heaven" (*zhitian* 知天) were at the core of the intellectual encounter between Chinese literati and the early modern West that the Catholic Church still represented. These classical Chinese terms were used by Chinese literati and Jesuits to accommodate both Western and Chinese views of practical studies, which included natural studies. In this cultural endeavor, we see an overlap between religious and scientific work on the part of the Jesuits and their Chinese converts and sympathizers.[3]

Previous estimates of Jesuit authors worldwide have suggested that between 1600 and 1773, they wrote more than 4,000 works, 600 journal articles (almost all after 1700), and 1,000 manuscripts dealing with the sciences. The vast majority were by Jesuit educators. More recent estimates indicate that some 450 works were translated or compiled by the Jesuits and their converts in China between 1584 and 1790. Over 26% of that total (120) was in the sciences, while 74% (330) were on religion. Usually the Latin texts were orally translated by a Jesuit and dictated to a Chinese in a style known as *kouyi* 口译 (oral translation) or *bishou* 笔受

---

[2] On the issue of *scientia* = *xuewen* in Chinese glosses of Latin terms, I have benefited from discussion concerning Latin-Chinese glossaries with Han Qi. Latin-Chinese word glossaries compiled by the Jesuits and their Chinese collaborators were forerunners of modern dictionaries of the Chinese language. See Federico Masini, "Using the works of the Jesuit missionaries in China to study the Chinese language: a research project", paper presented at the International Conference "Translating Western Knowledge into Late Imperial China", Göttingen University, December 6–9 (1999).

[3] Nicolas S.J. Standaert, "The Investigation of Things and the Fathoming of Principles (*Gewu qiongli*) in the Seventeenth-Century Contact Between Jesuits and Chinese Scholars", in *Ferdinand Verbiest (1622–1688): Jesuit Missionary, Scientist, Engineer and Diplomat*, ed. John W. Witek, S.J. Nettetal, 395–420.

(received writing). The Chinese collaborator then prepared a polished written version for review.[4]

For example, Matteo Ricci (1552–1610) translated Christopher Clavius's (1538–1612) 1607 *Elementorum* in this manner, as did Sabbathin de Ursis (1575–1620) for his 1612 *Taixi shuifa* 《泰西水法》 (*Western Techniques of Hydraulics*). Jean-Nicholas Smogolenski (1610–1656) and his collaborators introduced the European method for calculating eclipses in an astronomical work of circa 1656, which was also the first to introduce spherical trigonometry and logarithms. Later, Ferdinand Verbiest's (1623–1688) 1672 *Kunyu tushuo* 《坤舆图说》 (*Maps and Explanations of the Earth*) furnished further information on world geography beyond Ricci's earlier *mappa mundi*. By way of contrast, most 18th-century translations in China were theological works, and Jesuits turned instead to translating Chinese works into European languages.[5]

Interest in Natural Studies during the Ming Dynasty

The Jesuits in late Ming China saw the "investigation of things" and "exhaustively mastering principles" as a necessary way station to the doctrinal transmission of the experience of God to the Chinese they hoped to convert. For late Ming Chinese such as Fang Yizhi 方以智 (1611–1671), their recovery of the "concrete studies" (*shixue* 实学) of antiquity predisposed some literati to accept the Western learning brought by the Jesuits because it was an alternate form of the "investigation of things" and was presented by the Jesuits as a confirmation of Chinese ancient learning.[6]

---

[4] See R. Po-chia Hsia, "The Catholic Mission and Translations in China, 1583–1700," in Peter Burke and R. Po-chia Hsia, eds., *The Cultural History of Translation in the Early Modern World*, 39–51, for more recent estimates.

[5] See Tsien Tsuen-hsuin, "Western Impact on China Through Translation", *Far Eastern Quarterly* 13 (1954): 307–310, and Steven Harris, "Transposing the Merton Thesis: Apostolic Spirituality and the Establishment of the Jesuit Scientific Tradition", *Science in Context* 3.1 (1989): 29–65. See also *Siku quanshu zongmu* 《四库全书总目》, compiled by Ji Yun 纪昀 (1724–1805) *et al.*, 107.23b–24a. Compare Benjamin A. Elman, "Geographical Research in the Ming–Ch'ing Period", *Monumenta Serica* 35 (1981–1983): 1–18.

[6] Willard Peterson, "Fang I-chih: Western Learning and the 'Investigation of Things'", in Wm. Theodore de Bary *et al.*, eds., *The Unfolding of Neo-Confucianism*, 369–411.

Because of the physico-theology lurking in the Jesuits teleology of nature, however, the investigation of things was ultimately "to find God" for the Jesuits and "to fathom principles" of the *dao* 道 for the Chinese. Despite this theological twist to the Jesuit interpretation, the Jesuit conception and practice of *scientia* was ingeniously presented by some of the Chinese who collaborated with the Jesuits, such as Xiong Mingyu 熊明遇 (1579–1649), as roughly corresponding to the natural studies of the Chinese.[7]

Xiong noted in his late Ming "Personal Preface" to his *Gezhi cao* 《格致草》 (*Draft for Investigating Things and Extending Knowledge*) that "Literati took to heart the *Great Learning* and of course spoke of the necessity of first investigating things and extending knowledge". Unfortunately, they had failed to capture the unity of knowledge that the *Doctrine of the Mean* (*Zhongyong* 《中庸》) and *Mencius* (*Mengzi* 《孟子》) had extolled. What was required, Xiong explained, was a detailed examination of the heavens, earth, stars, constellations, the transformations of *qi* 气, plants and animals, and "one by one on the basis of what each phenomenon actually was, one could seek out the reason for why things are as they are and thereby illuminate the principle for why they could not be any other way".[8]

Both sides saw an order and purpose in the cosmos and on earth, which the Jesuits linked into a theology-informed geography to delineate God and nature as one. Most Chinese literati also saw the earth and heavens as a harmonious whole, but their teleological view of nature framed arguments for the design of the cosmos around an eternal and always changing *dao* rather than around the chronology of a divine providence informing the cosmic order in Christianity. In place of a cosmos made up of "four elements" (air-ether, fire, earth, water), the Chinese conceived of change in light of a "Supreme Ultimate" (*taiji* 太极), which through the medium of *yang* 阳 and *yin* 阴 forces set in motion the five evolutive phases (*wuxing* 五行, earth, fire, metal, water, and wood) of cosmic

[7] See Xu Guangtai, "Mingmo Qingchu xifang gezhixue de chongji yu fanying" 〈明末清初西方格致学的冲击与反应〉, in *Shibian chunti yu geren* 《世变群体与个人》, 236–258.

[8] See Xiong Mingyu's "*Zixu*" 自序 to the *Gezhi cao*, 1a–5a, and elsewhere for examples of efforts to inscribe Chinese views of nature with the teleologies of the Jesuits.

evolution and yielded the concomitant production and destruction cycles of the "myriad things" (*wanwu* 万物) in the world.[9]

Indeed, Alphonso Vagnoni's 1633 *Kongji gezhi* 空际格致 ("Investigation of the Atmosphere") was in part a refracted presentation of the theory of the four elements from the Conimbricenses edition of Aristotle's *Meteriologica*, which was then used in the Jesuit University of Coimbra in Portugal where many missionaries were trained before leaving for Asia. In his translation, for example, Vagnoni tried vainly to convince the Chinese of the error of their ways for including wood and metal and excluding air-ether as the building blocks of things in the world.[10] Rather than building blocks of the universe, Chinese literati perceived in the five phases evidence for the successive evolutive changes in all things. Chinese also perceived in *qi* a more fundamental material (*zhi* 质) and spiritual (*shen* 神) unity, which pervaded all things in the cosmos and undergirded the space–time evolution of *yin* and *yang*, rather than the lifeless air-ether element enunciated by Vagnoni, following Aristotle, as one of the substrates of matter.[11]

In light of the Jesuit use of the "investigation of things" to present their *scientia* to the Chinese, it is important to note that "natural studies" in China had also been classified under the phrase *gezhi* 格致 (investigating into and

---

[9] On the five phases, see Nathan Sivin, "The Myth of the Naturalists", in *Medicine, Philosophy, and Religion in Ancient China: Researches and Reflections*, ed. Nathan Sivin, 1–29, who notes that "by the first century B.C.E., a coherent theory of yin–yang and the Five Phases as aspects of *ch'i* [*qi*] emerged." For discussion, see Clarence J. Glacken, *Traces on the Rhodian Shore: Nature and Culture in Western Thought from Ancient Times to the End of the Eighteenth Century*, and Keith Thomas, *Man and the Natural World: Changing Attitudes in England, 1500–1800.*

[10] See the critique of Alphonso Vagnoni by 18th century Chinese literati in *Siku quanshu zongmu* 125.34a–b. For the Jesuit critique of *qi*, see Qiong Zhang, "Demystifying *Qi*: The Politics of Cultural Translation and Interpretation in the Early Jesuit Mission to China", in *Tokens of Exchange: The Problem of Translation and Interpretation in the Early Jesuit Mission to China*, ed. Lydia Liu, 74–106.

[11] See Vagnoni's *Kongji gezhi* A.1a-13b. See also Willard Peterson, "Western Natural Philosophy Published in Late Ming China", *Proceedings of the American Philosophical Society* 117.4 (August 1973): 295–322. Compare the account of the five phases in *Xingli daquan* 《性理大全》, compiled in 1415 by Hu Guang 胡广 *et al.*, 27.18a–19a, which remained orthodox for the civil examinations during the Ming and Qing dynasties. On the *Xingli daquan*, see my *A Cultural History of Civil Examinations in Late Imperial China*, 113–119.

extending knowledge, *gewu zhizhi* 格物致知). At other times, particularly in the medieval period, and often simultaneously with *gezhi* after the Yuan dynasty, such interests were expressed in terms of *bowu* 博物 (broad learning concerning the nature of things). For instance, *bowu*, not *gezhi*, was one of the six major classification categories for the 32 section headings in the *Gujin tushu jicheng*《古今图书集成》(*Synthesis of Books and Illustrations Past and Present*), the largest *leishu* 类书 encyclopedia in imperial Chinese history, under which over 6,100 items were organized. The full mapping out of the asymmetrical conceptual categories associated with these two potential candidates in Song and Ming times for natural studies (*gezhi*) and natural history (*bowu*) respectively remains incomplete.[12]

In addition, other technical terms in ancient and medieval bibliographic classifications, such as *shuji* 术技 (skills and techniques), were used to demarcate what we today refer to as science or technology.[13] In Han times, the *liuyi* 六艺 (Six technai) of the *History of the Former Han Dynasty* (*Hanshu* 《汉书》) focused on the six classical teachings, i.e., the "Six Classics" (*Change, History, Poetry, Annals, Rituals, Music*), and not the more ecumenical "Six Arts" (also *liuyi*) of rites, music, archery, charioteering, calligraphy, and mathematics. This interesting division of meanings for the same term (*yi* 艺) was followed thereafter in the official dynastic bibliographies.[14]

Although already in use during the Eastern Jin dynasty, the "four classifications" (*sibu* 四部) system became under the Sui dynasty the orthodox classification system for all knowledge in the Imperial Library. The bibliography section of the *History of the Sui Dynasty* (*Suishu jingji zhi* 《隋书・经籍志》) was compiled by Wei Zheng 魏徵 (580–643), then Director of the Palace Library. Wei applied the *sibu* sequence as orthodox historiography, i.e., the divisions according to which all documents in the Palace Library would be classified, for the first time.

In this framework, works on natural studies did not have their own main category, as in Han dynasty catalogs of books, and instead were

---

[12] Lionel Giles, *An Alphabetical Index to the Chinese Encyclopedia, Ch'in Ting Ku Chin T'u Shu Chi Ch'eng*. Giles translates *bowu* here as "Arts and Sciences".

[13] For discussion, see Cary Liu, "The Qing Dynasty Wen-yuan-ko Imperial Library", 83–124.

[14] *Hanshu* 30.1765-1775. Compare Cary Liu, 482–483.

included in a number of subcategories under the main four divisions. The *Liuyi* (Six technai) were further institutionalized as a classical canon equivalent to the *jingji* 经籍 (classics) or *jingdian* 经典 (classical canon) under the Tang dynasty, and the imperial classification scheme of "four divisions" was further legitimated politically by linking it to correlations with the four seasons, four directions, the four colors, and the four virtues.[15]

Later the Southern Song historian Zheng Qiao 郑樵 (1104–1162) in his *Tongzhi* 《通志》(*Comprehensive Treatises*) encyclopedia included a treatise on bibliography, which expanded the four classifications scheme to 12 main categories. Of the 12 main categories (*lue* 略), several were linked to natural studies:

1. Classics	5. History	9. Arts
2. Rituals	6. Pre-Han & Later Masters	10. Medicine
3. Music	7. Astrology	11. Encyclopedia
4. Philology	8. Five Phases	12. Literature[16]

Although Zheng Qiao's categories did not become normative, the encyclopedia was widely read and emulated by private scholars such as Sun Xingyan 孙星衍 (1753–1818), who in 1800 presented a new scheme of 12 major divisions of knowledge, which resembled Zheng's categories:[17]

1. Classics	5. Geography	9. Encyclopedias
2. Philology	6. Medicine & Law	10. Poetry
3. Masters	7. History	11. Arts
4. Astrology	8. Epigraphy	12. Fiction

---

[15] See Tsien Tsuen-hsuin, "A History of Bibliographical Classification in China", *The Library Quarterly* 22.4 (October 1952): 311–313, and Cary Liu, "The Qing Dynasty Wen-yuan-ko Imperial Library", 101–110.

[16] Tsien Tsuen-hsuin, "A History of Bibliographical Classification in China", *The Library Quarterly* 22.4 (October 1952): 314–315, and Ssu-yü Teng and Knight Biggerstaff, *An Annotated Bibliography of Selected Chinese Reference Works*, 109–110.

[17] See Benjamin A. Elman, *From Philosophy to Philology: Social and Intellectual Aspects of Change in Late Imperial China*, 163–168.

Hence, based on this brief summary of the classification terms and bibliographic locations for natural studies used in the Imperial Library since medieval times, when the "four classifications" were usually in effect, we cannot simply assume that there was a single and unified traditional field of natural studies in China known as *gezhixue* 格致学 before the Jesuits arrived in China.[18] Nonetheless, it appears that among late and post-Ming literati elites *gezhi* was becoming a common epistemological frame for the accumulation of knowledge. *Bowu* on the other hand carried with it a more common and popular notion of curiosities.[19] For example, the *Taiping yulan* 《太平御览》 (*Encyclopedia of the Taiping Era*, 976–983), compiled during the early years of the Northern Song dynasty, included earlier texts dealing exclusively with unusual events, strange objects, things, birds, spirits, and anomalies to provide a contemporary lexicon of textual usages in antiquity and medieval times that denoted the scope of *bowu* within classical writings, which the 18th-century *Gujin tushu jicheng* (see above) emulated.[20]

In the late 18th century, the *Siku quanshu* (*Complete Collection in the Imperial Four Treasuries*), which represented the climax of the classical scheme of disciplines, incorporated medicine and calendrical studies as subcategories under the Masters category (see Table 1). Similarly the mathematical aspects of music were subsumed under the Classics, while chronography and geography were listed under History. We thus find no single unified category for natural studies in the imperially authorized bibliographies after the ancient Han bibliographies organized such studies under the general category of "calculating skills" (*shushu lue* 数术略) or "skills and techniques" (*shuji* 术技). What we do find, however, is that the compilers of the Imperial Library catalog had linked mathematics and astrology under the same framework, which represented a Chinese response to the Jesuit impact on the new methods for successfully

[18] Xu Guangtai. "Mingmo Qingchu xifang gezhixue de chongji yu fanying", 236–258. Xu assumes that *gezhixue* was a widely used traditional designation for natural studies without showing its uses or scope before the Jesuits arrived.

[19] See Robert F. Campany, *Strange Writing: Anomaly Accounts in Early Medieval China*, 49–52, and Qiong Zhang, "Nature, Supernature, and Natural Studies in Sixteenth- and Seventeenth-Century China".

[20] *Taiping yulan*, 612.4a–10a.

Table 1. Forty-four Subdivisions of the *Siku quanshu* 《四库全书》 (*Complete Collection in the Imperial Four Treasuries*). Fields associated with Natural Studies have been underlined.

Classics	**History**
Change(s)	Dynastic Histories
Documents	Annals
Poetry	Topical Records
Rituals	Unofficial Histories
Spring & Autumn Annals	Miscellaneous Histories
Filial Piety	Official Documents
General Works	Biographies
Four Books	Historical Records
Music	Contemporary Records
Philology	Chronography
Masters	Geography
Scholars	Official Registers
Military Strategists	Institutions
Legalists	Bibliographies and Epigraphy
Agriculturalists	Historical Criticism
Medicine	**Literature**
Astrology & Mathematics	Elegies of Chu
Calculating Arts	Individual Collections
Arts	General Anthologies
Repertories of Science Miscellaneous Writers	Literary Criticism Songs & Drama
Encyclopedias	
Novels	
Buddhism	
Daoism	

reforming the calculations associated with the Qing calendar. The role of *gezhi* as a term for *scientia* was making inroads among Ming and Qing literati, however.[21]

Investigating Things & Extending Knowledge

When the Jesuits first arrived in Ming China, there was already considerable discussion among literati about an appropriate theory of knowledge. The debate often took the form of claims that "honoring morality" (*zundexing* 尊德性) took precedence over "formal knowledge" (*daowenxue* 道问学). Earlier, the Southern Song philosopher Zhu Xi 朱熹 (1130–1200), who became the core interpreter of the late imperial classical canon, argued that "investigating into and extending knowledge" presupposed that all things had their principle (*wanwu zhi li* 万物之理). Zhu therefore concluded: "One should in three or four cases out of ten seek principles in the outside realm." In most cases, however, moral principles should be sought within. Thereafter, the investigation of things became the key to opening the door of knowledge for literati versed in the Classics and Histories.[22]

Due to Zhu Xi's scholarly eminence after the Southern Song dynasty, *gezhi* became a popular *Daoxue* 道学 ("Way Learning", also called "Neo-Confucianism") term borrowed from the *Great Learning* (*Daxue* 《大学》; one of the Four Books) in the *Record of Rites* (*Liji* 《礼记》; one of the Five Classics) by literati to discuss the form and content of knowledge. There was considerable classical debate surrounding Zhu Xi's single-minded prioritizing of the *gewu* passage in the *Great Learning* to establish the epistemological boundaries for literati learning.[23] Yü Ying-shih has traced the 17th-century turn among

---

[21] See the editorial introduction to the "Tianwen suanfa lei" 天文算法类 in *Siku quanshu*, 106.1a–2a.

[22] See *Zhuzi yulei* 《朱子语类》, 18.14b–15a. See also Yamada Keiji, *Shushi no shizengagu*, 413–472, and Yü Ying-shih, "Some Preliminary Observations on the Rise of Ch'ing Confucian Intellectualism", *Tsing Hua Journal of Chinese Studies*, New Series 11.1 and 2 (December 1975): 105–146.

[23] Daniel Gardner, *Zhu Xi and the Daxue: Neo-Confucian Reflection on the Confucian Canon*, 27–59.

literati elites toward precise philology in classical studies back to 16th-century debates surrounding the Old Text version of the *Great Learning* (*Daxue guben* 《大学古本》).[24]

Wang Yangming 王阳明 (1472–1528), for instance, preferred the Old Text version of the *Great Learning* to gainsay Zhu Xi's "externalist" views of the investigation of things in the Four Books. One of the later stories associated with the youthful Wang Yangming related that in 1492, when he was in Beijing at the age of 21 *sui* 岁 (Chinese added a year at the first lunar new year after a child was born) to take the spring 1493 metropolitan examination, Wang had "investigated" (*ge* 格 i.e., observed intently as in meditation) the bamboo in his official residence with a friend to search for its principles until both became seriously ill; thereafter he questioned Zhu Xi's views on the investigation of things as a naive enterprise and also pointed to the vagueness in the application of *gewu* to natural studies.[25] The official account in Wang's *Chuanxi lu* 《传习录》 (*Instructions for Practical Living*), recorded between 1521 and 1527 while Yangming was in retirement, described the episode:

> From morning till night I was unable to find the principles of the bamboo. On the seventh day I also became sick because I thought too hard. In consequence [my friend and I] sighed to each other and said it was impossible to be a sage or a worthy, for we do not have the tremendous energy to investigate things that they have. After I had lived among barbarians for three years, I understood what all this meant and realized that there is really nothing in the things themselves to investigate, that the effort to investigate things is only to be carried out in and with reference

[24] See Yü Ying-shih, "Some Preliminary Observations on the Rise of Qing Confucian Intellectualism", *Tsing Hua Journal of Chinese Studies*, New Series 11.1 and 2 (December 1975): 125, for discussion of Wang Yangming's critique of Zhu Xi's elucidation of the *Great Learning*, which created a textual crisis in the 16th century. Concerning the authenticity of new versions of the *Great Learning* in the late Ming, see Lin Qingzhang, *Qingchu de chunjing bianweixue* 《清初的群经辨伪学》, 369–386.

[25] See "Wang Yangming nianpu", 3. Xu Guangtai, "In the Name of 'Gewu Qiongli': The Transmission of Western Learning in Late Ming and Early Qing" stresses the vague relation between natural studies and Zhu Xi's agenda for learning.

> to one's own body and mind, and that if one firmly believes that everyone can become a sage, one will naturally be able to take up the task.[26]

Wang failed the 1493 examination, not taking his *jinshi* 进士 (palace graduate) degree until 1499. Wang's alternate agenda for *gewu*, which he clarified in later years while serving as military plenipotentiary in Jiangxi province, dismissed tedious facts about things such as bamboo in favor of a unifying introspective knowledge of the self (*liangzhi* 良知), which could be harnessed for moral cultivation. Wang had demonstrated that principles were not so easily grasped by following Zhu Xi's methods.

Wang Yangming's claim that the principles of things existed only in the mind (*xin ji li* 心即理) occurred at a time when literati views of the economy, commodities, and objects and their significance were changing. As China's population grew from approximately 65/100 to 150/250 million between 1450 and 1600, the reach of the relatively static imperial bureaucracy declined. Similarly, anxious Ming literati wondered if the Song–Ming Cheng–Zhu 程朱 classical orthodoxy (*lixue* 理学) could still represent universal principles of knowledge at a time when domestic goods and things were financially converted into objects of wealth paid for by using imported silver. Late Ming literati such as Yuan Huang 袁黄 (1533–1606) worked out the tensions between morality and affluence by creating a new moral calculus for measuring private wealth by keeping track of good and bad deeds in "ledgers of merit and demerit" (*Gongguo ge* 功过格).[27]

Although literati after Wang Yangming still placed human understanding within a classical theory of knowledge, the quantity and exchange

---

[26] See *Chuanxi lu*, in *Wang Yangming quanji* 《王阳明全集》, 93, for the official account, translated in Wing-tsit Chan, *Instructions for Practical Living and Other Neo-Confucian Writings by Wang Yang-ming*, 249. The "bamboo" story may be apocryphal. See Schorr Adam, "The Trap of Words: Political Power, Cultural Authority, and Language Debates in Ming Dynasty China".

[27] Peterson, Willard. "'Chinese Scientific Philosophy' and Some Chinese Attitudes Towards Knowledge about the Realm of Heaven-and-Earth", *Past and Present* 87 (May 1980): 29. See also, Cynthia Brokaw, *The Ledgers of Merit and Demerit: Social Change and Moral Order in Late Imperial China*, 17–27.

velocity of things in the marketplace had multiplied exponentially. Ming elites were living through a decisive shift away from the traditional ideals of sagehood, morality, and frugality. Within an inter-regional market economy of exceptional scope and magnitude, gentry and merchant elites transmuted the impartial investigation of things for moral cultivation into the consumption of objects for emotional health and satisfaction. Ming painters presented the contemporary fondness for and connoisseurship of antiquities as a genre known as "Broadly Examining Antiquities" (*Bogu tu* 博古图). The paintings valorized the literatus as a collector of exquisite things.[28]

Subsequently, the delicate issue of the late Ming appearance of an even more ancient "stone inscribed version of the *Great Learning*" (*Daxue shiben* 《大学石本》), which was later determined a forgery, reopened for many late Ming and Qing literati Wang Yangming's infamous claim that Zhu Xi had in Song times manipulated the original text of this key passage on *gewu* to validate and make canonical his personal interpretation of the investigation of things. In particular, Wang Yangming gainsaid Zhu Xi's emphasis on *gezhi* ahead of morality (*chengyi* 诚意, "making one's intentions sincere"). For Wang and his late Ming followers the investigation of things and the extension of knowledge took a backseat to first making one's will sincere. In their attacks on Zhu Xi, the Jesuits never raised the issue of Zhu's changes to the *Liji* version of the *Daxue*.[29]

Such concerns continued, however, among Qing literati, when evidential scholars also challenged Zhu Xi's emendation of the *Daxue*. For example, at the age of ten *sui* Dai Zhen 戴震 (1724–1777) was studying Zhu Xi's standard version of the *Great Learning* when he asked his teacher about one of Zhu Xi's comments, which read:

> "The preceding chapter of classical text is in the words of Confucius, handed down by Master Zeng [Zengzi 曾子]. The ten chapters of

[28] Craig Clunas, *Superfluous Things: Material Culture and Social Status in Early Modern China*, 91–115, and Tim Brook, *The Confusions of Pleasure: Commerce and Culture in Ming China*, 190–228.

[29] See Wang Yangming, *Chuanxi lu*, 32–35, and Zhu Yizun 朱彝尊, *Jingyi kao* 《经义考》, 159.1a–7b, 160.1a–8a, 161.1a–12a. See also Wang Fan-shen, "The 'Daring Fool' Feng Fang (1500–1570) and His Ink Rubbing of the Stone-inscribed *Great Learning*", *Ming Studies* 35 (August 1995): 74–91.

> explanation that follow contain Zeng's views and were recorded by his disciples. In the old copies of the work [in the *Liji*], there appeared considerable confusion from the disarray of the [original] tablets. But now, availing myself of the decisions of Master Cheng [Cheng Yi 程颐, 1033–1107], and having arranged anew the classical text, I [Zhu Xi] have rearranged it in order as follows."[30]

Dai asked about Zhu Xi's rearrangement of the *Daxue* into: (1) a classical text by Confucius; and (2) ten commentarial chapters by Zengzi:

> "How does one know in this case that these are the words of Confucius recorded by Zengzi? Moreover, how does one know that Zengzi's intentions were recorded by his followers?"
>
> The teacher replied: "That is what the earlier literatus Zhu Xi said in his annotation."
>
> Dai Zhen asked another question: "When did Zhu Xi live?"
>
> The teacher answered: "Southern Song."
>
> Dai asked again: "When did Confucius and Zengzi live?"
>
> The teacher replied: "Eastern Zhou."
>
> Dai asked again: "How much time separates the Zhou [dynasty] from the Song?"
>
> Reply: "About two thousand years."
>
> Dai queried again: "Then how could Zhu Xi know it was so?" The teacher could not reply.[31]

In effect, Dai Zhen like Wang Yangming defied Zhu Xi's classical authority. As a *Daoxue* master, Zhu had elevated his own commentary on the *gewu* passage in the *Great Learning* chapter of the *Record of Rites*,

---

[30] See the translation of Zhu Xi's commentary in James Legge, *The Four Books*, 360, which I have modified. For discussion, see Daniel Gardner, *Zhu Xi and the Daxue: Neo-Confucian Reflection on the Confucian Canon*, 27–59.

[31] Wang Chang 王昶 (1725–1806), *Chunrongtang ji*《春融堂集》55.6b. This account was later included in "Dai xiansheng xingzhuang"〈戴先生行状〉, in Dai Zhen, *Dai Zhen wenji*《戴震文集》, 251–260.

one of the Five Classics, by putting it into the mouth of one of Confucius's direct disciples. Zhu thereby had made his own commentary one of Zengzi's ten commentarial chapters. Dai Zhen exposed Zhu Xi's claim that Zengzi had written this important chapter, which Zhu included as one of the Four Books, as historically undocumented, as was Zhu's claim that the *Daxue* version in the *Liji* was corrupt because it gave no gloss prioritizing the *gewu* passage as the starting point for classical learning.

When Cheng Yi's and Zhu Xi's views on all the Classics were declared orthodox for the Yuan civil examinations that were restarted in 1313,[32] *gezhi* as a *Daoxue* term was already used by the medical writer Zhu Zhenheng 朱震亨 (1282–1358) to denote technical learning. In Zhu's most famous work entitled *Gezhi yulun* 《格致馀论》 (*Views on Extending Medical Knowledge*), which was included in the *Siku quanshu* in the late 18th century,[33] Zhu opposed Song medical prescriptions, but he made a strong appeal to Yuan literati that they should include medical learning in their "Way Learning". In his view, medical learning was one of the key fields of study that not only complemented the moral and theoretical teachings of *Daoxue*, but it was also a key to the practical uses of the latter. The *Siku quanshu* editors cited Zhu's preface as arguing that medicine was one of the concrete fields that informed the "investigation into and extension of knowledge".[34]

In addition to its central epistemological place in literati classical learning since 1200, the notion of *gewu* was also applied to the collection, study, and classification of antiquities, as in Cao Zhao's 曹昭 (fl. 1387–1399) *Gegu yaolun* 《格古要论》 (*Essential Criteria of Antiquities*), which was published in the early Ming and enlarged several times thereafter. The work originally appeared in 1387/1388 with important accounts of ceramics and lacquer, as well as traditional subjects such as calligraphy, painting, zithers, stones, bronzes, and ink-slabs. The 1462 edition prepared by Wang Zuo 王佐 (*jinshi* of 1427) was considerably enlarged and included findings from the official Ming dynasty naval expeditions led by

---

[32] Benjamin A. Elman, *A Cultural History of Civil Examinations in Late Imperial China*, 29–38, 56–61.

[33] Zhu Zhenheng, *Gezhi yulun*, Vols. 746–638.

[34] See the "*Tiyao*" 提要 (Abstract) of Zhu Zhenheng's study prepared by the editors of the *Siku quanshu*, compiled by Ji Yun, Vols. 746–637.

Zheng He 郑和 (1371–1433) to Southeast Asia and the Indian Ocean from 1405 to 1433. Wang also added the subjects of imperial seals, iron tallies, official costumes, and palace architecture. In his "Preface", Wang added: "Whenever you see an object, you must look it all over, trace its appearance, and examine its history and origins. You should investigate its strengths and weaknesses, and distinguish its accuracy." He was particularly interested in ancient bronzes, calligraphic specimens, and curiosities.[35]

Similarly Li Shizhen's 李时珍 (1518–1593) *Bencao gangmu* 《本草纲目》(*Compendium of Materia Medica*), published in 1596, reclassified the entire *materia medica* according to a new logic, which revealed Li's concern with the "investigation of things." Li Shizhen's "Outline" ("Fanli" 凡例) for his work, for instance, stressed the problem of nomenclature, or *zhengming* 正名 ("rectification of names"), as the main entry. All other sub-entries explained the names of things themselves. Accordingly, Li Shizhen's enterprise was also a scholarly project:

> I have actually practiced what we literati scholars call the "study of the investigation of things". This can fill the hiatus in the commentaries on the *Erya* 《尔雅》(*Approaching What is Correct*) [dictionary] and the *Shijing* 《诗经》(*Poetry Classic*).

Li Shizhen's explicit work of medicine thus also belonged to the long tradition of *gewu*.[36]

Other late Ming *leishu* also encompassed a wide variety of books, including household manuals, quotation dictionaries, and collections of anecdotes. Their common feature was their topical arrangement. Those known as *Riyong leishu* 《日用类书》(household encyclopedias or

---

[35] See the abridged version of the *Gegu yaolun*, in Hu Wenhuan, *Gezhi congshu*, Vol. 25. See also Wang Zuo's "*Xu*" 序 1a–b, and Sir David Percival, *Chinese Connoisseurship, the Ko Ku Yao Lun: The Essential Criteria of Antiquity*. The new information from other parts of Asia, however, did not challenge the existing frameworks of knowledge in Ming China, which differs from the wider impact of 16th-century oceanic discoveries in early modern Europe. See Donald F. Lach, *Asia in the Making of Europe. Volume II. A Century of Wonder, Book 3: The Scholarly Disciplines*, 446–489.

[36] See Georges Métailié, "The *Bencao gangmu* of Li Shizhen — An Innovation for Natural History?" In Elisabeth Hsu, ed., *Innovation in Chinese Medicine*, 221–261, and Nathan Sivin, "Li Shih-chen", in Charles Gillispie, ed., *Dictionary of Scientific Biography*, Vol. 8, 390–398.

"daily-use compendia") by contemporary Japanese historians actually represented nouveau riches manuals for stylish living that were widely printed in South China, particularly by printers in Jianyang, Fujian, whose lower printing costs made cheaper editions accessible beyond the usual classically literate elite.

Seven types of "daily-use compendia" emerged: (1) those oriented toward general topics; (2) those oriented toward civil office; (3) reference works (phrase dictionaries, etc.); (4) literary digests; (5) names and people; (6) stories/anecdotes; and (7) childhood primers. Among the common classification divisions included in such collections were: the astrology section), geography/topography), human chronicles, official goods, penal laws and regulations, marriage protocols, burial rites, medical studies, nourishing life, etc. The audience for such quotidian works tended to be urbanites and commoners.[37]

On the other hand, the Ming scholar-merchant and Hangzhou bookseller Hu Wenhuan 胡文焕 (fl. c. 1596) prefigured the Sino-Jesuit dialogue concerning *scientia* in the 1630s when he compiled and published his widely circulated *Gezhi congshu* 《格致丛书》 (*Collectanea of Works Investigating into and Extending Knowledge*) in the 1590s as a late Ming repository of classical, historical, institutional, and technical works from antiquity to the present in China. The *Gezhi congshu* presented a cumulative account of all areas of textual knowledge important to a literati audience in the 17th century. Because Hu also had wide ranging interests in medicine, Buddhism and Daoism, the collectanea contained a broad range of classical texts and esoteric writings. Like many Ming printers, Hu mixed and matched his editions and changed several original works to conform to his own formats. Based on works from these earlier editions, the *Baijia mingshu* 《百家名书》 (*Famous Works of the Hundred Schools*) collectanea was later published in 1603 and included many overlapping works.[38]

---

[37] See Sakade Yoshinobu, "Kaisetsu — Mindai nichiyō ruisho ni tsuite", in Sakade Yoshinobu *et al.*, eds., *Chūgoku nichiyō ruisho shūsei* , Vol. 1, 7–30. See also Lucille Chia, "Printing for Profit: The Commercial Printers of Jianyang, Fujian (Song–Ming)", 242–245.

[38] See Wang Baoping, "Mindai no kakushoka Ko Bunkan ni kan suru kosatsu", *Kyūko* 36 (1999): 47–57, and Wang Baoping, "Zhongguo Hu Wenhuan congshu jingyan lu" 〈中国胡文焕丛书经眼录〉, *Zhong Ri wenhua luncong* 《中日文化论丛》 (1991): 6–25.

Hu Wenhuan spent a considerable amount of time in his printing shops in Hangzhou and Nanjing. The latter was another very important Ming publishing center in the Yangzi delta, along with Suzhou, and Hu seems to have relied on the Nanjing book market for his editions. The *Hanlin* 翰林 senior compiler Zhu Zhifan 朱之蕃 (1564–1624), who was famous empire-wide for finishing as the *optimus* on the 1595 civil palace examination, added prestige to Hu's reputation as a Hangzhou scholar-printer by preparing a long 1603 preface to the overlapping *Baijia ming-shu* collectanea. They probably met in Nanjing. Zhu's preface made clear that Hu's collectanea was intended as a comprehensive library of works stressing astrology, calendrics, flora and fauna, medicine and longevity, in addition to the more classical themes of poetry, literature, philology, and technical glosses of things, objects, and affairs. Zhu Zhifan used the term *gezhi* in his preface for the *Baijia mingshu* to suggest the overarching unity Hu Wenhuan applied to such works, which linked the collection to the *Gezhi congshu*.[39]

Morohashi Tetsuji has provided us with a probable catalog of the supposedly 346 total works collected by Hu for his dual collectanea. Only 181 were apparently extant in the *Gezhi congshu* by the late 18th century, according to the compilers of the *Siku quanshu*, who did not think very highly of Hu's pastiche of works, many of which he himself had written.[40] The editors attacked Hu's role as compiler by accusing him of simply lifting chapters out of several collections and presenting them as separate books, not his own to be sure, when they were really part of another, larger work. In addition, numerous different collections of both the *Baijia mingshu* and the *Gezhi congshu* are extant in libraries in China, Japan, Taiwan, and the United States.[41] According to Morohashi, the *Gezhi congshu* collectanea was divided into 37 categories, such as classical

[39] See Zhu Zhifan, "Xu" 序 to the *Baijia mingshu* 《百家名书》, 5b–6a. Compare as in L. Carrington Goodrich *et al*., eds. *Dictionary of Ming Biography*, 304–305, and Yu Weigang, "Hu Wenhuan yu Gezhi congshu" 〈胡文焕与《格致丛书》〉, *Tushuguan zazhi* 《图书馆杂志》4 (November 1982): 63–65.

[40] *Siku quanshu zongmu* 124.14a–b. See also Morohashi Tetsuji, *Dai kanwa jiten*, 15113.61.

[41] For the *Gezhi congshu*, I have relied primarily on the version housed in the Rare Books Collection of the National Library of Taiwan, which contains 46 works.

instruction, philology, phonology, historical studies, rituals and regulations, legal precedents, geography, mountains and streams, medicine, Daoism, Buddhism, preserving life, agriculture, stars, physiognomy, poetry and literature, painting, and epigraphy, among others, but these categories were not apparent in all versions printed.

The portions that can be unquestionably tied to the main themes of the *Gezhi congshu* reveal Hu Wenhuan's efforts to republish and cumulatively build on previous works that focused on natural phenomena and "names and their referents" (*mingwu* 名物) and "affairs and things" (*shiwu* 事物). Starting with works annotating the *Erya* dictionary and the *Shiming*《释名》(*Explication of Names*), the focus was on a comprehensive account of etymologies and word definitions that would shed a collective light on the golden ages of antiquity.[42] The work contained over 1,500 lexical entries arranged according to 27 major semantic categories beginning with heaven and earth. The *Shiming* linked names with their referents based on a correspondence theory of language.[43]

The "rectification of names" (*zhengming*) for all things became a passionate classificatory agenda in Hu Wenhuan's collectanea. Overall, the *Gezhi congshu* collectanea emphasized a broad learning of phenomena (*bowu*), one of Morohashi's 37 categories, which encompassed natural and textual studies within a humanist and institutional agenda. Within the collection, the medieval *Bowu zhi* 《博物志》(*A Treatise on Curiosities*), and the Song dynasty continuation titled *Xu bowu zhi* 《续博物志》(*Continuation to a Treatise on Curiosities*) were subsumed under the general category of *gezhi*. Other works included in the *Gezhi congshu* were the *Shiwu jiyuan* 《事物纪原》(*Record of the Origins of Things and Affairs*) compiled circa 1078–1085, the *Wuyuan*《物源》(*Origins of Things*) of the Ming dynasty, and the *Gujin shiwu kao* 《古今事物考》(*Examination of Ancient and Contemporary Things and Affairs*) in the late Ming.

In general these works stressed associating each human event, object/implement, or natural phenomena in terms of a teleology of their usefulness to humans and presented a genealogy of discovery that traced each

[42] On the *Erya*, see Michael Loewe, *Early Chinese Texts: A Bibliographical Guide*, 94–99.

[43] Liu Xi, "Preface" to the *Shiming*, in the *Gezhi congshu*, 1a–b. See also Michael Loewe, *Early Chinese Texts: A Bibliographical Guide*, 424–428.

item back to the appropriate sage, ruler, or scholar. The "Preface" to the *Shiwu jiyuan*, dated 1448 (with an internal date of 1444), opened by linking all myriad things and affairs to their principles which can be investigated by studying their origins:

> The myriad things, which pervade heaven and earth and extend from antiquity to the present, always have matters changing within them. Things have myriad variations, affairs have myriad transformations, and for every matter and every thing, none is without principle or an origin. If one does not fathom the principle [involved], then there is no way to complete the knowledge in our minds. If we do not research the origins [of a thing or matter], then how can we unravel and fathom its principle? Therefore, sagely studies give priority to investigating things and extending knowledge. Literary scholars value broad inquiry and wide learning so that not even one thing will be unknown. [Such lack of knowledge] would be an embarrassment for literati.[44]

The *Gegu yaolun* account of early Ming antiquities, mentioned above, was also included in the collection, but it was abridged by Hu Wenhuan to include only the key parts and titled *Gegu lunyao* 《格古论要》 (*On the most Important Items in the Investigation of Antiquities*). Hu noted:

> Antiquity must be investigated. When antiquity is investigated it always penetrates to the present. When things are investigated it always penetrates to humanity. Timely investigations are very meaningful.[45]

The inventory of organized knowledge in the works that were included in the *Gezhi congshu* ranged from heaven and earth to birds, animals, insects, fish, grasses, foodstuffs, architecture, and tools. Such knowledge presupposed a Chinese frame of reference for the systematic collection of data from a wide variety of native sources about China's natural resources,

---

[44] See these works in Hu, *Gezhi congshu*. For the "Preface" to the *Shiwu jiyuan*, see 1a–b. Compare Robert F. Campany, *Strange Writing: Anomaly Accounts in Early Medieval China*, 51–52.

[45] See Hu Wenhuan's "*Xu*" 序 to the *Gezhi congshu* edition of the *Gegu lunyao*, in Vol. 25, 1a–2a.

the arts, and manufactures. For example, in the *Poetry Classic* studies, Hu Wenhuan's own work, which was included in the *Gezhi congshu*, emphasized that knowledge of "the names of birds, animals, herbs, and trees" was very important for understanding the *Poetry Classic*. Hu's expertise was applied to detailing such information in the *Poetry Classic* by building on other works he had included in the *Gezhi congshu*.

In his "Preface", Hu Wenhuan replied to critiques that his own work was redundant or too nitpicking: "If one says that Master Zhu [Xi] brought together [all knowledge about the *Poetry Classic*] and that I have split it all up, then what good is erudition?"[46] Moreover, Hu Wenhuan included other relevant works to drive home his point that earlier literati had relied on mastering poetry for the civil examinations to attain fame. The result was that before 1056 literati had "a great deal of knowledge of birds and animal, herbs and trees, insects, fish, and dragons". When the examination essay took precedence, however, everyone mastered classical skills (*jingshu* 经术) and no longer bothered with lyric poetry or rhyme-prose. Ironically, because of the focus on classical skills, literati knowledge of things in the natural world diminished, which pointed to the importance of the *Shijing* and *Erya* for natural knowledge in both the *Gezhi congshu* and *Baijia mingshu* versions of Hu Wenhuan's collectanea.[47]

In addition to Hu Wenhuan's Ming "*Gezhi* studies", the *Guang bowu zhi*《广博物志》(*Expansion of a Treatise on Curiosities*) also paid more attention to natural history. Such works on *bowu* suggest that as a term it needs to be conceptually mapped asymmetrically with *gezhi*. Sometimes the former was included under the latter, sometimes not. In both *gezhi* oriented and *bowu* framed late Ming works, the transformation of objects into artifacts, antiquities, and art objects was attempted.

---

[46] See the "*Xu* 序" by Hu Wenhuan dated 1593, in the *Gezhi congshu*, 1a–3b.

[47] In the *Lunyu*《论语》, 17/9, Confucius tells students that through study of the *Poetry Classic* they can become acquainted with names of birds, beasts, and plants. See James Legge, *The Four Books*, 261: "The Master said, 'My children, why do you not study the *Poetry*? The *Poetry* stimulates the mind, enables one to view things, how to form groups, how to vent resentments. You learn the duty of serving one's father, and further the serving of one's lord. One becomes more knowledgeable with the names of birds, animals, herbs, and trees.'"

Indeed, the editors of the *Siku quanshu* made clear in their 1780s critique of the *Gezhi congshu* that they disapproved of Hu's inclusion of works on *bowu* because such accounts represented literary writings, which previously had been classified under *Xiaoshuo* 小说 ("idle chatter and gossip," i.e., fictional accounts). The editors nevertheless included Hu's collectanea under the *sibu* classification of "Masters" in the subcategory of *zajia* 杂家 (miscellaneous writers), rather than placing it under literature.[48]

It is intriguing that the term *gezhi* was also chosen by Ming literati in the 17th century as one of the native categories of specialized learning (*xuewen*), with the latter equivalent for many Ming literati to early modern European *scientia*. Early Jesuit translations of Aristotle's theory of the four elements (*Kongji gezhi* 《坤舆格致》, "investigation of space", 1633) and Agricola's *De Re Metallica* (*Kunyu gezhi*, "investigation of the earth", 1640) into classical Chinese, for example, had used the term *gezhi* in light of the Latin *scientia* (= "organized or specialized knowledge", or *xuewen*, as *scientia* was translated in Chinese in the 16th century).[49] Such titles suggest that Way Learning doctrine and natural studies, particularly medical and calendrical learning, were not mutually exclusive.[50]

Willard Peterson has noted how late Ming views of the *Daoxue* doctrine of the "investigation of things" had changed from a type of moral endeavor, purely, to an additional stress on external things. Fang Yizhi's magnum opus entitled *Wuli xiaozhi* 《物理小识》 (*Notes on the Principles of Things*) stressed "material investigations" to "comprehend the seminal forces" underlying patterns of natural change. Fang generally accepted Western explanations of natural phenomena, such as a spherical earth, limited heliocentrism, and human physiology, brought to China in the 17th century by the Jesuits, but he was critical of them for leaving behind material investigations and ending in unverified religious positions. Fang Yizhi favored, instead, descriptive knowledge of the natural world, and he inscribed the Way Learning interpretation of the

---

[48] *Siku quanshu zongmu,* 134.15a.

[49] See Pan Jixing, "The Spread of Georgius Agricola's *De Re Metallica* in Late Ming China", *T'oung Pao* 57 (1991): 108–118, and James Reardon-Anderson, *The Study of Change: Chemistry in China, 1840–1949*, 30–36, 82–88.

[50] Roger Hart, "Local Knowledges, Local Contexts".

"investigation of things" with a new view of the accumulation of knowledge, which gainsaid both the introspective focus of Wang Yangming and the moralist focus of Zhu Xi.[51]

In the interaction with Western *scientia*, then, Chinese literati were drawn into a moderate transformation of their own traditions of natural studies.[52] We will see in the last chapters of this volume that in the late Qing, between 1865 and 1900, reformist Chinese officials and scholars reworked *gezhixue* to designate "modern science". Subsequently *gezhixue* was replaced in the early 20th century by the Japanese neologism of *kagaku* 科学 as *kexue* for the Chinese equivalent for post-industrial revolution science. This repeated use of *gezhi* — from the late Ming to late Qing — suggests that native terms for Western science were contested in different ways at different times.[53]

Ming Civil Examinations and *Gezhi*

Ming dynasty examination records reveal that civil examinations also tested knowledge of astrology, calendrical studies, and other aspects of the natural world, which at times were referred to as "natural studies" (*ziran zhi xue* 自然之学).[54] The preeminent position of the Four Books and Five Classics was left unchallenged in the orthodox curriculum, but Ming candidates for both the provincial and metropolitan examinations, unlike their Song counterparts, were expected to grasp many of the technicalities in calendrics, astrology, anomalies, and the musical pitch series. The latter was the basis for official weights and measures.

---

[51] Willard Peterson, "Fang I-chih: Western Learning and the 'Investigation of Things.'" In Wm. Theodore de Bary *et al*., eds., *The Unfolding of Neo-Confucianism*, 369–411.

[52] *Gezhi congshu*, *passim*. See also Ssu-yü Teng and Knight Biggerstaff, *An Annotated Bibliography of Selected Chinese Reference Works*, 105.

[53] For discussion, see Lydia Liu, *Translingual Practice: Literature, National Culture, and Translated Modernity — China 1900–1937*, 20–42. See also my article "From Pre-modern Chinese Natural Studies to Modern Science in China", in *Mapping Meanings: The Field of New Learning in Late Qing China*, edited by Professor Michael Lackner and Natascha Vittinghoff.

[54] See Benjamin A. Elman, *A Cultural History of Civil Examinations in Late Imperial China*, 461–485. Compare Nathan Sivin, "Introduction", in Nathan Sivin, ed., *Science & Technology in East Asia*, xi–xv.

During the Tang, Song and Yuan dynasties, works on calendrical studies and astrology were banned from publication for security reasons. Only dynastic specialists working on the calendar in the Astronomy Bureau were allowed such knowledge, even though in practice popularly printed calendars and almanacs were widely available. Such restrictive policies continued outside the precincts of the Ming civil service examinations.[55]

In the early Ming, for example, the Yongle emperor put calendrical and practical studies near the top of what counted for official, literati scholarship. He ordered the chief examiner for the 1404 metropolitan examination (on which 472 graduates drawn from over 1,000 candidates were selected and appointed to high offices) to include questions that tested a candidate's "broad learning". The examiners selected policy questions (*shiwu ce* 时务策) dealing with astronomy, law, medicine, ritual, music, and institutions, and the emperor was especially pleased with the top policy answer that year. More importantly, the emperor had legitimated natural studies. Thereafter such questions regularly appeared on Ming civil examinations.[56]

Late Ming concerns about the accuracy of the official calendar forced Chinese literati to apply specific Jesuit techniques to reform the Ming calendar. Derived from the Vatican's adoption of the Gregorian calendar in 1582 to replace the Julian calendar under Christopher Clavius's leadership, the technical prowess that some Jesuits such as Matteo Ricci had learned as a result of studying under Clavius at the *Collegio Romano* proved fortuitous in Ming China.[57] But such influences did not carry over to the civil examinations during the Qing dynasty. The Manchu throne sought instead to monopolize this potentially volatile area of expertise within the confines of the court. The contemporary calendrical debates between Jesuits and literati-officials, which challenged the Yuan–Ming calendrical system during the Ming–Qing transition, gave the imperial

---

[55] Thatcher E. Deane, "The Chinese Imperial Astronomical Bureau: Form and Function of the Ming Dynasty 'Qintianjian' from 1365 to 1627", 259, 322–324, 399–400.

[56] See *Huang Ming sanyuan kao*《皇明三元考》, 2.3b, and *Zhuangyuan ce* 状元策, 15a.

[57] Peter Dear, "Jesuit Mathematical Science and the Reconstitution of Experience in the Early Seventeenth Century", *Studies in History and Philosophy of Science* 18 (1987): 135–141.

court pause about allowing possibly divisive questions on the calendar to appear in civil examinations.[58]

The collapse of the Ming dynasty and its Qing successor under non-Han rule created opportunities until 1685 for Jesuit experts in astronomy-astrology and music to break out of their subordinate positions and to challenge discredited Ming elites for political power under a new Manchu bannermen elite. The increased cultural importance of astronomical expertise, when the new dynasty had to reformulate in expert terms its calendrical and musical raison d'être as quickly as possible, probably outweighed, or at least challenged for a time, the cultural distinction accumulated by literati via mastery of classical studies. Court scholars such as Li Guangdi 李光地 (1642–1718) actively patronized specialists in calendrical calculations and made the musical pitch series a high priority in their officially financed research.[59]

Consequently, in the 1680s, when the Manchu dynasty had mastered its political and military enemies, the intellectual fluidity of the early decades of the Qing began to disappear, leaving Han literati and Manchu elites in a precarious balance at the top (and Jesuit, Muslim, and Chinese calendar specialists again in the middle) of the political and social hierarchies. For example, after 1711 the Kangxi emperor knew of the Paris Académie Royale des Sciences, which the French Jesuit Foucquet translated into Chinese as the *Gewu qiongli yuan* 格物穷理院 ("Academy for the Investigation of Things and Fathoming Principle"). The emperor also modeled the court's *Suanxue guan* 算学馆 (Academy of Mathematics) after it, but these remained temporary organizations created by the emperor to solve his immediate needs. The new academies had no great impact outside the Kangxi court, where the French Jesuits worked mainly on calendar reform and the empire-wide land survey. The emperor did not encourage a broader focus on natural studies as was the case in early modern France.[60]

---

[58] Jonathan Spence, *Emperor of China: Self-portrait of Kangxi,* xvii–xix, 15–16, 74–75, and Chu Pingyi, "Scientific Dispute in the Imperial Court", *Chinese Science* 14 (1997): 7–34. See also Huang Yinong 黄一农, "Qingchu tianzhujiao yu huijiao tianwenjia de zhengdou"〈清初天主教与回族天文家的争斗〉, *Jiuzhou xuekan*《九州学刊》5.3 (1993): 47–69.

[59] Arthur Hummel, *Eminent Chinese of the Ch'ing Period*, 473–475.

[60] Han Qi, "Gewu qiongli yuan yu Mengyangzhai"〈格物穷理院与蒙养斋：十七十八世纪之中法科学交流〉, *Faguo hanxue*《法国汉学》4 (2000): 302–324.

In the process, policy questions on the Qing provincial and metropolitan examinations virtually ceased to include natural studies. In addition, the role of the French Jesuits sent by Louis XIV to the Kangxi court as "royal mathematicians" and as members of the Paris Academy of Sciences declined. Their combination of scientific and religious objectives had proved as problematical for the Jesuits and their Church critics, as for the Kangxi court.[61]

By 1715, the Kangxi emperor successfully banned focus in the civil examinations on study of astronomical portents and the calendar because they pertained to Qing dynastic legitimacy. He could not restrict such interest among the literati community outside the civil examination bureaucracy, however. The emperor, for example, decreed in 1713 that thereafter all examiners assigned to serve in provincial and metropolitan civil examinations were forbidden to prepare policy questions on astronomical portents, musical harmonics, or calculation methods. The latest works in Qing natural studies, court projects on which the Kangxi emperor had employed Jesuit experts, were put off limits to examiners and examination candidates. The ban on natural studies occurred within a general effort by the court to keep the mantic arts and discussion of auspicious versus inauspicious portents out of public discussion.[62] In place of the banned natural studies, historical geography in particular prospered as an acceptable examination field of Qing scholarship, although map-making was kept secret by the throne.[63]

"High Qing" Evidential Studies and the Scope of *Gezhixue*

Such bans, however effective in the civil examinations, did not carry over to literati learning, where a decisive sea change in classical learning was occurring. There were clear limits to imperial policies outside

---

[61] Florence Hsia, "Some Observations on the *Observations*: The Decline of the French Jesuit Scientific Mission in China", in Éric Brian, ed., *Revue de synthèse: Les Jésuites dans le monde moderne Nouvelles approches*, 305–333.

[62] See *Huangchao zhengdian leizuan* 《皇朝政典类纂》, 191.7b–8a. For discussion of these court compilations, see Benjamin A. Elman, *From Philosophy to Philology: Social and Intellectual Aspects of Change in Late Imperial China*, 79–80.

[63] See also *Qingchao tongdian* 《清朝通典》, 18.2131.

the government. In contrast to their Ming "Way Learning" (*Daoxue*) predecessors, 18th-century "evidential research" (*kaozheng* 考证) scholars stressed exacting research, rigorous analysis, and the collection of impartial evidence drawn from ancient artifacts, historical documents, and texts. Evidential scholars made verification a central concern for the emerging empirical theory of knowledge they advocated, namely "to search truth from facts" (*shishi qiushi* 实事求是), a Han dynasty expression used as a slogan for impartial scholarship in the 18th century.[64]

For the most part, "High Qing" evidential scholars preferred not to use the term *gezhi* to describe their research agenda, because in the mid-18th century *gezhi* was a "Song Learning" category long associated with Zhu Xi's "studies of principle" (*lixue*), which they opposed.[65] In the 17th century, however, *gezhi* and *kaoju* (考据, i.e., = 考证) had been used together.[66] Under the Kangxi emperor, when a reaffirmation of Cheng–Zhu learning occurred, Song Learning (*Songxue* 宋学) views of *gewu* and natural studies remained perennial issues.[67] The mid-Qing program, however, involved the placing of proof and verification at the center of the organization and analysis of the classical tradition in its complete, multidimensional proportions, which now included aspects of natural studies and mathematics.

In the 19th century, when scholars softened their positions on the distinction between *kaozheng* and Song Learning, evidential scholars such as Ruan Yuan 阮元 (1764–1849; see also Chapter 5) and the Cantonese literatus Chen Li 陈澧 (1810–1882), whose teaching career at the *Xuehai tang* 学海堂 (Sea of Learning) Academy in Guangzhou influenced many local students, including Liang Qichao 梁启超 (1873–1929) and

---

[64] See *Hanshu* 5.2410: "When [Liu] De took the throne as King Xian of Hejian in 155 BC, he restored scholarship and honored antiquity. He sought the truth in actual facts."

[65] On 18th-century evidential research, see Benjamin A. Elman, *From Philosophy to Philology: Social and Intellectual Aspects of Change in Late Imperial China*, 57–85.

[66] For a 17th-century example of the linkage between *kaoju* and *gezhi*, see Lu Shiyi 陆世仪, "Sibianlu lunxue" 〈思辨录论学〉, in Wei Yuan 魏源, ed., *Huangchao jingshi wenbian*《皇朝经世文编》, 3.7–3.9.

[67] See *Kangxi jixia gewu bian*《康熙几暇格物编》, which was translated by the French Jesuits in 1779.

Kang Youwei 康有为 (1858–1927), linked the empiricism associated with *shishi qiushi* to the "investigation of things". Ruan Yuan noted:

> Many earlier literati discussed *gewu*, but they associated it [with so many] empty meanings that they apparently failed to grasp that this was not the basic intent of the sages. I don't dare to be different in my discussion of *gewu*, yet [I say] it is the search for truth from facts and that's all.

Later in the 19th century, Chen Li equated the "investigation of things" with *shishi* 实事 (concrete affairs = facts) and identified the "extension of knowledge" with *qiushi* 求是 (search for truth).[68]

By the late 19th century, reflecting the scholarly trends of the Qianlong era, the policy questions for civil examinations, for instance, began to exhibit a common five-way division of topics, usually in the following order: (1) Classics; (2) Histories; (3) Literature; (4) Statecraft; and (5) Local geography. The primacy of classical learning in the policy questions was due to the impact of evidential research among literati scholars, first in the Yangzi delta, and then empire-wide via examiners from the delta provinces of Jiangsu, Zhejiang, and Anhui. What fueled the popularity of the 1756 revival of first a Tang-dynasty-style poetry question on session one and then philology in the policy questions in session three of the civil examinations was the close ties between the rules for rhyming in regulated verse and the field of phonology, which became the queen of philology during the Qianlong reign.[69]

Qing dynasty evidential scholars such as Dai Zhen had in mind a systematic research agenda that built on paleography and phonology to reconstruct the meaning (*yi yin qiu yi* 以音求义) of Chinese words. Later scholars extended Dai's approach and attempted to use the "meanings" of Chinese words as a method to reconstruct the "intentions" of the sages, the farsighted authors of those words. Moreover, technical phonology when applied to the study of the history of the classical language reached

---

[68] See Ruan Yuan, *Yanjingshi ji* 《揅经室集》, 2.23b–2.24a, and Chen Li, *Dongshu dushu ji* 《东塾读书记》, 9.14. See also Benjamin A. Elman, *From Philosophy to Philology: Social and Intellectual Aspects of Change in Late Imperial China*, 245–247.

[69] See Benjamin A. Elman, *A Cultural History of Civil Examinations in Late Imperial China*, 503–520, 546–562.

unprecedented precision and exactness. To achieve this end, evidential scholars chose philological means, principally the application of phonology, paleography, and etymology, to study the Classics.[70]

Qing philologists, because they were trained as classicists, were faced with numerous passages in the official (those associated with Confucius) and unofficial classics (for example, the medical classics associated with the Yellow Emperor, and the mathematical classics) from antiquity, which required technical training in mathematics, astronomy, geography, and calendrical studies to decipher. *Kaozheng* scholars made astronomy, mathematics, and geography in their research programs another by-product of the changes in classical studies then underway.[71] Animated by a concern to restore native traditions in the precise sciences to their proper place of eminence, which early in the 18th century was legitimated by the Jesuit accommodation policy toward Chinese learning, later evidential scholars successfully incorporated technical aspects of Western astronomy and mathematics into the literati framework for classical learning. Qian Daxin 钱大昕 (1728–1804), in particular, acknowledged this broadening of the literati tradition, which he saw as the reversal of centuries of focus on moral and philosophic problems:

> In ancient times, no one could be a literatus (*Ru* 儒) who did not know mathematical calculation. Chinese methods [now] lag behind Europe's because *Ru* do not know mathematics.[72]

The impact of evidential research made itself felt in the attention *kaozheng* scholars gave to the Western fields of mathematics and astronomy first introduced by the Jesuits in the 17th century. Such interest had built upon the early and mid-Qing findings of Mei Wending 梅文鼎 (1633–1721), who was sponsored by Li Guangdi and the Manchu court once his expertise in mathematical calculation (*lisuan* 历算) and calendrical studies was recognized. Mei had contended that study of physical

---

[70] Hamaguchi Fujiō, *Shindai kōkyogaku no shisō shi teki kenkyū.*

[71] Han Qi, "Bai Jin de Yijing yanjiu he Kangxi shidai de xixue zhongyuan shuo" 〈白晋的易经研究和康熙时代的西学中源说〉, *Hanxue yanjiu* 《汉学研究》16, 1 (1998): 185–200.

[72] Qian Daxin, *Qianyantang wenji* 《潜研堂文集》, 3.335 (*juan* 卷 23).

nature gave scholars access to the "principles" (*li* 理) undergirding reality. In essence, Mei saw Jesuit learning as a way to boost the numerical aspects of the notion of moral and metaphysical principle.[73] At the same time, however, the imperial court and Mei Wending prepared preliminary accounts stressing the "native Chinese origins" (*Zhongyuan* 中源) of Western natural studies. Such origins made it imperative for Mei (and his highly placed follower in the early Qing court Li Guangdi) to restore and rehabilitate the native traditions in the mathematical sciences to their former glory. Under Kangxi's imperial patronage, mathematical studies were upgraded from an insignificant skill to an important domain of knowledge for literati that complemented classical studies.[74]

The term *gezhi* subsequently reappeared in Chen Yuanlong's 陈元龙 (1652–1736) *Gezhi jingyuan* 《格致镜源》(*Mirror Origins of Investigating Things and Extending Knowledge*), which was published in 1735. Chen had received the Kangxi emperor's orders in 1704 to compile a comprehensive work on astrology/astronomy, geography, human affairs, plants and trees, and insects, but he held on to the manuscript for 20 years before publishing it. A repository of detailed information divided into 30 categories culled from a wide variety of sources, the *Gezhi jingyuan* represented a post-Jesuit collection of practical knowledge by a well-placed scholar in the Kangxi and Yongzheng courts.

In his collection, Chen Yuanlong narrowed the focus of Hu Wenhuan's late Ming *Gezhi congshu*, some of which had already been lost, to cover almost exclusively the arts and natural studies. For instance, Chen left out all poetry, rhyme-prose, and stories, which separated the *Gezhi jingyuan* in the eyes of the *Siku quanshu* compilers from *Gezhi congshu*, because

[73] See Wang Ping, "Qingchu lisuanjia Mei Wending" 〈清初历算家梅文鼎〉, *Jindaishi yanjiusuo jikan* 《近代史研究所集刊》 2 (1971): 313–324, and John Henderson, "The Assimilation of the Exact Sciences into the Ch'ing Confucian Tradition", *Journal of Asian Affairs* 5.1 (Spring 1980): 15–31.

[74] See Bai Limin, "Mathematical Study and Intellectual Transition in the Early and Mid-Qing", *Late Imperial China* 16.2 (1995): 23–61, and Catherine Jami, "Learning Mathematical Sciences During the Early and Mid-Qing", in Elman and Woodside, ed., *Education and Society in Late Imperial China, 1600–1900*, 223–256. On the "Chinese origins" theory, see Quan Hansheng, "Qingmo de 'xixue yuan chu Zhongguo' shuo" 〈清末的西学源出中国说〉, *Lingnan xuebao* 《岭南学报》 4.2 (June 1935): 57–102.

the former left out all fictional material. Special attention was given to the origins and evolution of printing and stone rubbings, in addition to topics dealing with geography, anatomy, flora and fauna, tools, vehicles, weapons and tools for writing, as well as clothing and architecture. The Imperial Library editors included Chen's work in the *Siku quanshu* in the 1780s because Chen's catalog of entries "all were [examples of] broad learning and thus the work was titled 'investigating things and extending knowledge.'" The editors, who were for the most part partisans of evidential studies, were by now able to separate *gezhi* as a term from Song Learning and associate it with their emphasis on ascertainable knowledge derived from empirically derived research.[75]

The 17th-century impact of Jesuit knowledge in China was not always so easily domesticated in the 18th, however. Literati scholars took a range of positions concerning natural studies. A private scholar, Jiang Yong 江永 (1681–1762), for instance, combined classical loyalty to Zhu Xi's *Daoxue* teachings with knowledge of Western Jesuit studies obtained through evidential studies. Conservative as a classical scholar, Jiang was radical in his critique of both Han Learning (*hanxue* 汉学) and Mei Wending in natural studies for exalting nativist ancient studies in all cases. Jiang Yong recognized the advantages Western astronomy had over native traditions, while at the same time he continued to uphold the cultural superiority of the *Daoxue* view of morality. Although Jiang preferred Western learning for understanding the "principles" of nature because they were more precise and consistent than native traditions, he maintained a clear distinction between astronomical methods and cultural values.[76]

Overall, Ruan Yuan's compilation of the *Chouren zhuan*《畴人传》(*Biographies of Astronomers and Mathematicians*) while serving as governor of Zhejiang province in Hangzhou from 1797 to 1799, reprinted in 1849 and later enlarged, marked the climax of the celebration of natural studies

---

[75] See Chen Yuanlong, "Fan-li" 凡例, Vol. 1031, 1–3. See also *Siku quanshu zongmu* 136.25a–26a.

[76] Chu Pingyi, "Ch'eng-Chu Orthodoxy, Evidential Studies and Correlative Cosmology: Chiang Yung and Western Astronomy", *Philosophy and the History of Science: A Taiwanese Journal* 4.2 (1995): 71–108.

within the Yangzi delta literati world of 18th-century evidential research, which had been increasing since the late 17th century. Containing biographies and summaries of the works of 280 *chouren* 畴人, including 37 Europeans, this work was followed by four supplements in the 19th century. Bai Limin has noted how the mathematical sciences had begun to grow in importance among literati beyond the reach of the imperial court in the late 18th century. They were now linked to classical studies via evidential research. Because Ruan Yuan was a well-placed literati patron of natural studies in the provincial and court bureaucracy, his influential *Chouren zhuan* integrated the mathematical sciences with evidential studies. Mathematical and natural studies remained dependent on classical studies.[77]

Via evidential studies, philology became one of the key tools that later scholars cum mathematicians and scientists, such as Xu Shou 徐寿 (1818–1882) and Li Shanlan 李善兰 (1810–1882), employed to build conceptual bridges between Western learning and the traditional Chinese sciences in the 19th century. In the process, modern Western science and mathematics were initially introduced in the 19th century as compatible with native classical and technical learning. The Song–Yuan "heavenly element notation" (*tianyuan shu* 天元术; also called "single unknown" techniques) and "four elements notation" (*siyuan shu* 四元术; also called "four unknowns" techniques) forms for expressing and solving quadratic and higher algebraic equations of several unknowns, once recovered in the late 18th century by Mei Juecheng 梅瑴成 (1681–1763) — Mei Wending's grandson — and others, were thought by Qing evidential scholars to be superior to the algebraic techniques introduced to China by the Jesuits. Not until the introduction of the differential and integral calculus in the mid-19th century, for which the Chinese could not find a precedent in China, did Li Shanlan and other Chinese mathematicians finally admit that although the "four elements notation" was perhaps superior to Jesuit algebra, the Chinese had never developed anything resembling the calculus.

[77] Arthur Hummel, *Eminent Chinese of the Ch'ing Period*, 402. See also Bai Limin, "Mathematical Study and Intellectual Transition in the Early and Mid-Qing", *Late Imperial China* 16.2 (1995): 23–30.

Chapter 3

Some Comparative Issues — Ming–Qing Border Defense and Jesuit Learning in Late Imperial China

For over a century, Europeans have heralded the success of Western science and assumed the failure of modern science to develop of itself elsewhere. Since 1954, Joseph Needham had stressed the unique rise of modern science in Europe, but at the same time he acknowledged the achievements in traditional Chinese science and technology until 1600. In the decades since Needham answered his provocative question — "why didn't the pre-modern Chinese develop modern science?" — we have increasingly acknowledged that our focus on the "failure" of Chinese science to develop into modern science is heuristically interesting but historiographically misguided. We are now forced to reassess how the history of science globally should be rewritten.

This chapter will focus at the end on why late imperial Chinese never learned about the "Newtonian Century" in Europe and its analytic style of mathematical reasoning until after the Opium War (1839–1842). Some still contend that the Qing state during the Macartney mission in 1793, for example, was too closed-minded to learn about the emerging early modern world. With hindsight, such views appear incontrovertible, but there were many external factors to China, such as the collapse of the Jesuit mission in China and Europe in the middle of the 18th century, which help explain why the Newtonian revolution came so late in Asia, and not in the 18th century.

Debates about Freedom and Curiosity

The contested nature of the interaction between late imperial Chinese and early modern Europeans over the meaning and significance of natural studies since 1550 is a little-known story. Narratives of the history of science worldwide from 1500 to 1800 have, until recently, been portrayed mainly through European frames of reference, even when comparative themes are stressed. These Eurocentric portraits of the rise of modern science, while not monolithic, mainly represent variations of a single-minded historical teleology of Western European scientific "success", and non-Western "failure". Usually the narrative plots of such accounts reproduce uncritically the story of the 17th century Protestant-based scientific revolution or return to the teleologies of the medieval, Catholic roots of modern science.

The basic argument presented in Toby E. Huff's *The Rise of Early Modern Science: Islam, China, and the West*, one of the most ecumenical of works in the genre of the comparative historiography of science, is that the fortuitous concatenation of Greek natural philosophy, Roman civil law, and rationalist Christian theology during the 12th century Renaissance created an intellectual climate conducive for the emergence of the European scientific revolution in the 17th century. By comparison, the promising development of Arabic science before the 14th century was subsequently aborted by the power of the Islamic clergy to islamicize foreign knowledge and exclude natural science from the education curriculum in universities. Similarly in China, where the Song dynasty was well advanced in technology before the 13th century Mongol conquest, the retarding influence of the imperial state and its civil service examination curriculum based on the Confucian Classics prevented autonomous institutions conducive to the social legitimation of men of science to emerge. Only in late medieval Europe did legal, social, and institutional revolutions occur that permitted rise of autonomous universities in which study of natural science was affirmed by both state and society. Only in Europe did the requisite "neutral spaces for public discourse" appear that "created structures of action and agency" conducive for growth of the impartial and skeptical "ethos of science".[1]

Adopting what Huff calls "civilization frames of reference", his book describes "cultural climates" in medieval Europe, the Islamic world, and imperial China that abetted or hindered emergence of "modern science" in each cultural region. The cultural trajectories of Christian Europe, Islamic Asia, and Confucian China are examined according to affinities with the

[1]Toby E. Huff, *The Rise of Early Modern Science: Islam, China, and the West*. See also my review of the first edition in *American Journal of Sociology*, (November 1994): 817–819.

roles of scientists in society, development of social norms of impartiality and legitimate skepticism, and elaboration of an autonomous reward system under control of professional members of scientific communities. Islamic breakthroughs in study of the natural world were initially advanced because of input from Greek and Indian natural philosophy and mathematics. In the fields of optics, astronomy, and medicine, Islamic scholars were important pioneers, but the fate of the Marâgha Observatory, which flourished in Tabiz between 1259 and 1305, is emblematic according to Huff of the short-lived success of Islamic science. Clerical power based on Islamic mysticism, fundamentalist appeals to sacred law (*shari'a*), and the inherent elitism and secrecy of Islamic elites over the long term created obstacles that effectively prevented emergence of modern science.

In China, according to Huff, social and political elites had access to the fruits of Islamic science in the Mongol era, and they had access to European science through the Jesuits in the 17th century. Despite their technological head start in the Song dynasty, Chinese were unable to break through roadblocks erected by the imperial state (required secrecy imposed on state astronomers and mathematicians) and the alternative and more appealing reward system (based on competitive examinations) to become a Mandarin official that diverted incentives toward study of the Confucian Classics and away from precise study of natural phenomena. In Europe, according to Huff, the late medieval "legal, social, and institutional revolution" singularly provided fertile ground for planting seeds for modern science in the 12th century that grew to maturity in the 17th.

Huff's history of European, Islamic, and Chinese science focuses on the intellectual history of natural philosophy in cultural and social context. By rejecting Joseph Needham's claim that there is no meaningful distinction between science and technology, Huff too conveniently concludes that experimentalism after Galileo was not the driving force in the birth of modern science and that in China technological superiority had little to do with Chinese natural knowledge. By sowing the seeds of scientific revolution in Catholic culture before the Protestant Reformation, Huff presents the "symbolic technology" of late medieval Europe as the key to that revolution without fully coming to grips with the problem of why it took over five centuries for European natural philosophy to begin to produce the industrial revolution.

In the final analysis, Huff's otherwise thoughtful comparative history of science essentializes science as a universal mental exercise practiced best in neutral institutions within an open society. The intellectual revolution — its philosophical roots and cultural context in Catholic Europe — is not properly contextualized historically in light of the later political and industrial significance of the scientific revolution in Protestant Northwestern Europe. By surgically eliding technology and industry from the problem of the scientific revolution, Huff thus avoided coming to grips with the more convincing "Protestant" history of science and industry as elaborated by Margaret C. Jacob,[2] or the actual practice of modern science as described by Bruno Latour.[3]

Moreover, without any discussion of technology, Huff's history of science in Catholic Europe, the Islamic world, and in imperial China was limited to the domains of Catholic, Islamic and Confucian educational institutions. Huff's comparison of Catholic Europe, Islamic Asia, and Confucian China in terms of natural philosophy and educational institutions was timely and rewarding, but his conclusions were insufficient to understand **why** elites and commoners in Europe, the Islamic world, and China were interested in the natural world and **how** they actually created and applied their natural knowledge before the scientific revolution.

To answer such criticism, Huff has recently completed a new book entitled *Intellectual Curiosity and the Scientific Revolution: A Global Perspective*. It leaves behind his earlier fetishes of the medieval world and mainly addresses curiosity about science in early modern Europe from 1600 to 1800. Part one focuses on the role of the telescope and other new instruments in leading Europeans to increased curiosity about the world, which climaxed with Newton (1642–1727) et al., and the making of the Protestant scientific revolution. Despite his intellectual progress in finally understanding the importance of early modern Europe in the history of science, Huff has mobilized the increasing factual evidence in his new volume for one major end: to rave about the superiority of European

[2]Margaret Jacob, *Scientific Culture and the Making of the Industrial West.*

[3]Bruno Latour, *Science in Action: How to Follow Scientists and Engineers Through Society.*

science due to their superior curiosity about the world as best exemplified by their use of the telescope (see also the discussion in Chapter 5).[4]

Adam Schall's (1591–1666) *On the Farseeing Optic Glasses* (*Yuanjing shuo* 《远镜说》, i.e., the telescope) was translated into Chinese and available in Ming China in 1626. In the early 1630s, Rome sent Beijing a telescope, the first of many that Huff overlooks. He also misses, however, the importance of the telescope in the massive hunting and topographical expeditions the Manchu emperors initiated in the 18th century, one of which was portrayed by Lang Shining 郎世宁 (Giuseppe Castiglione, 1688–1766) et al. in the far right-hand corner of their "Portrait [of the Qianlong Emperor] Troating for Deer, 1741" 弘历哨鹿图 (Palace Museum, Beijing), included above and below (see Figures 1 and 2).[5] After incorrectly reviewing how the telescope was received in Qing China, Moghul India, and the Ottoman Empire, Huff summarily concludes that Jesuit efforts to demonstrate the superiority of Western science in the court of the Manchu emperor failed because they were not in the least bit interested. Why then did Kangxi and Qianlong both include a large telescope on horseback in their expeditions (see Figure 2). As we will see further in Chapter 4, the telescope was also used to measure the extent of their empire and mark out its boundaries with the Russians at the treaties of Nerchinsk (1689) and Kiakhta (1727).[6]

Instead of analysis, Huff turns to "Orientalist" hubris to make his point for him. He summarizes the Ottoman lack of curiosity in light of a recorded instance in which the telescope was used to spy on the Sultan's harem in Constantinople. How ingenious! The point of course is that the telescope did not stimulate a new era of scientific research in other civilizations such as Qing China in the way it did in early modern Europe.

---

[4]Toby E. Huff, *Intellectual Curiosity and the Scientific Revolution: A Global Perspective*.

[5]Dorothy Berinstein, "Hunts, Processions, and Telescopes: A Painting of an Imperial Hunt by Lang Shining (Giuseppe Castiglione)", *RES: Anthropology and Aesthetics* 35, Intercultural China (Spring 1999):170–184. These images may be used freely under the Creative Commons licensing agreement for any non-commercial application.

[6]See Peter S. Perdue, *China Marches West: The Qing Conquest of Central Eurasia*. See also Eva-Maria Stolberg, "Interracial Outposts in Siberia: Nerchinsk, Kiakhta, and the Russo-Chinese Trade in the Seventeenth/Eighteenth Centuries", *Journal of Early Modern History* 4, 3–4 (December 1999): 322–336.

Figure 1. Troating for Deer, 1741.

What about Japan? We know, for example, that the writer Ihara Saikaku's 井原西鶴 (1642–1693) illustrated novel entitled *Life of a Sex-mad Man* (*Kōshoku ichidai otoko*《好色一代男》, 1682) included the playful use of the telescope by a Japanese "peeping Tom". But Huff conveniently ignores the other findings of Timon Screech and others who have stressed the role of the telescope, along with Jesuit learning and "Dutch studies", in forming the "scientific eye and visual wonders" in Edo Japan during the 18th century.[7] This lack of scientific curiosity in other lands is a given for

[7]Timon Screech, "Rethinking the Visual Revolution in Edo: Inoue Masashige and International Visual Culture in the Years after the Prohibition of the Seas", in *The Scientific Eye and the Visual Wonders in Edo*, 14–20. See pages 96–107 of the catalog on telescope viewing in Tokugawa Japan.

Figure 2. Lang Shining (Giuseppe Castiglione) and others, *Portrait [of the Emperor] Troating for Deer* (detail), 1741. Hanging scroll, ink and colors on silk, 267.5 × 319 cm. Palace Museum, Beijing.

Huff, but he curiously never tries to explain what might have caused the Europeans to be so much more curious than the Qing Chinese or Tokugawa Japanese. Can we agree with Huff, once again, that it was simply the working out of liberty and curiosity via "neutral institutions within an open society"? Does the Ottoman example hold for all of Asia in the same way? For all of Istanbul? Huff's view certainly doesn't hold for early modern China or Japan.

China, India, and Japan

Arguably, by 1600, Europe was already ahead of Asia in producing basic machines such as clocks, screws, levers, and pulleys that would be applied increasingly to the mechanization of production. Once harnessed to steam power, machines became the engines of the 19th-century industrial revolution. But in the 17th century, Europeans still sought the technological secrets for silk, textile weaving, porcelain, and tea production from the

Chinese.[8] Chinese literati in turn borrowed from Europe new algebraic notations (of Hindu-Arabic origins), geometry, trigonometry, and logarithms from the West.

The epistemological premises of modern Western science were not triumphant in China until the early 20th century. Until 1900, Chinese elites interpreted the transition in early modern Europe — from new forms of scientific knowledge to new modes of industrial power — on their own terms. Each side made a virtue out of the mutually contested accommodation project, and each converted the other's forms of natural studies into acceptable local conventions of knowledge. The accommodation project should be viewed as something more than an imaginary shared rationalism between the Jesuits and Chinese literati. As we will see below, the comparison between late imperial China and early modern Europe, as with India and Japan, should also take into account natural anomalies, monsters, supernatural events, and religious faith.[9]

If Europeans increasingly thought themselves scientifically and technologically superior after 1500, as Michael Adas has shown, neither the Chinese nor Japanese agreed with this perspective until the effects of the 19th-century industrial revolution were visible. In Japan, for instance, aristocratic and merchant elites domesticated Western learning using classical Chinese and Japanese frames of reference — despite the impact of "Dutch Learning" in the 17th and 18th centuries — until the Meiji period (1868–1911).[10] Unlike the

[8]Donald F. Lach, *Asia in the Making of Europe. Volume II. A Century of Wonder, Book 3: The Scholarly Disciplines*, 397–400. See also Han Qi, *Zhongguo kexue jishu de xichuan ji qi yingxiang* 《中国科学技术的西传及其影响》, 135–169, on printing, porcelain, metallurgy, and textiles, and Lydia Liu, "Robinson Crusoe's Earthenware Pot", *Critical Inquiry* 25 (Summer 1999): 728–757.

[9]See Donald Mungello, *Curious Land: Jesuit Accommodation and the Origins of Sinology*, 23–43, and Lionel Jensen, *Manufacturing Confucianism: Chinese Traditions and Universal Civilization*, 34–75. See Zhang Qiong, "About God, Demons, and Miracles: The Jesuit Discourse on the Supernatural in Late Ming China", *Early Science and Medicine* 4, 1 (February 1999): 1–36.

[10]See Michael Adas, *Machines as the Measure of Men: Science, Technology, and Ideologies of Western Dominance, passim.* See also Wai-ming Ng, "The *I Ching* in the Adaptation of Western Science in Tokugawa Japan", *Chinese Science* 15 (1999): 94–115, Grant Goodman, *Japan: the Dutch Experience*, 228, and Masao Watanabe, *The Japanese and Western Science*. Compare Adam Clulow, *The Company and the Shogun: The Dutch Encounter with Tokugawa Japan*.

colonial environment in India, where British imperial power after 1700 could dictate the terms of social, cultural, and political interaction between natives and Westerners, natural studies in late imperial China were until 1900 part of a nativist imperial project to master and control Western views on what constituted legitimate natural knowledge.[11]

In South Asia, on the other hand, the British colonial regime successfully set the epistemological agenda for natural studies in "India" by defining the body of Western knowledge that would be augmented in that colony through local collecting procedures. Such knowledge would in turn be ordered and classified according to the standards of authoritative British scientific practice. Colonial forms of knowledge, according to Bernard Cohn, were translated into reports, statistical records, histories, gazetteers, legal codes, and encyclopedias that induced elites in "India" to become part of Britain's project of political and cultural control. Through such knowledge formation within the Western structures of science, colonized natives acquired enough practical experience to understand how natural knowledge was acquired, studied, and interpreted.[12]

Gyan Prakash adds to Cohn's perspective an account of how the British regime actually "staged" science in India via museums, exhibitions, and governmental projects throughout the colony. Such stagings thereby presented science as a universalist sign of modernity, which augmented colonial rule by educating native elites in the acceptable forms of scientific knowledge and natural history. Western educated elites then portrayed modern science and technology as a preferred value system and useful technology, which could enrich indigenous traditions. Prakash also describes how native elites renegotiated the terms of their acceptance of the British regime of science and technology and created a hybrid discourse of science and nation that included the *Vedas* and inscribed Hindu philosophy with modernist claims of nationalism. The colonized thus

[11]On India, see Bernard Cohn, *Colonialism and Its Form of Knowledge: The British in India*, 5–56. See also Gyan Prakash, *Another Reason: Science and the Imagination of Modern India*, 3–14, which notes that the British "civilizing mission" in India initiated the cultural authority of modern science in South Asia. Prakash adds that nativists also identified a body of indigenous South Asian traditions consistent with Western science, but he focuses on colonialist discourses rather than pre-colonial traditions in natural studies in India.

[12]Bernard Cohn, *Colonialism and Its Form of Knowledge: The British in India*, 5–76.

were not passive agents in their participation in the production of modern science. Colonial power provided the intellectual space within which Indian intellectuals appropriated science and refashioned their own traditions in light of the ideals of universal science and not simply British colonial power. Prakash describes this process as the indigenization of the authority of science and the formation of a Hindu modernity out of the colonial experience. One ironic result was the eventual gainsaying of the British colonialist regime by the very Indian nationalists cum modernists the British Raj produced.[13]

Fascinating as the colonial case is in the rise of modern science in India, late imperial Chinese literati and the imperial government, whether under a Chinese or Manchu ruler, openly and effectively contested all European claims to scientific and religious superiority at every stage of interaction since 1550. One of the reasons why we have detailed accounts of the conditions in Chinese prisons in the 16th and 17th centuries, for instance, is that those accounts were prepared by overly aggressive religious proselytizers from the Domincan and Franciscan orders who were locked up by the imperial state as rabble-rousers.[14]

Because they were not and did not perceive themselves as subordinated to the West, until the late 19th century, Chinese and Manchus did not engage in the covert colonialist renegotiations that Hindu nativists in India carried out. At the same time, the imperial court induced Jesuit calendrical, military, and land mensuration experts to work as imperial minions in the state bureaucracy to augment first the Ming and then the Qing dynasty's own project of political and cultural control. Consequently, it would be a historiographical mistake to underestimate Chinese efforts to master on their own terms the Western learning (known as *Xixue* 西学 or *gezhixue* 格致学) of the Jesuits in the 16th, 17th, and 18th centuries.[15]

---

[13]Gyan Prakash, *Another Reason: Science and the Imagination of Modern India*, 17–85. See also Paula Findlin, *Possessing Nature: Museums, Collecting, and Scientific Culture in Early Modern Italy*, 1–11.

[14]Derk Bodde, "Prison Life in Eighteenth Century Peking", *Journal of the American Oriental Society* 89 (April–June 1969): 311–333, gives references to earlier accounts of prisons.

[15]See Xu Guangtai, "Ruxue yu kexue: yige kexueshi guandian de tantao" 〈儒学与科学：一个科学史观点的探讨〉, *Qinghua xuebao* 《清华学报》, New Series, 26, 4 (December 1996): 369–392.

Pre-Jesuit Cartography and Descriptive Geography

Since the Song dynasties, China had sometimes supported a massive navy, which the Mongols expanded to invade Japan in 1274 and again in 1281 and to attack Java. Subsequently, the early Ming navy carried out enormous excursions into Southeast Asia and the Indian Ocean from 1403 to 1434. Ming rulers also declined any political accommodation on land with the Mongols and their nomadic neighbors, a policy that lasted from 1368 until 1570. They saw in such accommodations, which included possible diplomatic and commercial relations, one of the reasons for the fall of the Song in 1280 to the steppe barbarians. As a result of its more narrow-minded defensive and punitive measures, the Ming faced continual threats from their Mongol neighbors on the steppe.[16]

During the first century of Ming rule, for example, the "perennial problem of the Mongol menace" became so acute that officials in Beijing feared an imminent reconquest by the Mongols. The infamous Tumu debacle of 1448–1449, for instance, began when the Oirats, who had reunified Western Mongol forces on the steppe, had captured the reigning Zhengtong emperor when he led a Ming military campaign against them. In the climactic battle, the Ming army lost half of its 500,000 soldiers. Rather than capitulate to Oirat demands for ransom or move the capital from Beijing back to Nanjing, the Ming court chose to replace the captured emperor with his younger brother, who was enthroned as the Jingtai emperor and led the successful defense of Beijing. The proposal to retreat to the south reminded Ming officials too much of the disastrous experiences of the Southern Song when it had moved its capital from Kaifeng to Hangzhou.[17]

The crisis continued when the Zhengtong emperor, a worthless hostage by then, was returned by the Oirats in 1450. Several years later he

---

[16]Henry Serruys, *Sino-Mongol Relations during the Ming*, in *Melanges Chinois et Bouddhiques, Vol. 1: The Mongols in China during the Hung-wu Period, 1368–1398*, and *Vol. 2: The Tribute System and the Diplomatic Missions*, 8. Compare David Robinson, *Martial Spectacles of the Ming Court*, Chapter 4.

[17]See Philip de Heer, *The Care-Taker Emperor: Aspects of the Imperial Institution in Fifteenth-Century China as Reflected in the Political History of the Reign of Chu Ch'i-yü*, and Frederick W. Mote, "The T'u-mu Incident of 1449", in Frank Kierman, Jr., and John Fairbank, eds., *Chinese Ways in Warfare*, 243–272.

retook the throne as the Tianshun emperor during the coup d'état of 1457 during which the officials who had sacrificed their emperor in 1449 were accused of treason and executed. After 1457, the Ming pulled its army back from the Inner Mongolian steppe, and the rebuilt Great Wall became in the 16th century the only immediate barrier separating the capital from Mongol tribes.[18]

In the 1460s, the Ming court terminated naval expeditions to Southeast Asia and the Indian Ocean, which had been especially prominent from 1403 to 1434, in favor of a strictly coastal navy. The Tianshun emperor, who had already suppressed the great "Treasure Ships" of the Ming navy before his capture by the Oirats, maintained a military vigil on his northern borders after returning to power. In 1474, only 140 warships of the 400 ships in the main fleet remained. Thereafter, all seagoing ships of more than two masts were scrapped and used as lumber, while the court attended to military campaigns against the Mongols and rebellious southeastern tribes. A coastal navy thereafter defended the China coast from Chinese and Japanese pirates in the 16th century but in vain. This northern border focus on steppe defenses lasted until the mid-16th century when the "wokou" 倭寇 pirate menace loomed in the Yangzi delta and the southeast.[19]

Chinese naval power revived briefly in the late Ming as a result of Ming efforts to aid Korea and halt Hideyoshi's (1536–1598) invasion of Korea. Ming loyalists subsequently defeated the Manchu Qing dynasty in major naval and land battles along the Fujian coast. This brief naval efflorescence lasted until the 1680s, but thereafter new types of sailing vessels were still developed, such as the Zhejiang junks first built in 1699. They were used in northern waters for the Ningbo–Nagasaki trade between Japan and China, which lasted into the 18th century despite Japan's alleged "closed door" policies.[20]

---

[18] Arthur Waldron, *The Great Wall of China: From History to Myth*, 79–86, notes how Ming forces withdrew from steppe garrisons as early as 1403–1410. See also Alastair Johnston, *Cultural Realism: Strategic Culture and Grand Strategy in Chinese History*, 195–197.

[19] On the Ming fleets, see Joseph Needham, *Science and Civilisation in China,* Vol. 4, Part 3 (hereafter SCC), 477–553. On Ming relations with Southeast Asia, see Wang Gungwu, *Community and Nation: China, Southeast Asia and Australia*, 77–146.

[20] See Lo Jung-pang. "The Decline of the Early Ming Navy", *Oriens Extremus* 5, 2 (1958): 147–168.

During the late Ming, native attempts at serious cartography and descriptive geography provided Chinese with a foundation that cumulatively helped bridge the gap between the orthodox symbolic geography popular since the Song period and the Jesuit techniques that stimulated the emergence of more precise cartography in the 17th century. Chinese achievements in geography climaxed with the production of schematic grid-maps, sailing charts, and relief maps. Although few maps from early dynasties have survived, two other extant maps, which demonstrate the remarkable development of map-making in China, were carved on stone in the 12th century. Informed by Tang models, the maps were entitled the *Map of China and Foreign Areas* (*Huayi tu*《华夷图》) and *Map of the Tracks of Emperor Yu* (*Yuji tu*《禹迹图》). The latter (see Map 1), carved

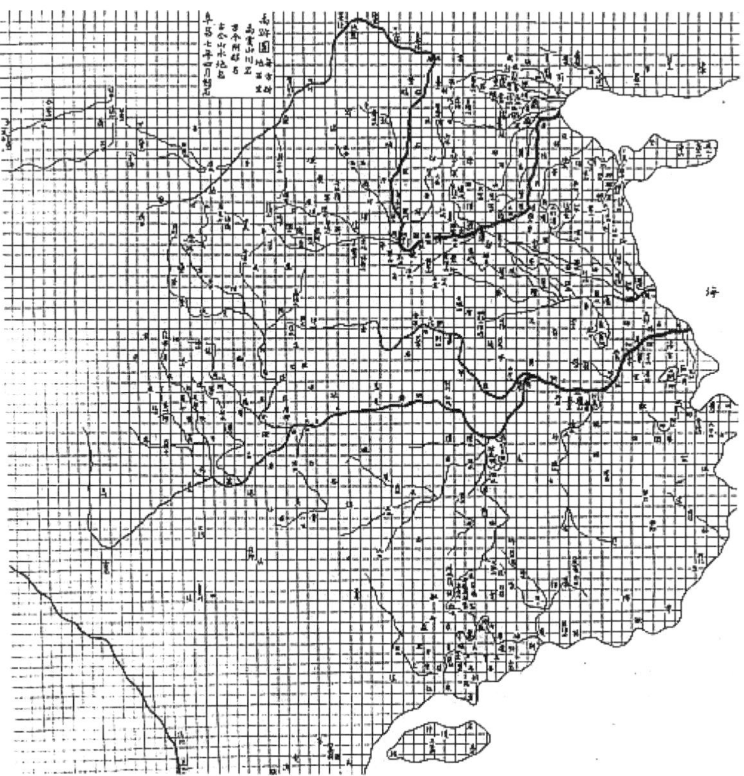

Map 1. *Map of the Tracks of Emperor Yu* (*Yuji tu*《禹迹图》). From Joseph Needham, *Science and Civilisation in China*, Vol. 3.

in 1137 as a grid based on a scale of approximately 1:1.5 million, is described by Needham as "the most remarkable cartographic work of its age in any culture".

The medieval cartographer Pei Xiu 裴秀 (224–271) first used the method of indicating distances by a rectangular grid system. Since then, cartography in China was often based on the grid tradition, although many important maps still followed the *Map of China and Foreign Areas* model and were not rendered using a grid. Jia Dan 贾耽 (730–805), the greatest Tang cartographer, constructed a map of Chinese and foreign regions for the emperor. The map, now lost, was 30 feet long, 33 feet high, and used a grid scale of one inch for 100 Chinese miles, i.e., equal to about 33 English miles, the same scale used in the *Map of the Tracks of Emperor Yu*. Jia Dan's map likely depicted all of known Asia because of its huge scale.[21]

Zhu Siben 朱思本 (1273–1337) inherited this grid format. Perhaps the best known Chinese cartographer until the 20th century, Zhu used grid-maps to summarize the large body of new geographical information that the Mongol conquests in Asia had added to the earlier fund possessed by Tang and Song geographers. Zhu's map, circa 1320, was known as the *Mongol Atlas of China* (*Yuditu*《舆地图》). An early Ming map of about 1390 replicated the imperial pretensions of Mongol maps and charted the empire's sway from Central Asia and Japan to the Atlantic via Southeast Asia and the Indian Ocean world.[22]

In the 16th century, Luo Hongxian 罗洪先 (1504–1564) discovered a manuscript copy of Zhu Siben's map of the known world, which Zhu prepared between 1311 and 1320 but never printed. Luo revised and enlarged Zhu's grid format in 1541. He also added new information, some derived from the early Ming naval explorations led by Admiral Zheng He 郑和 (1371–1433). Luo's map (*Guang yutu*《广舆图》, *Enlargement of a Map of the Earth*), printed in about 1555, included clear depiction of the

[21] See Richard Smith, "Mapping China's World: Cultural Cartography in Late Imperial Times", in Yeh Wen-hsin, ed., *Landscape, Culture and Power in Chinese Society*, 62–68. See also Alexander Wylie, *Notes on Chinese Literature*, 43–67, for pre-modern works on geography.

[22] Richard Smith, "Mapping China's World: Cultural Cartography in Late Imperial Times", in Yeh Wen-hsin, ed., *Landscape, Culture and Power in Chinese Society*, 67–68.

Cape of Good Hope and southern Africa, leading some to speculate that Zheng He was the "Vasco da Gama of China" and had traversed from the Indian to Atlantic Ocean and back some 70 years before da Gama did it in reverse.[23]

Ming Knowledge of Foreign Countries

Geographers and cartographers in Ming China were fairly knowledgeable of the southern regions (*nanyang* 南洋, i.e., what Chinese referred to as Southeast Asia) and foreign countries in the Indian Ocean and Arabian Peninsula (*xiyang fanguo* 西洋番国, i.e., "Western tributary states"). Information of this type peaked in China as a result of the early 15th-century voyages of the Ming fleet. Zheng He, the eunuch commander of seven Ming expeditionary fleets launched between 1405 and 1433, reached at one time or another Sumatra, Java, Ceylon, Hormuz, Aden, and East Africa.

The expeditions overlapped with the extent of the Ming maritime tributary system around which trade and diplomatic relations were organized in Ming China. Moreover, Ma Huan 马欢 (ca. 1380–1460) left a descriptive account of Zheng's fleets in 1433, which was based on his service as Muslim translator for three of the seven expeditions (1413–1415, 1421–1422, 1431–1433). Ma's *Captivating Views of the Ocean's Shores* (*Yingya shenglan*《瀛涯胜览》) described the countries the fleet visited and portrayed the customs, religions, and lifestyle in each place. He also depicted topography, geology and wildlife.[24]

Subsequently, other members of Zheng He's expeditions described their adventures. For example, Fei Xin's 费信 (1388–1436?) *Captivating Views from a Star Guided Vessel* (*Xingcha shenglan*《星槎胜览》) and Gong Zhen's 巩珍 *Gazetteer of the Western Tributary States* (*Xiyang fan'guo zhi*《西洋番国志》, 1434) were based on first-hand visits and helped fill the gap in Indian Ocean accounts between Marco Polo's *The Travels* and Portuguese reports in the late 15th century. References to

---

[23] SCC, Vol. 3, 533–551, Vol. 4, Part 3, 500–502, and *Dictionary of Ming Biography*, L. C. Goodrich, *et al.*, eds., 982. Compare such speculations with the claims presented as facts to unsuspecting audiences in Gavin Menzies, *1421: The Year China Discovered America*, passim.

[24] See Jan Julius Lodewijk Duyvendak, *Ma Huan Re-examined*, 8–9.

the expeditions and the exotic gifts brought back to China were added, for instance, to Cao Zhao's 曹昭 (fl. 1387–1399) *Essential Criteria of Antiquities* (*Gegu yaolun*《格古要论》), which was published in the early Ming and enlarged several times. The work originally appeared circa 1387–1388 with important accounts of ceramics and lacquer, as well as traditional subjects such as calligraphy, painting, zithers, stones, bronzes, and ink-slabs.[25]

Excerpts and extracts were subsequently included in the massive early 18th-century *Synthesis of Books and Illustrations Past and Present* (*Gujin tushu jicheng*《古今图书集成》), the largest encyclopedia in Chinese history. The Qianlong Imperial Library also included a 1520 annotation of Ma Huan's account by Zhang Sheng 张升 (1442–1517). In addition, the Imperial Library summarized the content of Gong Zhen's work and mentioned Huang Xingzeng's 黄省曾 (1490–1540) 1520 *Records of Western Tributaries* (*Xiyang chaogong dianlu*《西洋朝贡典录》), although neither was copied into the collection. Huang's work drew on and cited earlier sources.[26]

Within China, Ming dynasty explorers such as Wang Shixing 王士性 (1547–1598), Xie Zhaozhi 谢肇制 (1567–1624), and Xu Hongzu 徐宏祖 (also known as Xu Xiake 徐霞客, 1586–1641) traveled to the most remote borders of the empire. In their travels, they collected notebooks outlining in descriptive terms the river systems, topography, and cultural aspects of the places they visited. Xu Hongzu, for example, discovered the main source of the West River in his travels through much of Southwest China and was able to determine that the Mekong and Salween Rivers were different river systems.[27]

---

[25] *Hsing-ch'a-sheng-lan: The Overall Survey of the Star Raft by Fei Hsin*, translated by J. V. G. Mills and revised, annotated, and edited by Roderich Ptak. See also Sir David Percival, trans., *Chinese Connoisseurship, the Ko Ku Yao Lun: The Essential Criteria of Antiquity*.

[26] *Siku quanshu zongmu*《四库全书总目》, compiled by Ji Yun 纪昀 (1724–1805) *et al.*, hereafter SKQSZM, 78. 14b–19a.

[27] SKQSZM, 71.3b–4b, 78.14b–15a, SCC, Vol. 3, 511, 557–558. See also Ting Wen-chiang, "On Hsu Hsia-k'o (1586–1641): Explorer and Geographer", *New China Review* 3, 5 (October 1921), 325–337. Detailed discussion of Xu Xiake's travels can be found in Li Chi, *The Travel Diaries of Hsu Hsia-k'o*. Yi Zheng, "Xu Xiake's Travel Notes: Motion, Records and Genre Change," in Yi Zheng and Ofer Gal, eds., *Motion and Knowledge in the Changing Early Modern World: Orbits, Routes and Vessels*, 31–45.

Cartography and Ming Military Defense

Interest in cartography and geography in China was tied to problems of border or maritime defense. Luo Hongxian became interested in geography because of Sino-Japanese pirate raids on the maritime provinces in the Yangzi delta and along the Southeast Coast in the 16th century, which were aided by Chinese accomplices. The Ming government urgently needed maps and geographical advisors to cope with the situation. Luo spent three years collecting geographical materials for military defense and in the process discovered the manuscript of Zhu Siben's grid-map. Zheng Ruozeng 郑若曾 (1505–1580), a native of Suzhou prefecture, was also involved in coastal defense because of the pirate threat. In the 1540s, Zheng compiled a strategic atlas of China's coastal region extending north from the Liaodong Peninsula to southern Guangdong, which included maps and strategic information. More important, however, was a work on coastal geography compiled under then Zhejiang Governor Hu Zongxian's 胡宗宪 (1511–1565) auspices by Zheng Ruozeng.

First published in 1562, the resulting *A Maritime Survey: Collected Plans* (*Chouhai tubian* 《筹海图编》) was modeled after Luo Hongxian's work. The *Maritime Survey* was an ambitious survey of geographical minutiae about the China coast, which contained accurate accounts of China's neighbors, Japan and Korea. Hence, it was not simply an atlas. A major contribution to Chinese historical geography, the *Maritime Survey* marked a major turning point in Chinese geographic studies. Arguably its completion provided the impetus for later literati to incorporate Jesuit information into the Chinese tradition of geography and cartography.[28]

Prior to the 16th century, threats to China's security had mainly come from the north and northwest, e.g., the Jurchen, Mongols, and the Manchus. Hence, earlier geographers such as Zhu Siben focused attention on the northern frontier areas in their geographical accounts. In the 16th century, however, the primary military menace came from Japanese pirates and their Chinese collaborators along the South China Sea

[28]Originally completed in 1562, the *Maritime Survey* was re-edited by Hu Zongxian's descendants and reissued in 1624. The editors of the later editions dropped Zheng Ruozeng's name as the author and replaced it with Hu Zongxian, their revered ancestor. Compare the account in SKQSZM, 69.31a–32b.

coastline. Zheng Ruozeng's research on the maritime provinces in the southeast thus stimulated a shift in geographical research. A number of 16th-century geographical treatises on maritime defense were either inspired by or followed the pattern of the *Maritime Survey*.[29]

Because of their travels, European Jesuits could claim expertise in world geography when they arrived in Ming China in the late 16th century. Their charts of the globe were based on the New World discoveries of the Portuguese and Spanish explorers they often accompanied. The *mappa mundi* that Matteo Ricci (1552–1610), his collaborators, and others produced for literati in the late 16th century caused an uproar because the Ming imperial system was based on a cosmological view of its geographical centrality in the world, which the Jesuit maps seemed at first to challenge. It was difficult, for example, for literati to accept the Jesuit claim that the earth was round, because this suggested that the "vaulted heavens" (*gaitian* 盖天) were circular as well. Many literati considered the earth flat and the sky a finite vault overhead.[30]

Matteo Ricci's *Mappa Mundi*

Jesuits added to the geographical knowledge of Ming literati in the late 16th century. Produced with the help of Chinese converts, the first edition of Matteo Ricci's *mappa mundi* entitled *Complete Map of the Earth's Mountains and Seas* (*Shanhai yudi quantu* 《山海舆地全图》), for instance, was printed in 1584. A flattened sphere projection with parallel latitudes and curving longitudes, Ricci's world map went through eight editions in all between 1584 and 1608. The third edition was entitled the *Complete Map of the Myriad Countries on the Earth* (*Kunyu wanguo quantu* 《坤舆万国全图》) and printed in 1602 with the help of Li Zhizao 李之藻 (1565–1630; see Map 2 below). Li studied European mathematics and astronomy, in addition to geography, after meeting Ricci in Beijing in 1601.[31]

---

[29] SCC, Vol. 3, 514–517.

[30] Cordell Yee, "Traditional Chinese Cartography and the Myth of Westernization", in J. B. Harley and David Woodward, eds., *The History of Cartography. Volume 2. Book 2: Cartography in the Traditional East and Southeast Asian Societies*, 170–202.

[31] Ricci's world map is reproduced in SCC, Vol. 3, 582–583 (Plates XC and XCI). A 1608 color version is included in *Symbols of Power: Masterpieces from the Nanjing Museum*.

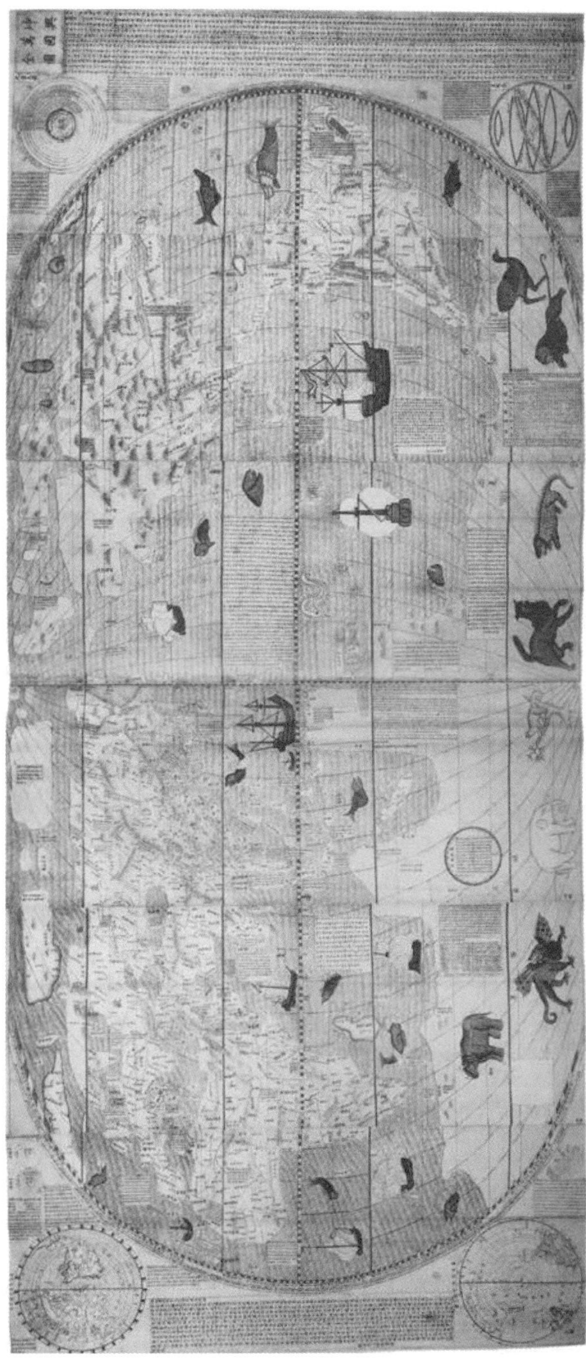

Map 2. Ricci's Map of 1602.

Source: http//geog.hkbu.edu.hk/GEOG1150/Chinese/Catalog//Catalog_main_11.htm.

Ricci's description of the form and size of the earth obliged many 17th-century Chinese literati to revise their views of the world. For the first time, Chinese became aware of the exact location of Europe in relation to their own country. In addition, Ricci's maps contained technical geographical lessons for Chinese geographers: (1) Ricci taught Chinese cartographers to localize places by means of circles of latitude and longitude; (2) he invented many geographical terms and names, including Chinese terms for Europe, Asia, America, and Africa; (3) Ricci's maps transmitted to China the most recent discoveries made by European explorers; (4) he described the existence of five terrestrial continents surrounded by large oceans; (5) the maps introduced the sphericity of the earth; (6) finally, Ricci spoke of five geographical zones and their location on the earth, i.e., the Arctic and Antarctic Circles, the zone between the Arctic Circle and the Tropic of Cancer, etc.[32]

An indication of the initial impact of Ricci's maps, especially the 1608 edition printed in Beijing, was its inclusion in geographical works produced by literati scholars in the closing years of the Ming dynasty. For example, Zhang Huang 章潢 (1527–1608), who had met Ricci in Jiangxi in 1595, added Ricci's *Complete Map* to his own illustrated collection, which Zhang called the *Compendium of Maps and Materials* (*Tushu bian* 《图书编》, 1613). In addition to Ricci's 1584 map of the world, Zhang also printed European depictions of the northern and southern hemispheres along with traditional maps of the four seas in his massive collection.[33]

The only known copy of the first Chinese map of the world produced in 1593 was entitled the *Comprehensive Map of Heaven and Earth and the Myriad Countries and Ancient and Modern Persons and Artifacts*

---

[32] Helen Wallis, "The Influence of Father Ricci on Far Eastern Cartography", *Imago Mundi* 19 (1965), 35–45, and Kenneth Ch'en, "Matteo Ricci's Contribution to, and Influence on, Geographical Knowledge in China", *Journal of the American Oriental Society* 59, 3 (September 1939), 325–359. Compare Theodore Foss, "A Western Interpretation of China Jesuit Cartography", in Charles Ronan, S.J., and Bonnie Oh, eds., *East Meets West: the Jesuits in China, 1582–1773*, 209–251.

[33] Zhang Huang, *Tushu bian*, 19.33a–40a. See also Kenneth Ch'en, 343–347. For the initial reaction of Chinese literati, see Pasquale M. d'Elia, "Recent Discoveries and New Studies (1938–60) on the World Map in Chinese of Father Matteo Ricci S.J.", *Monumenta Serica* 20 (1961), 82–164.

(*Qiankun wanguo quantu gujin renwu shiji* 《乾坤万国全图古今人物事迹》) and included geographical information brought by the Jesuits. It was based on Ricci's first world map of 1584, which is now lost. The map offered a traditional representation of China with foreign lands arranged around the periphery. It served chiefly as an administrative map for officials and thus included statistical information such as population (based on families) and locally produced commodities.

Topological rather than topographical, the 1593 Chinese version fit European lands in along its edges without affecting traditional cartography. The New World was shown as a series of small islands surrounding the "Central Kingdom". Methodologically, however, the compiler stressed the "achievements of investigating things and extending knowledge", which by now served as a native trope for the accumulation of knowledge. The linkage of the world map of 1593 to classical studies based on investigating things and extending knowledge heralded the application of such native terminology to European *scientia*.[34]

The 1602 *Complete Map of the Myriad Countries on the Earth* corresponded to one of the first issues of Ricci's third world map and is the earliest version that survives. This 1602 edition followed the *Typus Orbis Terrarum* by Abraham Ortelius (1527–1598), a Flemish scholar and geographer, which was first published in a 1570 European atlas called the *Theatrum Orbis Terrarum*, which was itself based on Gerardus Mercator's (1512–1594) prominent 1569 world map. The chief alteration for the Chinese version was that Ming China was placed at the center of the map to appeal to Chinese dynastic sensibilities. The New World was located on the eastern borders. A fourth edition of the map was prepared in 1604.[35]

Ricci's mapmaking was continued by later Jesuits who in 1623 produced a lacquered wooden globe, which updated Ricci's map and also stressed the sphericity of the earth. Giulio Aleni (1582–1649) and Yang

---

[34] *The Library of Philip Robinson*, Part II, 77–95, especially 76–77, 78–79. The 1593 world map was engraved by Liang Zhou 梁辀, an education official in Wuxi, and printed in Nanjing. See also John Day, "The Search for the Origins of the Chinese Manuscript Copies of Matteo Ricci's Maps", *Imago Mundi* 47 (1995).

[35] *The Library of Philip Robinson*, Part II, 79–80. See also Lionel Giles, "Translations from the World Map of Father Ricci", *Geographical Journal* 52, 6 (1918), 367–385, and 53, 1 (1919), 19–30.

Tingyun 杨廷筠 (1562–1627) quickly followed with their *Account of Countries not listed in the Records Office* (*Zhifang waiji* 《职方外纪》), a treatise on world geography, which was later included in the 1628 *First Collection of Celestial Studies* compiled by Li Zhizao. Aleni's *Account* grew out of notes on Ricci's 1601 *mappa mundi* that Pantoja and Ursis had prepared for the Ming court.[36]

Aleni's translation represented the first detailed exposition in Chinese of a world geography that drew on Renaissance traditions of local lore and Ptolemaic cosmography to describe Asia, Europe, Africa, and the Americas. Another section focused on the oceans. In addition to the *First Collection of Celestial Studies*, other Chinese collectanea reproduced Aleni's treatise in the 18th century, thus making it more influential during the Qing dynasty than Ricci's maps, which were quickly forgotten by literati. In the 19th century, Aleni's work was reprinted three times in 60 years after Euro-American geography became de rigueur for literati trying to understand the military consequences of the Opium War.[37]

A world map composed by the Naples Jesuit Father Francesco Sanbiasi in 1648 was designed as simplified version of the large maps by Ricci. It was drawn as an oval projection with China at the center. Later, Michael Piotyr Boym, a Polish Jesuit who had served the embattled Southern Ming in the 1650s, produced the *Map of the Middle Kingdom* (*Zhongguo tu* 《中国图》), circa 1652, which he brought back to Rome in 1656. The original manuscript was compiled to convey Jesuit knowledge of China to Europe. This version was superseded by Martinus Martini's (1614–1661) *Novus Atlas Sinensis* published in an Amsterdam atlas series in 1655.[38]

Although Matteo Ricci introduced the system of longitude and latitude to Ming China, the grid system employed by Luo Hongxian and his

---

[36] Alexander Wylie, *Notes on Chinese Literature*, 58.

[37] See Bernard Hung-kay Luk, "A Study of Giulio Aleni's *Chih-fang wai-chi*", *Bulletin of Oriental and African Studies* 40 (1977), 58–84, and Eugenio Menegon, *Un solo Cielo: Giulio Aleni S.J. (1582–1649)*, 141–146. See also Richard Smith, *Chinese Maps: Images of "All Under Heaven"*, 49–54.

[38] *The Library of Philip Robinson*, Part II, 83, 85–86, 89, and 95. Martini had been in China from 1643 to 1646. See also Boleslaw Szczesniak, "The Seventeenth Century Maps of China: An Inquiry into the Compilations of European Cartographers", *Imago Mundi* 13 (1956), 116–136.

predecessors still exercised a dominant influence on Chinese cartography throughout the late Ming and Qing periods. For example, an admired genre known as "Complete Maps of All under Heaven" (*Tianxia quantu* 天下全图) was initiated by the Ming loyalist Huang Zongxi 黄宗羲 (1610–1695) in 1673 and continued by a number of talented literati scholars, interested in mapmaking. Cao Junyi's ambitious "Complete Map" of 1644, for instance, mixed inaccurate classical geographical lore with precise recognition of Europe, Africa, India, and Central Asia by providing longitudes and degrees for estimating their distances from Nanjing, the Southern Ming capital.[39]

In the early Qing, Ferdinand Verbiest (1623–1688), with the help of others, produced two works that dealt with world geography. His 1674 world atlas included a comprehensive map in two hemispheres with gazetteer information about each part of the globe. Later abridged, it was based largely on Aleni's *Account of Countries not listed in the Records Office*. Similarly, Verbiest's *Main Records about the West* (*Xifang yaoji* 《西方要记》) drew on the topical organization in Aleni's 1637 *Answers about the West* (*Xifang dawen* 《西方答问》), which compared China to the Europe in light of geographical lore. These works were considered important enough by the compilers of the Imperial Library during the Qianlong reign to be copied into the collection in the 1780s.[40]

The final stage in the development of traditional Chinese map-making came in the 18th century when French Jesuits on behalf of the Kangxi emperor conducted systematic surveys of the entire Manchu realm between 1708 and 1718. They drew up a series of maps of the Qing empire and its border areas, which became known as the 1718 *Kangxi Atlas* (*Huangyu quanlan tu* 《皇舆全览图》). Along with succeeding maps

---

[39]Richard Smith, "Mapping China's World: Cultural Cartography in Late Imperial Times", in Yeh Wen-hsin, ed., *Landscape, Culture and Power in Chinese Society*, 71–83, introduces the "Complete Maps" genre.

[40]See Bernard Hung-kay Luk, "A Study of Giulio Aleni's *Chih-fang wai-chi*", *Bulletin of Oriental and African Studies* 40 (1977), 76–77. See also Chen Minsun, "Ferdinand Verbiest and the Geographical Works by Jesuits in Chinese 1584–1674", 123–164, and Lin Tongyang "Ferdinand Verbiest's Contribution to Chinese Geography and Cartography", 135–164, both in John Witek, S.J., ed., *Ferdinand Verbiest, S.J. (1623–1688): Jesuit Missionary, Scientist, Engineer, and Diplomat*.

in the Yongzheng and Qianlong periods, the *Kangxi Atlas* surpassed earlier Jesuit surveys completed in the 17th century. Although the Manchu court restricted access to and reproduction of these surveys and maps, they remained the chief sources of foreign geographical information concerning China until the 20th century.[41]

Rather than switching to Jesuit cartography, late 18th-century literati such as Ma Junliang马俊良 sometimes presented both a traditional picture of a China-centered world and global maps that loosely duplicated Ricci's *mappa mundi*. Ma's *Capital Edition of a Complete Map Based on Astronomy* (*Jingban tianwen quantu*《京板天文全图》) circulated widely in the 1790s and included two smaller global maps of the Euro-Asian-African and Pacific hemispheres above his detailed map of the Qing empire that roughly corresponded to Ricci's *Map of the Myriad Countries on the Earth* rendered in a late Ming encyclopedia.[42]

Despite displaying substantial variations, a prominent feature of the "Complete Maps" was the mingling of traditional and newer techniques, such as the hybrid overlap between the grid and latitude-longitudinal approach for large-scale land maps. Indeed, Li Zhaoluo's 李兆洛 (1769–1841) famous 1832 atlas of the Qing empire showed both grid and latitude and longitude lines on the same map, indicating the reluctance of native geographers to give up the traditional system long after the Jesuit *mappa mundi* were introduced. This nativist pattern for domesticating European learning was repeated in other new fields of learning such as mathematical astronomy during the late 18th century.[43]

The Jesuits in Late Imperial China

Literati scholars and Chinese and Muslim specialists in the government's imperial calendrical bureau interpreted early modern Western achievements

---

[41] Walter Fuchs, "Der Jesuiten-Atlas der Kanghsi-Zeit", *Monumenta Serica*, Monograph Series 4 (1943), 60–75. See also James A. Millward, *Beyond the Pass: Economy, Ethnicity, and Empire in Qing Central Asia, 1759–1864*, 70–72.

[42] Richard Smith, "Mapping China's World: Cultural Cartography in Late Imperial Times", in Yeh Wen-hsin, ed., *Landscape, Culture and Power in Chinese Society*, 77–81.

[43] SCC, Vol. 3, 551–586. Maps based on latitudinal and longitudinal degrees were essential for ocean going vessels, which the Chinese did not prioritize for their coastal fleet.

in calendrical reform and natural studies in light of nativist traditions of scholarship. Derived from Christopher Clavius's (1538–1612) success in moving the Papacy and Western Christendom from the Julian to the Gregorian calendar in 1582, the technical prowess that some Jesuits, such as Matteo Ricci, had learned as a result of studying under Clavius proved fortuitous in Ming China. Ming concerns about the accuracy of the official calendar forced Chinese literati to evaluate and apply specific Jesuit techniques to reform the Ming calendar.[44]

This "local" research agenda among the court's calendrical specialists represented neither an indigenous modernization process nor the beginnings of a modest scientific revolution, at least by Western standards.[45] And in not searching for a Western form of modernity until the late 19th century, late imperial Chinese and Manchus were not acting out a purely anti-Western ideological agenda either, although at times court politics in Beijing interceded and the Jesuits as bearers of Western tidings were faced with the political animosities such "new" learning produced among those in power who were satisfied with the "old" learning.

Unfortunately, one of the most common generalizations scholars make today concerning the role of "science" (= "natural studies") in late imperial China is that, after about 1300, studies of astronomy and mathematics were in steady decline there until the arrival of Jesuit missionaries in the 16th century.[46] When Matteo Ricci described the scientific prowess of Chinese during the late Ming dynasty, he noted that they "have not only made considerable progress in moral philosophy but in astronomy and in many branches of mathematics as well. At one time they were quite proficient in arithmetic and geometry, but in the study and teaching of these branches of learning they labored with more or less confusion." Ricci concluded: "The study of mathematics and that of medicine are held in

---

[44] Peter Dear, "Jesuit Mathematical Science and the Reconstitution of Experience in the Early Seventeenth Century", *Studies in History and Philosophy of Science* 18 (1987): 135–141, and Willard Peterson, "Calendar Reform Prior to the Arrival of Missionaries at the Ming Court", *Ming Studies* 21 (1986): 45–61.

[45] But see Nathan Sivin, "Why the Scientific Revolution did not take place in China — or didn't it?" reprinted in Sivin, *Science in Ancient China: Researches and Reflections*, VII, 45–66.

[46] Keizō Hashimoto, *Hsu Kuang-ch'i and Astronomical Reform*, 17.

low esteem, because they are not fostered by honors as is the study of philosophy, to which students are attracted by the hope of the glory and the rewards attached to it."[47]

Chinese mathematics and astronomy, according to this view, had reached their pinnacle of success during the Song and Yuan dynasties but had declined precipitously during the Ming.[48] This longstanding perspective has been challenged by recent studies that indicate: (1) the Jesuits and their Chinese collaborators knew little of Chinese natural studies; (2) mathematics and calendar reform were important concerns among Ming literati before the arrival of the Jesuits in China.[49] Others have demonstrated that the Jesuits misrepresented their knowledge of contemporary European astronomy to suit their religious objectives during the late Ming and early Qing period. Such self-serving tactics, which produced contradictory information about new, Copernican trends in European astronomy, lessened their success in transmitting the European sciences to late Ming dynasty literati.[50] From this new perspective, late Ming scholars were not lifted out of their scientific "decline" by contact via the Jesuits with European astronomy. Rather, they themselves reevaluated their astronomical legacy and its current inadequacies, successfully taking into account pertinent features of the European sciences introduced by the Jesuits.[51]

Placing natural studies in China within its own internal and external contexts enables us to reconstruct its communities of interpretation.[52] Simply understanding Chinese views of the natural world in light of our

---

[47] *China in the Sixteenth Century: The Journals of Matteo Ricci: 1583–1610*, translated into Latin by Father Nichola Trigault and into English by Louis J. Gallagher, S.J., 31–33.

[48] For the conventional perspective, see SCC, Vol. 3, 173, 209, and Ho Peng Yoke, *Li, Qi, and Shu: An Introduction to Science and Civilization in China*, 169.

[49] See Roger Hart, *Imagined Civilizations: China, the West, and Their First Encounter*. See also Thatcher E. Deane, "The Chinese Imperial Astronomical Bureau: Form and Function of the Ming Dynasty 'Qintianjian' from 1365 to 1627".

[50] Nathan Sivin, "Copernicus in China", in *Colloquia Copernica II: Études sur l'audience de la théorie héliocentrique*, 63–114.

[51] Jacques Gernet, *China and the Christian Impact*, 15–24. See also Nathan Sivin's biography of "Wang Hsi-shan (1628–1682)", in Charles Gillispie, ed., *Dictionary of Scientific Biography*, Vol. 14, 159–168, and Thatcher E. Deane, "The Chinese Imperial Astronomical Bureau: Form and Function of the Ming Dynasty 'Qintianjian' from 1365 to 1627", 401–441.

[52] Compare Stanley J. Tambiah, *Magic, Science, Religion, and the Scope of Rationality*, 154.

own modern Western distinctions between science and non-science has not gotten us very far beyond the usual narratives of Western "success" in science and the concomitant Chinese "failure". The issue of "superiority" or "inferiority" of natural studies in late imperial China vis-à-vis early modern Europe should be "bracketed" as an unnecessary value judgment, because by today's standards both the pre-Newtonian, Aristotelian framework for natural studies (dominant in Europe to 1600) and the organicistic system of classical explanations (used in imperial China) have each been discredited.

Choosing between the "four elements" of the Greeks and Romans, which the Jesuits transmitted to Ming China, and the "five phases" of the Chinese, which the Jesuits discredited, is now a historical — not a cutting-edge scientific — problem.[53] Indeed, the exact premises for the European monopoly of modern science since 1700 (some would mistakenly say since 1500) must be unraveled before we can compare and contextualize the bicultural historical dialogue from 1550 to 1800 between late imperial Chinese elites (who in no way felt inferior) and early modern Europeans (many of whom recognized in learned Chinese literati a powerful intellectual tradition) concerning fathoming and measuring the natural world.[54]

Civil Examinations, Natural Studies, and Anomalies

Views that late imperial literati, unlike their Song and Yuan predecessors, were participants in a strictly humanist civilization, whose elite participants were trapped in a literary ideal that eschewed interest in the natural world, have been common since the Jesuits.[55] Historians have typically appealed for corroboration to the infamous Chinese civil examination

---

[53]See Zhang Qiong, "Demystifying *Qi*: The Politics of Cultural Translation and Interpretation in the Early Jesuit Mission to China", in Lydia Liu, ed., *Tokens of Exchange: The Problem of Translation and Interpretation in the Early Jesuit Mission to China.*

[54]Gyan Prakash, *Another Reason: Science and the Imagination of Modern India*, 17–120, describes how modern science in colonial India initially was a British monopoly that was translated intellectually and transmitted materially downward to India's social and political elites and later was indigenized in nationalist ideology.

[55]See Michael Adas, *Machines as the Measure of Men: Science, Technology, and Ideologies of Western Dominance*, 41–68, 79–95.

system. Matteo Ricci, for example, wrote: "The judges and the proctors of all examinations, whether they be in military science, in mathematics, or in medicine, and particularly so with examinations in philosophy, are always chosen from the senate of philosophy, nor is ever a military expert, a mathematician, or a medical doctor added to their number."[56] Catholic scholars were aware of the role played by political and social institutions in Chinese cultural matters, and the Jesuits realized that the civil recruitment system achieved for Chinese education a degree of standardization and importance unprecedented by early modern European standards.[57]

The examination ethos had carried over for a time into the domains of medicine, law, fiscal policy, and military affairs during the Northern Song dynasty. For example, Shen Kuo 沈括 (1031–1095) wrote that during the Huangyou reign 皇祐 (1049–1053) civil examination candidates were asked to comment on a *fu* 赋 (rhyme-prose) on astronomical instruments. (A similar *fu* was used in the 1679 special examination administered by the Kangxi emperor). The answers were so confused about the celestial sphere, and the examiners were themselves so ignorant of the subject, however, that all candidates were passed with distinction.[58] In addition, we have assumed that the classical curriculum for Ming civil examinations had refocused elite attention on a classical (*Daoxue* 道学, i.e., "Way Learning") orthodoxy stressing moral philosophy and literary values and away from earlier more specialized or technical studies. Conventional scholarship still contends that technical fields such as law, medicine, and

---

[56] *China in the Sixteenth Century: The Journals of Matteo Ricci: 1583–1610*, 41.

[57] Donald F. Lach, *China in the Eyes of Europe: The Sixteenth Century*, 780–783, 804. Compare Kiyosi Yabuuti, "Chinese Astronomy: Development and Limiting Factors", in Shigeru Nakayama and Nathan Sivin, eds., *Chinese Science: Explorations of an Ancient Tradition*, 98–99. See also George H. Dunne, S.J., *Generation of Giants: The Story of the Jesuits in China in the Last Decades of the Ming Dynasty*, 129–130, and Benjamin A. Elman and Alex Woodside, eds., *Education and Society in Late Imperial China, 1600–1900, passim.*

[58] See SCC, Vol. 3, 192. See also Robert Hartwell, "Historical-Analogism, Public Policy, and Social Science in Eleventh- and Twelfth-Century China", *American Historical Review* 76, 3 (June 1971): 690–727, and his "Financial Expertise, Examinations, and the Formulation of Economic Policy in Northern Song China", *Journal of Asian Studies* 30, 2 (1971): 281–314. On the 1679 examination, see my *A Cultural History of Civil Examinations in Late Imperial China*, 548.

mathematics, common in Tang and Song examinations, were not replicated in late imperial examinations.[59]

When faced with foreign rule (first under the Mongols, 1240–1368, and later under the Manchus, 1644–1911) significant numbers of literati, in addition to the usual number of candidates who failed, turned to occupations outside the civil service such as medicine. In the 18th and 19th centuries, when demographic pressure meant that even provincial and metropolitan examination graduates were not likely to receive official appointments, many literati turned to teaching and scholarship as alternative careers.[60] Moreover, examiners used policy questions on natural events and anomalies to gainsay the widespread penetration of popular religion and the mantic arts among examination candidates and to keep such beliefs out of politics.[61]

Careful scrutiny of Ming dynasty examination records reveals that civil examinations also tested the candidates' knowledge of astrology (*tianwen* 天文), calendrics (*lifa* 历法), and other aspects of the natural world, which were referred to as "natural studies" (*ziranxue* 自然学).[62] The preeminent position of the Four Books and Five Classics was left unchallenged in the

---

[59]See, however, Zhang Hongsheng, "Qingdai yiguan kaoshi ji tili"〈清代医官考试及体例〉, *Zhonghua yishi zazhi* 《中华医史杂志》 25, 2 (April 1995): 95–96, on Qing examinations to choose a limited number of medical officials, which were based on Ming precedents. See also Liang Jun 梁峻, *Zhongguo gudai yizheng shilue* 《中国古代医政史略》. Calendrical and cosmological questions were required in Ming examinations administered for candidates applying for positions in the Astronomical Bureau. See Thatcher E. Deane, "The Chinese Imperial Astronomical Bureau: Form and Function of the Ming Dynasty 'Qintianjian' from 1365 to 1627", 197–200.

[60]Robert Hymes, "Not Quite Gentlemen? Doctors in Song and Yuan," *Chinese Science* 7 (1986): 11–85, Jonathan Spence, *To Change China: Western Advisers in China, 1620–1960*, and Joseph Levenson, "The Amateur Ideal in Ming and Early Qing Society: Evidence from Painting", in John Fairbank, ed., *Chinese Thought & Institutions*, 320–341. See also Benjamin A. Elman, *From Philosophy to Philology: Social and Intellectual Aspects of Change in Late Imperial China*, 67–137.

[61]See Benjamin A. Elman, *A Cultural History of Civil Examinations in Late Imperial China*, 295–370.

[62]See Benjamin A. Elman Elman, *A Cultural History of Civil Examinations in Late Imperial China*, 461–485. Compare Nathan Sivin, "Introduction," in Sivin, ed., *Science & Technology in East Asia* xi–xv.

orthodox curriculum, but Ming candidates for both the provincial and metropolitan examinations, unlike their Song counterparts, were expected to grasp many of the technicalities in calendrics, astrology, anomalies (*zaiyi* 灾异) and the musical pitch series (*yuelü* 乐律). The latter was the basis for official weights and measures. Indeed, during the Tang, Song, and Yuan dynasties, works on calendrics and astrology had been banned from publication for security reasons. Only dynastic specialists working on the calendar in the Astronomy Bureau were allowed such knowledge, even though in practice popularly printed calendars and almanacs were widely available. Such restrictive policies continued outside the precincts of the Ming civil service examinations.[63]

Typical of the naturalistic outlook favored in written examinations, policy questions and answers were often pervaded by a distancing orientation to natural calamities and were clearly opposed to what were called "disjointed" (*buhe* 不和) interpretations of nature. Limitations in the human understanding of the natural operations of the cosmos, according to both examiners and candidates, had to be recognized. Otherwise, to impute human meaning and intent to calamities, as prognostication and the mantic arts presumed, was to humanize heaven and translate human knowledge into human fear and ignorance. Furthermore, the usual appeal to the personal agency of the sage-kings as men who confronted the events of their time and rectified them indicated that notions of fate that implied resignation in the face of calamities were unacceptable for orthodox literati operating in the public domain. What mattered was not the supernatural origins of floods or droughts but rather what concrete policies were followed to deal with them as natural events. Governance by men took precedence in a world in which the complete workings of heaven were beyond one's understanding.[64]

---

[63] See Lucille Chia, "*Mashaben*: Commercial Publishing in Jianyang, Song–Ming", in Paul Jakov Smith and Richard von Glahn, eds., *The Song–Yuan–Ming Transition in Chinese History*. See also Thatcher E. Deane, "The Chinese Imperial Astronomical Bureau: Form and Function of the Ming Dynasty 'Qintianjian' from 1365 to 1627", 259, 322–324, 399–400.

[64] See Benjamin A. Elman, *A Cultural History of Civil Examinations in Late Imperial China*, 346–360.

We see in the incomplete efforts by Ming literati to naturalize anomalies and supernatural events by appealing to the operation of yin and yang, the five phases, and *qi* 气, which carried over to the Qing, an interesting parallel to the early modern European transformation of supernatural miracles and monsters as prodigies into naturalized phenomena. The treatment of monsters and attitudes toward them in Europe evolved in the 16th and 17th centuries, and their religious significance was gradually diminished. Changing attitudes towards anomalies and monsters thus can serve as a barometer for different outlooks concerning natural and supernatural causes, which the Jesuits used in discussions with Ming literati about God and miracles to defend Christianity's superiority. In contrast to the monism of *qi* that most Chinese affirmed as psycho-physical reality, the Jesuits presented a numinous world of a personal God, angels, and the eternal souls of the dead.[65]

Originally treated by Europeans as divine prodigies contrary to nature, monsters became after 1500 more and more natural wonders. Subsequently in the 17th century, Francis Bacon (1561–1626) in his *Novum Organon* set the research agenda for including them in natural history. Prodigies as anomalies served Bacon as a means to explain the secrets of nature. Monsters eventually lost their autonomy as prodigies and were integrated by French specialists in the Parisian Académie des Sciences into the medical disciplines of comparative anatomy and embryology. Instead of prodigies, monsters became counter-examples to normal embryological development, and thus examples of medical pathology, in which nature's uniformity undergirded the exceptionalism of monsters.[66]

Similarly with regard to the treatment of miracles in Europe, Lorraine Daston describes how "the debate over the evidence *of* miracles became a

---

[65] See Zhang Qiong, "About God, Demons, and Miracles: The Jesuit Discourse on the Supernatural in Late Ming China", *Early Science and Medicine* 4, 1 (February 1999): 6–23, about Chinese demonology and Jesuit exorcism. For the Jesuit critique of *qi*, see Zhang Qiong, "Demystifying *Qi*: The Politics of Cultural Translation and Interpretation in the Early Jesuit Mission to China", in Lydia Liu, ed., *Tokens of Exchange: The Problem of Translation and Interpretation in the Early Jesuit Mission to China*.

[66] Katharine Park and Lorraine Daston, "Unnatural Conceptions: The Study of Monsters in Sixteenth-Century France and England", *Past and Present* 92 (August 1981): 20–54.

debate over the evidence *for* miracles". In moving from "signs" to "facts", miracles as preternatural (what rarely happens) or supernatural (God's unmediated actions) phenomena eventually lost much of their religious meaning. By focusing on the natural causes for deviating instances, Bacon and others made the anomaly and the miraculous central to scientific discourse. The goal in 17th-century naturalism was to explain whenever possible the natural causes of marvels and prodigies as parts of natural philosophy and not as pure miracles.

In addition, in those cases where miracles seemed a plausible explanation for an anomaly, the European intellectual community began to require reliable "evidence *of* miracles". Experts were now needed to distinguish fraud in sifting through "evidence *for* miracles". According to Daston, faith in the pure evidence of a miraculous event was short-lived when the evidentiary problem was to distinguish a true miracle from a false one. Ignorant enthusiasts or purposive charlatans rather than Satan could now be blamed for such claims, and even the Church began to distinguish divine signatures from forgeries, which now required a panel of Church leaders to sift through the evidence. In the process, miracles themselves lost their evidentiary power and their autonomy when the Council of Trent stiffened the requirements to distinguish between religion and superstition. The Church in effect strengthened the hands in Europe of political and religious authorities as arbiters of miracles.[67]

Just as the 16th and 17th centuries in Europe undermined the categories of the supernatural and miracles among Protestants and Catholics, so orthodox Chinese literati were also setting limits in official circles to the use of anomalies as political or religious signs. In Europe among the learned and powerful elites, however, the evidentiary requirements there, as Daston notes, implied far less confidence in the authenticity of supernatural events. Yet, both China and Europe saw any emotionally charged enthusiasm for the marvelous as a possible sign of deception and ignorance. For both, the dangers of false prodigies among the intellectually heterodox or politicized rabble-rousers were very real.

But among late imperial Chinese elites, and the Jesuits in China, the supernatural never diminished among them as much as among more

[67] Lorraine Daston, "Marvelous Facts and Miraculous Evidence in Early Modern Europe", *Critical Inquiry* 18 (1991): 93–124.

secular, Protestant elites in Northern Europe as a counterpart to the natural world. The 1558 civil examination demonstrates that the naturalization of unusual phenomena among Ming literati, as for the Catholic Jesuit missionaries still tied to a medieval vision of the natural world, remained incomplete.[68] By way of contrast, among Protestant men of science the boundary between the natural and the artificial dissolved into the notion of God as a great artisan, who could be mimicked by the great artist or man of science. The mechanization of nature was the result. Nevertheless, even in Protestant Europe the "non-natural" persisted as a legal category governing familial and sexual relationships when moral sanctions were used for the most outrageous and heinous crimes such as patricide.

The new mechanical philosophy interrupted the development of naturalism because this view of matter included an inert and barren view of matter, which was infused with the supernatural activity of God. Newton observed that stability of the solar system was maintained by divine activity in mechanical philosophy; it could not be accounted for by the laws of mechanics alone. The miraculous nature of gravity depended on theological considerations as an integral part of early Newtonianism.[69]

Divine Activity in Scholastic Philosophy thus continued into the Newtonian era. Indeed, Aristotle (384–322 B.C.) had excluded divine providence from his naturalism, and Acquinas (1225–1274) had added a notion of the supernatural process to resolve the conflict between Aristotle and Christianity. The Protestant reaction against naturalism led by Luther (1483–1546) and Calvin (1509–1564) had maintained the principle of immanent divine causation, while at the same time attacking the Scholastic distinction between the natural and supernatural. Calvin in particular rejected occultism and stressed that inert matter required the spiritual excitation of natural impulses. But Newton's theory was preferred because it brought God into continuous activity with creation, unlike Leibniz's naturalism, which was infused with godless monads.[70]

---

[68] On the Jesuits, see Zhang Qiong, "About God, Demons, and Miracles: The Jesuit Discourse on the Supernatural in Late Ming China", *Early Science and Medicine* 4, 1 (February 1999): 35–36.

[69] Keith Hutchison, "Supernaturalism and the Mechanical Philosophy", *History of Science* 21 (1983): 297–299.

[70] Ibid., 299–325.

The new naturalism evolved in association with mechanical philosophy. We cannot view the scientific revolution as a purely naturalistic movement. The legacy of the new mechanical philosophy yielded important parallels between natural and political philosophy, such that the new divine right of kingship became enmeshed in question of God's participation in running the world.[71] Finally, as Daston tells us, naturalization as an intellectual and cultural process does not explain the late 17th century shifts in the meanings of nature in Europe because the process was not a product of specific achievements of the scientific revolution. She maintains that the keys to the imposition of a rationalized order on natural concepts came from theology and jurisprudence, and that the authority of nature alone would not have sufficed to justify European political regimes in the 17th and early 18th century. The reinterpretation of nature, that is, the naturalization of anomalies and miracles, still required divine props — a physico-theology — for its authority. Until nature was universalized as neutral and amoral, portents and prodigies could not be reduced to non-issues.[72]

So too in China, as long as the literati vision of a rational and orderly cosmos within which the macrocosmic interactions of the heavens and earth informed the microcosmic patterns of differentiation and organization in the creation and evolution of all things in the world, the naturalization of anomalies in Ming and Qing times still required the cultivation of moral perfection of all learned scholars and officials. Only the sage could truly "investigate things and extend knowledge", just as in England only the early Fellows of the Royal Society possessed the circumstances, education, cultural heritage, and moral equipment of the early modern English gentleman to engage rightfully in the new practice of empirical and experimental science.[73]

---

[71] Keith Hutchison, "Supernaturalism and the Mechanical Philosophy", *History of Science* 21 (1983): 326–333.

[72] Lorraine Daston, "The Nature of Nature in Early Modern Europe", *Configurations* 6 (1998): 149–172.

[73] Steven Shapin, *A Social History of Truth: Civility and Science in Seventeenth-Century England*, 122–123.

Final Comments

Such evidence from the 18th century gainsays the usual conclusions that Western and Chinese scholars have drawn concerning the "failure" of the Macartney mission to open China to European trade and science in 1793. Joanna Waley-Cohen has reevaluated the Qing dynasty's so-called "blindness" to world developments in the 18th century and revealed how this erroneous assessment grew out of Western technological superiority after the 19th-century industrial revolution, which was then read back into the 1793 Macartney mission to China by later historians and diplomats. This misassessment in Western attitudes toward China, Waley-Cohen argues, was also due in part to the Qing court's need under the Qianlong emperor to reassert the "public Chinese attitude of superiority toward foreigners" in the factionalized court politics of 1793, even though the emperor at the same time avidly employed Jesuit experts in the arts of warfare for late 18th-century military campaigns against rebels within the empire. Moreover, the Qing court welcomed the advice of Jesuits in their midst concerning cannon-building and empire-wide cartography.[74]

The earlier "modernization narrative" that described British imperial expansion colliding with a "sinocentric" Qing state unsympathetic with the needs of scientific knowledge should be amended.[75] Furthermore, the Qianlong emperor's famous letter to George III gainsaying Western gadgets should not be read as the statement of a Manchu empire completely out of touch with historical reality. As Waley-Cohen also shows, the emperor's famous letter to the British king was not a categorical rejection of Western technology, which has become the standard interpretation. Coming before the industrial revolution, the scientific trinkets the Macartney mission brought to China were contested by the court.

When contextualized, the emperor's reaction to the Macartney mission can be understood in light of the mutual misunderstandings that swelled from the overstated claims Macartney made about the

---

[74] Waley-Cohen, "China and Western Technology in the Late Eighteenth Century", *American Historical Review* 98, 5 (1993): 1525–1544.

[75] See the still "Orientalist" account in Alain Peyrefitte's *The Immobile Empire*, translated by Jon Rothschild, which is based on his *L'Empire immobile ou Le Choc des mondes*.

pre-industrial revolution gadgets cum gifts — a replica of the solar system, for example — he had brought for the emperor (who did not think the planetarium so fabulous). Later emperors who found English military firepower irresistible in the 1839–1842 Opium War and thereafter were dealing with a different set of technological circumstances.[76] That literati scholars had incorporated mathematical study into evidential research and made natural studies a part of classical studies is another piece to the puzzle concerning the fate of natural studies and technology in late imperial China since the Jesuits first made their presence felt in the 17th century.

During the transition from the last dynasty to the Republic of China, new political, institutional, and cultural forms emerged that challenged the creedal system of the late empire and refracted the latter's cultural forms of knowledge, such as traditional Chinese medicine. Just as the emperor, his bureaucracy and literati cultural forms quickly became symbols of political and intellectual backwardness, so too traditional forms of knowledge about the natural world were uncritically labeled as "superstition" (*mixin* 迷信, "confused belief"), while "modern science" in its European and American forms was championed by new intellectuals as the path to objective knowledge, enlightenment, and national power. Even those who sought to maintain Chinese traditional medicine by modernizing it according to Western standards of rigor, however, also played a part in the denigration of past medical practices.[77]

The dismantling of the native traditions of natural studies, among many other categories, that had linked natural studies, natural history, and medicine to classical learning from 1370 to 1905 climaxed during the cultural and intellectual changes of the New Culture Movement. When their iconoclasm against classical learning and its traditions of natural studies climaxed after 1915, New Culture advocates helped replace the imperial tradition of *gezhixue* with modern science and medicine. This concluded a millennium of elite belief in literati values and 500 years of an empire-wide classical orthodoxy that had encompassed the Chinese natural studies and local technologies. Socially, classical credentials no

[76] James Hevia, *Cherishing Men From Afar: Qing Guest Ritual and the Macartney Embassy of 1793.*

[77] See Peter Buck, *American Science and Modern China, 1879–1936.*

longer confirmed gentry status or technical expertise, so sons of gentry turned to other avenues of learning and careers outside officialdom, particularly the sciences, modern medicine, and engineering. Literati increasingly traveled to Shanghai, Fuzhou, and other treaty ports to seek their fortunes in arsenals and shipyards as members of a new gentry-based post-imperial Chinese intelligentsia that would become the seeds for modern Chinese intellectuals.

Culturally, the longstanding affinity between literati learning and natural studies was also severed between 1905 and 1915. In other words, the linguistic monopoly of that official, classical knowledge by cultural elites no longer mattered as much socially or politically. As elites turned to Western studies and modern science, fewer remained to continue the traditions of classical learning that had been the basis for imperial orthodoxy and literati statuses before 1900. The millennial hierarchy of literati learning, based on the Four Books and Five Classics, study of the Dynastic Histories, mastery of poetry, and traditional natural studies was demolished in favor of modern science and its impact via Darwinism on social and historical studies.[78]

After 1911 a remarkable intellectual consensus emerged among Chinese and Western scholars that imperial China had failed to develop science before the Western impact. Even the Chinese protagonists involved in the 1923 "Debate on Science and Philosophy of Life" accepted the West as the repository of all scientific knowledge and only sought to complement such knowledge with moral and philosophical purpose (see Chapter 7).[79] The consensus until the 1960s then drew on heroic accounts of the rise of Western science to demonstrate that imperial China had no science worthy of the name. Both Western scholars and many — but not all — Westernized Chinese scholars and scientists so essentialized European natural studies into a universalist ideal that when Chinese

---

[78] James R. Pusey, *China and Charles Darwin*, *passim*. Benjamin A. Elman, "Towards a History of Modern Science in Republican China", in Benjamin A. Elman and Jing Tsu, eds., *Science and Technology in Modern China*: *1880s to 1940s*, 15–38.

[79] See Wang Hui, "From Debates on Culture to Debates on Knowledge: Zhang Junmai and the Differentiations of Cultural Modernity in 1920's China". See also Charlotte Furth, *Ting Wen-chiang: Science and China's New Culture*.

studies of the natural world, her rich medieval traditions of alchemy, or pre-Jesuit mathematical and astronomical achievements in China were discussed, they were usually treated dismissively and tagged with such epithets as "superstitious", "prescientific", or "irrational" to contrast them with the triumphant objectivity and rationality of the modern sciences. Because China had had no industrial revolution and had never produced capitalism, therefore the Chinese never produced science either. This "frame" of analysis carried over to India and Japan.

Chapter 4

The Jesuit Role as "Technical Experts" in "High Qing"

Earlier accounts have generally overvalued or undervalued the role of the Jesuits in Ming–Qing intellectual life. In many cases the Jesuits were less relevant in the ongoing changes occurring in literati learning. In the medical field, for example, before the 19th century few Qing literati-physicians (*ruyi* 儒医) took early modern European "Galenic" medicine seriously as a threat to native remedies. Instead, most were determined to pierce the veil of Song metaphysical and cosmological systems that since the Jin and Yuan periods had been inscribed in the theory and practice of traditional Chinese medicine. They sought to recapture the pristine meanings formulated in the medical classics of antiquity. They called into question the dominant framework of analysis, which Jin–Yuan–Ming physicians, following Song precedents, had enshrined as the theoretical norm in medical texts and casebooks (see further in Chapter 7).[1]

On the other hand, this chapter will show that the Kangxi revival of interest in mathematics was closely tied to the introduction of Jesuit algebra (*jiegen fang* 借根方), trigonometry (*sanjiao xue* 三角学), and logarithms (*duishu* 对数). As a result, Mei Juecheng 梅瑴成 (1681–1763) and others realized that they no longer had access to many of the works originally included in the medieval Ten Computational Classics (*Shibu suanjing* 十部算经). Moreover, in addition to the *Sea Mirror of Circular Measurement* (*Ceyuan haijing* 《测圆海镜》) of 1248, which was the oldest extant work on the "single unknown" (*tianyuan shu* 天元

[1]Asaf Goldschmidt, *The Evolution of Chinese Medicine: Song Dynasty, 960–1200*. See also Chu Pingyi, "Tongguan tianxue, yixue yu ruxue: Wang Honghan yu Ming-Qing zhi ji zhongxi yixue de jiaohui" <通贯天学、医学与儒学：王宏翰与明清之际中西医学的交会〉, *Lishi yuyan yanjiu suo jikan*《历史语言研究所集刊》70, 1 (1999): 165–201, for an example of the impact of European medicine in the 17th century.

术; also translated as "heavenly element notation") technique for algebra, the seminal works of Song–Yuan literati mathematicians on polynomial algebra (*siyuan shu* 四元术) were unavailable in Mei Wending's 梅文鼎 (1633–1721) time.

After the relatively "closed door" policies of the Yongzheng emperor and his successors, a large-scale effort to recover and collate the treasures of ancient Chinese mathematics were prioritized in the late 18th and early 19th century. In addition to famous scholars such as Dai Zhen 戴震 (1724–1777), Qian Daxin 钱大昕 (1728–1804), Ruan Yuan 阮元 (1764–1849), and Jiao Xun 焦循 (1763–1820) who stressed mathematics in their research, the editing of ancient mathematical texts and the continued digesting of European mathematical knowledge was carried out by a series of literati mathematicians who were also active in evidential studies (*kaozhengxue* 考证学). Through the recovery and collation of ancient mathematical texts, the alleged superiority of Jesuit mathematics was increasingly disparaged by Qing scholars who appealed to the "Chinese origins of Western learning" (*Xixue Zhongyuan* 西学中源) as a historical reality and not just a political tactic to justify calendrical reform.

Despite setbacks during the early 18th century Rites Controversy, the Jesuits in China remained important "experts" (*zhuanjia* 专家) in the Astro-calendric Bureau (*Qintian jian* 钦天监) and supervisors in the Qing dynasty's imperial workshops. The technical expertise of the Jesuits in the China mission during the 18th century also ranged from translating Western texts and maps, introducing surveying methods to producing cannon, pulley systems, sundials, telescopes, water pumps, musical instruments, clocks, and other mechanical devices. Their European enemies accused the Jesuits of making themselves useful to local rulers for their personal advantage rather than in the name of Christianity. Adam Schall (1592–1666) and Ferdinand Verbiest (1623–1688) not only championed the role of mathematics in Christianizing literati elites, but they also produced instruments and weapons at the behest of both the Ming and Qing dynasties.[2]

While in Beijing, for example, the French Jesuits Michel Benoist (1715–1774) and Jean-Joseph-Marie Amiot (1718–1793) interested

[2] Li Bin, "Xishi wuqi dui Qingchu zuozhan fangfa de yingxiang" 〈西式武器对清初作战方法的影响〉, *Ziran bianzheng fa tongxun*《自然辩证法通讯》24. 4 (2002): 45–53.

themselves in electricity. In 1755 they sent a report of their tests to St. Petersburg. Such experiments were kept secret. In 1773, Benoist demonstrated an air pump to the Qianlong emperor, who decided that the proper name for the device would be a "pipe to wait for the *qi*" (*houqi tong* 候气筒), a reference to the perennial efforts by the Astro-calendric Bureau to determine the onset of spring by measuring the *qi* emanating from the earth, which the Kangxi emperor had ridiculed early in his reign.[3]

In 1645, for instance, Schall had gone through the motions of submitting a report about the "waiting for the *qi*" (*houqi* 候气) procedures used by the Bureau to determine the onset of spring within a fortnightly period, whose precise timing was one of the Bureau's charges. "Waiting for the *qi*" was a technique for measuring the earth's emanations that had changed over time. It had switched from a method to establish the correct dimensions of the musical pitch pipes, whose ratios were used for weights and measures, to one measuring the onset of the fortnightly periods. The procedures associated with "waiting for the *qi*" involved burying 12 musical pitch pipes of graduated lengths in a sealed chamber and filling the pipes with ashes produced by burning the pith of a reed. In antiquity it had been believed that when the sun entered the second fortnight of any month, the earth's *qi* as a seminal force would rise and expel the ashes from the pipes.[4]

When Yang Guangxian 杨光先 (1597–1669) brought his famous 1664 suit against the Jesuits, their attitude toward "waiting for the *qi*" as a means to determine auspicious dates (*jixiong* 吉凶= "hemerology") became a focus of Yang's attacks. Yang accused Schall of relying on his own calculations and ignoring the old system to determine the onset of spring. When Schall replied that system had been in decline and was no longer used by his predecessors in the Ming Astro-calendric Bureau,

---

[3] J. L. Heilbron, *Electricity in the 17th and 18th Centuries: A Study of Early Modern Physics*, 121, 352, 405. See also Joseph Needham and others, *Science and Civilisation in China* (hereafter SCC), Vol. 3, 450–451. Compare Stephen Shapin and Simon Schaffer, *Leviathan and the Air-Pump: Hobbes, Boyle, and the Experimental Life*.

[4] Derk Bodde, "The Chinese Cosmic Magic known as Watching for the Ethers", in Soren Egerod and Elise Glahn, eds., *Studia Serica Bernhard Karlgren Dedicata, Sinological Studies Dedicated to Bernhard Karlgren on His Seventieth Birthday October Fifth 1959*, 14–35, argues that the practice was in disrepute by the 16th century.

however, the court noted that he had earlier ascertained the onset of spring by using the "waiting for the *qi*" method. The Kangxi regents were unhappy with Schall's pretences, because they meant that the timing of the onset of spring had never really been verified based on the old system.[5]

The political background for Yang Guangxian's anti-Christian attacks makes it clear that the role of the Astro-calendric Bureau in determining auspicious days had continued under the Jesuits, much to the dissatisfaction of other Catholic orders in China who were opposed to astrology and hoped to Christianize imperial "time". Schall had underestimated how the intertwining of astronomy, "waiting for the *qi*" traditions, and yin-yang numerology in telling fortunes and determining auspicious days might ensnare the Jesuit mission in a calendrical battle where predicting eclipses would not be the key issue. These claims suggested that the Jesuits were organizing converts to plot against the dynasty.

In late 1668, Verbiest, as Schall's successor in the Bureau, was able to make the case that "using reed pipes and flying ashes" was no way to measure the 24 fortnightly periods of the year. Verbiest argued that the onset of spring was determined by the intersection of the sun with the equatorial, a completely different orientation from Yang's focus on "waiting for the *qi*". The Kangxi emperor endorsed Verbiest's calculations, but the Jesuits still had to follow traditions associated with the cosmological and ritual aspects of the calendar. We will see that the Kangxi emperor also relied on Verbiest and the Jesuits to mediate between the Qing and Russian empires when their geographic borders expanded in the 1670s and 1680s.[6]

After 1670, the Jesuits depended almost exclusively on Manchu imperial patronage of their technical expertises and not on the support of

---

[5] Huang Yi-long, "Court Divination and Christianity in the K'ang-hsi Era", *Chinese Science* 10 (1991): 1–20, and Huang Yi-long and Chang Chih-ch'eng, "The Evolution and Decline of the Ancient Practice of Watching for the Ethers", *Chinese Science* 13 (1996): 82–85, 90–96, 104.

[6] Huang Yi-long and Chang Chih-ch'eng, "The Evolution and Decline of the Ancient Practice of Watching for the Ethers", *Chinese Science* 13 (1996): 95–97, and Chu Pingyi, "Scientific Dispute in the Imperial Court: The 1664 Calendar Case", *Chinese Science* 14 (1997): 7–34.

Chinese literati. The opposite had been the case in the late Ming when literati had been the order's stalwarts and sanctioned the quest for converts. Literati scholars who were leaders in the 18th-century turn toward evidential studies recognized the importance of European studies, particularly mathematics and astronomy, but remained suspicious of the Jesuits' motives in coming to China.[7]

After 1735, few classical scholars had any contact with the Jesuits who remained in China, even though many were still prominent in the Qianlong court as architects and glassmakers. Most intellectual issues addressed by Han Learning (*hanxue* 汉学) scholars, for instance, revolved around the complexities of the native classical heritage and its misrepresentations in accounts by followers of Song and Ming "Way Learning". The Jesuits were reduced from the intellectual equals of classically trained literati to foreign experts whose special skills entitled them to a lower status in the bureaucracy as official minions of the emperor.

Mensuration and Cartography in the 18th Century

In geography, for instance, Qing scholars during the late 17th and 18th century reacted to the Jesuit contributions in world geography by domesticating such new knowledge in the midst of Manchu empire building and the court's use of Jesuit surveying techniques to measure its domains. At the same time that the dynasty took advantage of international changes in Central Asia, literati took an internalist turn intellectually by focusing on native topics in their new geographic works.[8]

The late Qianlong compilers of the 1787 edition of the *Comprehensive Analysis of Civil and Military Institutions during the Qing Dynasty* (*Huangchao wenxian tongkao* 《皇朝文献通考》), which included documents and materials covering the period 1644–1785, demoted the mention of Europe to a minor section on Italy within the category of the "Four Frontiers" (*siyi* 四裔). Twenty-four chapters in the traditional

[7] Huang Yi-long and Chang Chih-ch'eng, "The Evolution and Decline of the Ancient Practice of Watching for the Ethers", 99.

[8] See Benjamin A. Elman, *From Philosophy to Philology: Social and Intellectual Aspects of Change in Late Imperial China*, 72–122.

category of "Geography and Lands" (*yudi* 舆地) dealt with imperial domains, while only eight covered the borderlands. The compilers of the "Four Frontiers" focused on Korea, the Ryukyu Islands, Vietnam, and other neighboring tributary countries.

The section on Italy mentioned that "what the Italians [i.e., Matteo Ricci] had said about the division of the world into five continents followed from Zou Yan's [邹衍] Warring States theory of the Sacred Ocean [*shenhai* 神海], although the Italians dared to add that the land of China was but one of the five continents [*wudazhou* 五大洲]". Such claims infuriated the compilers who dismissed the Italians as too grandiose. They could not be taken seriously because they were simply trying to impress Chinese with European customs, goods, governance, and education. By citing Zou Yan on the question of continents, the *Comprehensive Analysis* was drawing on a long tradition of classical interpretation that the ancient "nine regions" (*jiuzhou* 九州) were surrounded by oceans and thus the geographic term for "regions" (*zhou* 州) could also serve as a native term for "continent".[9]

Still, the *Comprehensive Analysis* included detailed geographic discussions of Europe and Russia. Moreover, the dramatic impact that European surveying methods had in China early in the 18th century continued to pique the interest of the Qianlong emperor and his literati in court when the Kangxi map of the empire (*tianxia quantu* 天下全图) was updated using European surveying methods. European geographical content may have been overlooked, but European methods were still admired and copied in geography as well as astronomy.[10]

Xue Fengzuo 薛凤祚 (d. 1680) studied astronomy and mathematics under the Jesuit Jean-Nicholas Smogolenski. Xue then applied the techniques of spherical trigonometry and logarithms to surveying, which was appreciated by the Qianlong Imperial Library editors in their review of Xue's compendium on the Yellow River and Grand Canal (*Lianghe qinghui*《两河清汇》). The editors noted that Xue's mathematical expertise was an invaluable aid in analyzing problems related to flood control and canal upkeep. His use of European trigonometry was recognized as a clear

---

[9] *Huangchao wenxian tongkao*《皇朝文献通考》, in *Shitong*《十通》(The ten comprehensive encyclopedias of civil and military governance), Vol. 2, 298.7469–7470.

[10] *Huangchao wenxian tongkao*, Vol. 2, 298.7469–7474, 298.7481–7489.

improvement over the native forms of trigonometry known as "double application of proportions" (*chongcha* 重差, i.e., properties of right triangles expressed as a function of angles), which had dominated Chinese surveying techniques until that time.[11]

Although the internal turn in Qing geographic research did not produce a cartographic reconceptualization of foreign lands, we should not underestimate Jesuit and European cartographic influence in China. Through systematic gathering of materials that they would then critically scrutinize and in some cases quantify, Qing scholars combined evidential research methods with data collection and organization. As research pushed forward in the 18th century, Asian and Chinese continental geography became a key discipline.[12]

Despite the internal turn of this research away from concern with maritime lands far from China, achievements in geographical knowledge during this period were still evident in military defense and historical and descriptive geography, particularly along the Qing borders with Russia, Zungharia, and Kashgaria in Siberia and Central Asia. In addition, the cultural construction of Mukden (Shenyang 沈阳) and its environs as the exclusive homeland of the Manchus was achieved in part through the mapping of the area under Jesuit direction. The mapping of Manchuria began circa 1690 in the aftermath of negotiations with the Russians to determine the boundaries of the Amur River in northeast Asia.

Such achievements lent themselves to the cumulation of geographical knowledge. To be sure, the Americas were still depicted as parts of the Asian land mass north of Great Wall in a Chinese world map circa 1743, which was based on Liang Zhou's 梁辀 1593 world map. Moreover, Phillipe Foret's account of the planning for the Manchu summer capital in Chengde and for other imperial sites demonstrates how European cartographic technologies could coexist with earlier Chinese geographic practices such as the geomancy informing cartography, landscape architecture, and urban arrangements. The geography of the Qing empire was

---

[11] *Siku quanshu zongmu* 《(四库全书总目)》, compiled by Ji Yun 纪昀 (1724–1805) *et al.*, hereafter SKQSZM, 69.22a–23a.

[12] See Benjamin A. Elman, "Geographical Research in the Ming-Ch'ing Period", *Monumenta Serica*, 35 (1981–1983): 1–18. See also Matthew W. Mosca, *From Frontier Policy to Foreign Policy: The Question of India and the Transformation of Geopolitics in Qing China*.

intertwined with Tibetan Buddhism, and this relationship was an essential component of Manchu expansion into Central Asia. That expansion, as we will see below, also required a substantial investment in state-of-the-art mapping techniques from Europe to delineate accurately the Russo-Chinese border in the 18th century.[13]

Cartography, Sino-Russian Relations, and Qing Imperial Interests

Early Manchu rulers recognized the need for better records for land use and taxation. In 1646, a cadastral survey was undertaken, but its geographic inadequacies were recognized. When the Russians appeared in force along the northern frontier in the 17th century, the Manchu court and Chinese officials required more accurate geopolitical information to deal with the latest threats to the empire. After the Tungusic chief Gantimur defected from the Qing in 1670 to allegedly Russian territory, the Russians quickly took advantage.

A crisis in Sino-Russian relations ensued from the 1670s to 1690s, when the Manchus learned that the Russians had already built a fortress in 1654 along the Amur River at Nerchinsk in Gantimur's native region. The Kangxi emperor refused any further trade or diplomatic relations with Russia until the deadlock was resolved. Meanwhile, the Russians and Zunghar Mongols both expanded their interests in the northwest while the Qing was preoccupied in the south and southwest during the Revolt of Three Feudatories from 1673 to 1681. Much like late Ming calendar reform, Qing recognition of its geographic needs preceded the European contributions to Chinese cartography. The changing borders threatened the Manchu homeland.[14]

---

[13] See Mark Elliot, "The Limits of Tartary: Manchuria in Imperial and National Geographies", *Journal of Asian Studies* 59, 3 (August 2000): 603–646, and Richard Smith, *Chinese Maps: Images of "All Under Heaven"*, 54–59. Compare Phillipe Foret, *Mapping Chengde: The Qing Landscape Enterprise*, Chapter 6.

[14] Peter Perdue, "Boundaries, Maps, and Movement: Chinese, Russian, and Mongolian Empires in Early Modern Central Eurasia", *The International History Review* 20, 2 (June 1998): 267–268, and Laura Hostetler, *Qing Colonial Enterprise: Ethnography and Cartography in Early Modern China*, 66–71. Compare Mark Elliot, "The Limits of Tartary: Manchuria in Imperial and National Geographies", 619–620.

During the Rites Controversy, the Manchu court was embroiled simultaneously in military threats from Zunghars and Russians along the borders of the empire in Central Asia, which introduced new elements into the storm over the Jesuits and their loyalty to the Qing dynasty. For example, when the Russian mission was allowed into Beijing in 1676 to negotiate trading agreements and population movements, Ferdinand Verbiest was involved. The lack of a clear boundary in the Amur River area and the ambiguous claims to sovereignty in the area later led to the treaty of Nerchinsk in 1689, negotiated by the French Jesuit Gerbillon, which demarcated the frontiers between the Qing and Russia.

Jesuits and Mapping the Qing Empire

The Jesuits and others were commissioned to provide the necessary data that would enable Qing leaders to stem the tide of Russian infiltration into Manchu and Mongolian homelands. The geographical knowledge that accrued during this time was an important addition to earlier information of foreign lands. In the process, the Kangxi court's awareness of the actual geographical divisions of the Sino-Russian frontier slowly came apace of their long-standing knowledge of and interest in Southeast Asia.

Verbiest, however, seems to have secretly provided Russian missions with maps and descriptions of the border region with Siberia, which included the locations of Manchu forces obtained from Russian deserters. Because Russian expansion in Siberia challenged Qing power, the Jesuits thereafter had limited access to the most sensitive frontier areas of the Qing empire when the survey for the *Kangxi Atlas* was carried out. Territorial claims and dynastic security compelled the Qing court to hire only those Jesuits who did not intend to return to their native lands. The dynasty avoided circulating such information too widely in and outside of China. By 1727, Qing knowledge of the region of Amuria was seen in light of the realities of Russian penetration into Siberia.[15]

---

[15] Mark Mancall, *Russia and China: Their Diplomatic Relations to 1728*, 98–109. See also Peter Perdue, "Boundaries, Maps, and Movement: Chinese, Russian, and Mongolian Empires in Early Modern Central Eurasia", 269–271, and Laura Hostetler, *Qing Colonial Enterprise: Ethnography and Cartography in Early Modern China*, 77–78. Compare the account in John Witek, ed., *Ferdinand Verbiest (1623–1688): Jesuit Missionary, Scientist, Engineer, and Diplomat*, passim.

The geographical ambitions of the empires of the Qing dynasty, Imperial Russia, and the Mongol Zunghars in Central Asia led to the redrawing of the frontier boundaries between Russia and the Qing in the region and the crushing of the autonomous state of Zungharia in 1760 by Qing armies. Peter Perdue has noted how 18th-century Central Asian borders were constructed through three stages: (1) military confrontation; (2) negotiated treaties; and (3) symbolic representation on maps. The Qing dynasty, Zungharia, and Russia each produced important new maps of unprecedented scale and accuracy as political and ideological weapons in their struggle for control of Central Asia.

When Russia and China defined their mutual borders in the treaties of Nerchinsk in 1689 and Kiakhta in 1727, new surveying techniques for cartography were applied by both to the newly defined borders. In addition, new classification systems and ethnographic atlases to control the movements of refugees, nomads, tribes, traders, soldiers, and other mobile groups across the borders were compiled. Both sides used tax and land registers, censuses, border patrols, passports, and visas to keep people from moving freely across the borders. Each also applied 17th-century European technical knowledge that was transmitted through the Jesuits to survey their new territories.[16]

The Kangxi emperor, like Louis XIV, mapped his entire empire out of strategic concern. The Jesuits produced their first survey of Beijing in 1700, which the emperor checked. Later, he asked for a survey of portions of the Great Wall in 1707. In 1710 further surveys along the Amur River helped mark the strategic bases on the border with Russia. The *Kangxi Atlas* attempted to systematize the Qing dynasty's knowledge of its imperial territories and rationalize its claims vis-à-vis the Zunghars and Russians.

The Manchu homelands were surveyed between 1709 and 1712 and a complete map of greater Mukden, i.e., "Manchuria" (*Manzhou* 满洲), was produced. The text and maps that were included in the 1733 edition

[16] See Peter Perdue, "Boundaries, Maps, and Movement: Chinese, Russian, and Mongolian Empires in Early Modern Central Eurasia", 263–286, and Laura Hostetler, "Qing Connections to the Early Modern World: Ethnography and Cartography in Eighteenth-Century China", 623–662. See also Mark Elliot, *The Manchu Way: The Eight Banners and Ethnic Identity in Late Imperial China*.

of the *Collected Statutes and Precedents of the Great Qing* (*Daqing huidian*《大清会典》) were concerned with military deployments and garrison towns. They were compiled under the auspices of the Ministry of Military Personnel (*Bingbu* 兵部). The maps that the Jesuits prepared for the Manchu homelands became the starting point for later Japanese and European maps of the region in the 18th and 19th century.[17]

The atlas and its subsequent Qianlong era revisions shared features that were consistent with contemporary European maps. They left out pictorial elements and drew on astronomical observations to calculate longitude and latitude based on a precise scale. Hostetler interprets such developments in light of Qing evidential studies and the change in research epistemologies that affected scholarly views of geography in the late 17th century, a time when government interests increasingly focused on internal military defense and historical and descriptive geography. In the original maps, however, "China" (Zhongguo 中国) and the "Qing empire" (Qingchao 清朝) were not coterminous. "China" was presented as one distinct part of the Qing empire, and the Manchus homelands were another. Two other versions of the map from the same surveys, however, were entirely in Chinese with no Manchu script, perhaps to avoid offending Han Chinese cultural sensibilities. These Chinese language maps elided the Manchu view that the maps included distinct administrative and cultural spheres, to which Zungharia and Tibet would later be added.[18]

French and Russian Imperial Cartography

France was a leader in cartographic activity under Louis XIV after he appointed Jean-Baptiste Colbert (1619–1683) as minister for home affairs.

[17] See Walter Fuchs, "Materialen zur Kartographie der Mandju-Zeit, part 1", *Monumenta Serica*, 1 (1935–1936): 395–396, and Fuchs, "Materialen zur Kartographie der Mandju-Zeit, part 2", *Monumenta Serica*, 3 (1938). Compare Mark Elliot, "The Limits of Tartary: Manchuria in Imperial and National Geographies", 621–632.

[18] Peter Perdue, "Boundaries, Maps, and Movement: Chinese, Russian, and Mongolian Empires in Early Modern Central Eurasia", 274–275, and Laura Hostetler, *Qing Colonial Enterprise: Ethnography and Cartography in Early Modern China*, 17–18, 76. Compare David Turnbull, "Cartography and Science in Early Modern Europe: Mapping and the Construction of Knowledge Spaces", *Imago Mundi*, 48 (1996).

Colbert made France a center for science and solidified that role in Europe by founding the Parisian Academy of Sciences in 1666. Louis XIV also promulgated topographical surveys for territories based on astronomical observations that were initiated under his chief of astronomy, the Italian Gian Domenico Cassini (1625–1712). While in the employ of Pope Clement IX, Cassini published a series of tables in 1668 based on the eclipses of Jupiter's moons. Colbert invited Cassini to France in 1669 to make astronomical observations crucial to improved navigation and mapping. Because Cassini communicated with the Jesuits in China, he sought such observations globally. Sent to China in 1685, the French Jesuit John Baptist de Fontenay (1643–1710) had been the only well-trained mathematician in his group. He was responsible for astronomical observations and communicated regularly with Cassini in the Paris Academy of Science until his return to Paris in 1703. Fontenay's return effectively marked the end of French Jesuit scientific work in China.[19]

Enlarged in 1676, Cassini's *Ephemerides* (*Rili* 《日历》) permitted astronomers to determine the latitude and longitude of the point from which they made their observations. In 1679, France began a national survey relying on Cassini's tables for accurate measurements. The French Academy required observations from around the globe, which in part led to Louis XIV sponsoring the French Jesuits in China under the Missions Etrangères in 1663. Once mapmaking became a vital component of imperial expansion, the cartographic technology to carry out accurate geodetic surveys spread quickly. France, Russia, and the Qing employed experts regardless of their origins. Colbert had recruited the Italian Cassini, Kangxi employed the French Jesuits, and Russia engaged the Swedish officer Strahlenberg, who was taken prisoner by Russians in 1711, to collect information about Siberia, Mongolia, and neighboring regions.

Peter the Great, like the Kangxi emperor, used maps to measure the growth of the Russian empire and to legitimate its claims. In 1698, Peter had already commissioned a survey of his new territories, and new maps for an atlas were completed in 1701, although full surveys of the empire

[19] See Roger Hahn, *The Anatomy of a Scientific Institution: The Paris Academy of Sciences, 1666–1803*, 66–67, 90, 96–97. Compare Florence C. Hsia, *Sojourners in a Strange Land: Jesuits and their Scientific Missions in Late Imperial China*.

were not formally initiated until 1727 by Catherine the Great. For France, the Qing, and Russia, the requirement of better maps was tied to imperial expansion. New surveying and mapmaking techniques were essential. In the midst of Russian expansion into Siberia, the Qing empire more than doubled in size from 1660 to 1760 in a global context of population growth and colonial exploration and expansion.[20]

The Kangxi surveys were quickly completed by 1717, while French surveys took until 1744 to accomplish. The Qianlong revisions of the 1717 survey were finished in 1755, while the second edition of the French survey appeared in 1788. Similarly, the Russian imperial atlas, which followed the French national survey, appeared in 1745. Peter the Great used cartography and his European experts, who were also hired to explore the North Pacific, to put Russia on the map of 18th-century Europe. The *Kangxi Atlas*, for instance, had decisively changed European mapmaking when the Jesuit maps first arrived in France in 1725 and the new information was digested in Paris, London, and elsewhere.

Similarly, the latest mapping technology was effective for the Qing in legitimating and consolidating the empire, and became the basis for China's modern territorial claims in the 20th century. The Kangxi emperor's gift of his survey to Peter in 1721 indicated a desire to apprise Russia of Qing sovereignty and cartographic sophistication. It did not record all the strategic information the Qing had about the northern border areas, however.[21]

More importantly, the Kangxi emperor sought peace with the Russians to free his hand in wars with the Zunghar Mongols in Central Asia. By neutralizing Russia, the Qing court prevented a possible Russo-Zunghar alliance against them. Hence, when the Russians demanded the principle of equality at the Nerchinsk peace negotiations, the Manchus did not allow ceremonial difficulties to interfere with their primary diplomatic task.

---

[20] Laura Hostetler, *Qing Colonial Enterprise: Ethnography and Cartography in Early Modern China*, 71–75. See also Peter Perdue, "Military Mobilization in Seventeenth and Eighteenth-Century China, Russia and Mongolia", *Modern Asian Studies*, 30, 4 (1996): 757–793.

[21] Laura Hostetler, *Qing Colonial Enterprise: Ethnography and Cartography in Early Modern China*, 74–79, and Mark Elliot, "The Limits of Tartary: Manchuria in Imperial and National Geographies", 626.

The emperor relented on the usual Qing ceremonial claim of imperial superiority when dealing with bordering states. Accordingly, the Treaty of Nerchinsk represented a compromise in which the marking out of the frontier was more favorable to the Manchus, while the Russians kept Nerchinsk. In addition, the Manchus conceded that trade could be initiated by either side, and each could cross the border with passports. Furthermore the problem of the repatriation of fugitives was settled.

As an instrument of diplomacy, the economic concessions made by the Qing government in the 1689 treaty proved their political worth when the leader of Zungharia, Galdan, proposed an alliance with the Russians in 1690. Joint military action against the Manchus was now impossible, however, because the Russians were bound by treaty with the Qing. The Kangxi emperor was left free to eliminate the Zunghar threat, which he did in 1696. Galdan's death in 1697 eliminated the Mongols as a potentially divisive third force in Central Asia.

In 1718, the Russians contemplated full normalization of Sino-Russian relations during Rites Controversy, which was damaging Jesuit and Catholic interests in China. Peter the Great, for example, expelled the Jesuits from Russia in 1719 and tried to install a Russian "bishop" in Beijing in 1722. Russian authorities unsuccessfully kept this effort secret from the Manchus and the Jesuit enemies of the Greek Orthodox Church in Beijing, but Qing suspicions prevented the appointment. Subsequently in 1728, the Zunghar threat against the Manchus in Turkestan and Tibet revived. Again, the Manchus eliminated the threat through the Zunghar wars in the 1750s, which were facilitated by the Treaty of Kiakhta that ended Russian interference. Since 1727, the Kiakhta treaty had established officially supervised trade in Amuria that stabilized the Russian-Qing frontier until the 19th century.

The Qianlong reign brought a complete victory over the Zunghars and the incorporation of Ili in the far northwest by the Qing in 1755. Manchu military victories led to Qing overconfidence vis-à-vis the Russians, which generated a ban on trade caravans to Beijing from Russia after 1755. Qing success in Central Asia in the 18th century thus occurred in the context of Russian expansion into Amuria. Through compromise and accommodation, Russian interests in trade and Manchu interests in Central Asia were negotiated. Diplomacy, warfare, and timely mapping of strategic frontiers enabled the Qing dynasty to incorporate major portions of Amuria, Zungharia, and Kashgaria at the expense of the Mongols, Uighurs, Kazaks, Tajiks, and Russians.[22]

The Jesuit Role in High Qing Arts, Instruments, and Technology

Before the Napoleonic Wars were resolved in Britain's favor, the Qing court and its literati continued to co-opt certain aspects of European learning and artistic expertise. For example, Jesuits remained as court clockmakers and geometers, and a Jesuit, usually Swiss, was always in charge of the imperial clock collection in the 18th century. When compared to the industrializing tendencies in Britain and France, the imperial workshops in Qing China, like the Academy of Mathematics under the Kangxi emperor, were never breakthroughs to European-style science and technology.

Such workshops were, however, evidence for the important role of Jesuits working with Chinese artisans in manufacturing clocks, porcelain, glass, and building pavilions and gardens during the late empire. Interestingly, when the Macartney mission presented clocks and watches to the court to demonstrate British inventiveness, the Qianlong emperor was not particularly impressed and rejected any official trade with Britain. His haughty imperial tone belied his extensive collecting and reproducing European manufactures and clockwork at home.[23]

Clock making in the Kangxi Era

The French and Swiss Jesuit's expertise in mathematics, astronomy, cartography, and clock making was connected to the unprecedented

---

[22] Mark Mancall, *Russia and China: Their Diplomatic Relations to 1728*, 149–159, 209–210. The translation of the treaty is on 280–283. See also Eric Widmer, *The Russian Ecclesiastical Mission in Peking during the Eighteenth Century*, 45–58, 174–178.

[23] Catherine Pagani, *"Eastern Magnificence & European Ingenuity": Clocks of Late Imperial China*, 39–57, 70–74. On earlier clockwork in China, see SCC, Vol. 4, 220–266. Compare Yulia Frumer, "Clocks and Time in Edo Japan" (Princeton University Ph.D. dissertation in the History of Science, 2012).

degree of precision in their time and space measurements. When the Kangxi emperor commissioned the Jesuits to survey the entire empire between 1708 and 1717, the required mapping was accomplished in collaboration with official Chinese astronomers and Manchu officials. Chinese clockmakers were trained because the Kangxi emperor also thought clock making skills were worth appropriating. Chinese who worked under Jesuits were expected eventually to make native clocks.[24]

According to Catherine Pagani, mechanical clocks in early modern Europe were a metaphor for God's maintenance of the universe. In an age when a mechanistic view of nature was emerging, Europeans believed that the mastery of clock mechanics would lead to better understanding of God's design of the world. The Jesuits accepted this view, but they also recognized that when mechanical clocks were presented to potentates at home and abroad this act of gift giving gained them access to high places in the papal court as well as in the Chinese empire. Their skills as architects and glass makers paralleled their careers as calendricists and clock makers in late imperial China.

Matteo Ricci (1552–1610) believed, for example, that he had gained admission to the Wanli court by presenting a clock and repeating watch to the Ming emperor, a tactic which followed established European gift-giving practices. When Johann Terrenz Schreck and Wang Zheng 王徵 prepared their 1627 translation entitled *Diagrams and Explanations of the Marvelous Devices of the Far West* (*Yuanxi qiqi tushuo lu zui*《远西奇器图说录最》), it represented the first work in Chinese that provided information about European escapement techniques for delivering power at regular intervals to move the train of wheelwork inside a timepiece. Subsequently, the Wanli emperor had his eunuchs work with the Jesuits to master the art of repairing the clocks he had been presented.

In the late 17th century, the Kangxi emperor established a number of workshops that manufactured luxury items under the auspices of the Office of Manufacture in the imperial household. These were modeled

[24] Catherine Jami, "Western Devices for Measuring Time and Space: Clocks and Euclidian Geometry in Late Ming and Ch'ing China", in Chun-chieh Huang and Erik Zürcher, eds., *Time and Space in Chinese Culture*, 169–200. See also Nicolas S.J. Standaert, *Handbook of Christianity in China, Vol. One: 635–1800*, 840–850.

after the French Academy. Mentioned in palace documents of 1689 and 1692, the workshops were at the outset staffed by Jesuits and workmen from Guangdong province. After the imperial workshops were formally established in 1693, there may have been as many as 31 shops. Most clocks in the 18th century were completed in shops inside the Forbidden City. An Office of Clock Manufacture was subsequently created in 1723, which lasted until 1879 when the last list of clocks in the court was compiled.[25]

Three major sites were operating in 1756 under the Office of Manufacture within the Imperial Household Department. There, clock-makers, painters, and engravers worked in the decorative arts using enamels and glass for the clocks. The Lofty Pavilion gardens (*Yuanming yuan* 圆明园) also served as venue for clock making. In 1752, for instance, Amiot mentioned that the Jesuits prepared a hemicyclical mechanical theater clock with three scenes, which was built for the empress dowager's 60th birthday and remained in the Lofty Pavilion. Manchu and Chinese interest in elaborate mechanical clockwork, as in Europe, clearly represented the linkage of imperial power and prestige.[26]

Five European horologists were working in Beijing in 1701 during the Kangxi reign. Under the Qianlong emperor, 11 Jesuits built clocks in the imperial workshops. By 1800, Chinese artisans were proficient enough to make mechanical clocks themselves, and the role of missionaries in the production process was limited. The end of the Jesuit order was also an important reason. Outside of Beijing, for instance, imperial clock making workshops were also established in Suzhou and Hangzhou.

The native industry for clocks served as an effective form of tribute among Qing officials. Such modest levels of industrial production by native artisans were monopolized by the court and its literati elites. Production in the provinces was controlled by Chinese merchant guilds. The popularity of mechanical watches among Chinese elites mirrored imperial taste in the 18th century. Mechanical clocks were mentioned in

---

[25] Catherine Pagani, "*Eastern Magnificence & European Ingenuity*": *Clocks of Late Imperial China*, 26–57, 181–184, and Craig Clunas, "Ming and Qing Ivories: Useful and Ornamental Pieces", in William Watson, ed., *Chinese Ivories from the Shang to the Qing*, 122.

[26] Catherine Pagani, *"Eastern Magnificence & European Ingenuity": Clocks of Late Imperial China*, 58–98.

novels such as *Dream of the Red Chamber* (*Honglou meng*《红楼梦》), where owning one reinforced the high status of the Jia 贾 family.[27]

Imperial Factories for Glassware

Despite the damage their influence suffered in the Manchu court after the Rites Controversy, The French Jesuits remained heavily involved in the glass workshop that the Kangxi emperor had established in 1696 as one of his palace workshops. The German Father Bernard-Kilian Stumpf (1655–1720) in particular worked with the French Jesuits to produce decorative glass under imperial auspices. To please the emperor the Jesuits arranged to have two glassworkers sent to Beijing in 1699 to work in the imperial workshop.[28]

The Beijing workshop was producing high-quality glassware from the beginning of the 18th century. Decorative snuff bottles were fabricated in a wide variety of colors and shapes. The emperor presented such glassware as gifts to high officials on his southern tour to the Yangzi delta in 1705, for instance. He also presented the papal legate Maillard de Tournon (1668–1710) with an enameled glass snuff bottle at the outset of what proved to be a disastrous series of meetings in 1706 to discuss the rites controversy. Matteo Ripa (1682–1746), an Italian Jesuit who arrived in China in 1710, noted that by 1715 the glass workshop consisted of several furnaces for glassmaking, which required a large number of skilled craftsmen under Stumpf's supervision. In 1719, Jean-Baptiste Gravereau (1690–1762), a French Jesuit expert in enamel, arrived in Beijing. The enameled glassware produced in the workshop was of high enough quality that in 1721 the Kangxi emperor sent the Pope two cases of enamel ware, in addition to 136 pieces of Beijing glass.[29]

---

[27] Ibid., 76–78, 91–93.

[28] Emily Curtis, "Plan of the Emperor's Glassworks", *Ars Asiatiques* (Paris) 56 (2001): 81–90.

[29] Peter Lam, "The Glasswork of the Qing Imperial Household Department", in *Elegance and Radiance: Grandeur in Qing Glass*, The Arthur K. F. Lee Collection, 46–47, and Yang Boda, "Qingdai boli gaishu" 〈清代玻璃概述〉, *Gugong bowuyuan yuankan*《故宫博物院院刊》(1983): 13–16. See also Yang Boda, "An Account of Qing Dynasty Glassmaking", in *Scientific Research in Early Chinese Glass*, 144.

After the Yongzheng emperor's anti-Jesuit policies took effect, the imperial workshop, like the Kangxi era Academy of Mathematics (*Suanxue guan* 算学馆) increasingly relied on native glassmaking talent, particularly after Stumpf died in 1720. The workshop also was encouraged to manufacture enamel colors independently of the Jesuits. In addition, enamel colors were sent to the imperial pottery kilns in Jingdezhen, where new, enamel-decorated porcelains appeared for the first time. The overglaze blue of the Kangxi era had been introduced to Jingdezhen in 1700, and under the Yongzheng reign a translucent pink enamel made from colloidal gold was used in porcelain. The opaque pink glaze with specks of metallic gold developed by the imperial workshop drew on a recipe for ruby glass in a clear blue matrix. The source for the ruby glass derived from an ancient Venetian formula, rediscovered and developed in Germany.[30]

Despite his animosity towards Christianity and the Jesuits, the Yongzheng emperor established a branch of the glass workshop at the workshops in the Lofty Pavilion gardens in the northern suburbs of Beijing, where since 1712–1713 astronomical and mathematical work had also been carried out. Production also continued at the Jesuit glass workshop into the Qianlong era, when two additional Jesuits joined the imperial glass workshops in 1740. Glass production reached its high point when the Jesuits became involved in constructing European-style palaces and gardens for the Lofty Pavilion gardens in the 1750s.

European-style glassware was produced for display in the elaborate buildings that the Qianlong emperor asked Giuseppe Castiglione (1688–1766) and others to design for him. Castiglione's designs for Gabriel-Leonard de Brossard's (1703–1758) carved glass and enamels were also prominently displayed in the palaces of the Forbidden City (*Gugong* 故宫). Chinese decorative themes were connected with European illustrative skills for shading and perspective to adorn glassware. By 1766, the Qianlong emperor had turned the Lofty Pavilion gardens into a treasure house of gardens, pavilions, paintings, glassware,

---

[30] Zhang Rong, "Imperial Glass of the Yongzheng Reign", in *Elegance and Radiance: Grandeur in Qing Glass, The Arthur K. F. Lee Collection*, 64. See also Rosemary Scott, "Eighteenth Century Overglaze Enamels: The Influence of Technological Development on Painting Style", in *Colloquies on Art & Archaeology in Asia*, 156–158.

porcelains, and furniture, whose artistic merits and technical prowess were of the highest standards.[31]

Jesuits and Garden Architecture

Due to the Qianlong emperor's patronage, the Jesuits found new life in the Qing cultural world of the 1750s through their expertise in painting, designing, and building the lavish European-style palaces and interiors that the emperor sought. Initially impressed with the grandeur of European-style fountains, the emperor asked Castiglione to draw up plans for such fountains in the Lofty Pavilion gardens. Castiglione in turn sought the help of Michel Benoist, who had arrived in China in 1744 and later presented the court with an accurate account of Copernican cosmology. Because Benoist was knowledgeable in mathematics and hydraulics, as many Jesuits were, he was able to present the emperor with a model fountain, which the court quickly authorized Castiglione to build alongside the Baroque-style palatial buildings.[32]

The garden designers and builders were entirely Jesuits, with Castiglione playing the most prominent role. They were well prepared for their task in Beijing because in contemporary Rome the popes had glorified themselves and the Church via an ambitious scale of building palaces, piazzas, and fountains throughout the old city, which had been supervised by the clergy. The mid-18th century was also a time when the impact of chinoiserie-style architecture, gardens, and lakes on European rulers and aristocrats led to the construction in Europe of Chinese-style rooms, gardens and pagodas, such as the 1762 garden and pavilion in Wales designed by the architect Sir William Chambers (1723–1796) for the Princess Dowager Augusta (1719–1772).[33]

---

[31] Zhang Rong, "Imperial Glass of the Yongzheng Reign", 63. See also Peter Lam, *Elegant Vessels for the Lofty Pavilion: The Zande Lou Gift of Porcelain with Studio Marks*, 33–36. Compare Emily Curtis, "Glass for the Qing Court: The Jesuit Workshop", paper presented at the colloquium on "Art Brokering for China: The Missionary Connection", sponsored by the UCLA Center for Chinese Studies in conjunction with the Southern California China Colloquium, May 4, 2002.

[32] See the special publication of *Yuanming yuan* 圆明园 (Lofty Pavilion), 1 (November 1981).

[33] Richard Rudolph, "Early China and the West: Fertilization and Fetalization", in Rudolph and Schuyler Cammann, *China and the West: Culture and Commerce*, 4–5.

When the first pavilion was completed in 1747, the Qianlong emperor indicated his pleasure, according to Benoist. Located on 65 acres, the European section (*Xiyang lou* 西洋楼) of the Lofty Pavilion gardens was 750 meters long and 70 meters wide when completed and stood at the northern end of the Long Spring Garden (*Changchun yuan* 长春园) inside the Lofty Pavilion's massive grounds. The second phase of the European section inside the Lofty Pavilion took another eight years to complete.[34]

A three-story palatial building designed by the Italian Jesuit architect Ferinando Moggi (1684–1761) was known as the Belvedere (*Fangwai guan* 方外观) when completed in 1759. Built in a crescent shape evoking Islam, it had marble balustrades enclosed by a moat. After entering from the bronze outdoors stairway that rose up to the second floor, one encountered two tall stone tablets with Arabic inscriptions. The front door resembled that at St. Andrea Al Quirinale in Rome and faced a marble bridge with flamboyant balustrades across from a moat that lead to a smaller garden.

To the east a 36-room pavilion was named the Sea Calming Hall (*Haiyantang* 海宴堂) when completed in 1781. The largest structure built in the European section, it was likely modeled on Court of Honor at Versailles. It housed a reservoir with goldfish under a glass ceiling. A room on each side contained hydraulic machines to pump the water in the reservoir and feed the outside fountains and cascades. Nearby stood a triple gateway each with triumphal arches that resembled the Triumphal Arch in Paris.[35]

The pavilions, fountains, and arches designed by the Jesuits comprised about one-fifth of the Lofty Pavilion. For such a large-scale project, they were required to supervise many Chinese artisans, builders, and masons, whose architectural talents were indispensable for constructing the European section using Chinese materials and tools. Hence, the

---

[34] See Wong Young-tsu, *A Paradise Lost: The Imperial Garden Yuanming Yuan*, 59–65, and the drawings included in Régine Thiriez, "The Qianlong Emperor's European Palaces", in *The Delights of Harmony: The European Palaces of the Yuanmingyuan & the Jesuits at the 18th Century Court of Beijing.*

[35] See Victoria Siu, "Castiglione and the Yuanming Yuan Collections", *Orientations* (Hong Kong) 19 (November 1988): 72–79.

Baroque models that the Jesuits designed melded with Renaissance, French, and Rococo styles that the imperial laborers also reproduced. Elements of Chinese garden architecture, such as the inclusion of "Lake Tai stones" (*Taihu shi* 太湖石) and bamboo pavilions, were added to the mix of buildings in the European section, which included roof tiles in native yellow, green, and blue colors.[36]

Overall, the Lofty Pavilion represented a syncretism of architectural styles and tastes that the Manchu court favored as part of their efforts to create a universal vision of their power first in Asia and then in the world. The most representative building of this syncretic style was the small Belvedere, which combined the design of a Florentine building with a classic double Chinese roof. The interior designs for the pavilions were also inspired by European models. Glass windows, wood plank floors, handrails, flower terraces overlooking lawns, mechanical clocks, hanging lamps, and oil paintings. Even Gobelins tapestries of French beauties presented by Louis XV were placed throughout the European section. Large nude figures were not included, however. Chinese garden elements outside and traditional scroll paintings inside added to an ambience of Manchu pretensions as global style setters.[37]

Final Comments

The leaders of the 1793 Macartney mission defined their historical role vis-à-vis their Jesuit predecessors by presenting Great Britain, and thus themselves, to the Manchu court and Chinese literati as the mechanical leaders of Europe and as enthusiastic teachers of their new scientific knowledge. Lord Macartney believed that the gifts that he had brought — chief among them a solar planetarium synchronized by "the most ingenious mechanism that had ever been constructed in Europe" — were more sophisticated than the armillary spheres, mechanical clocks, and

[36] Wong Young-tsu, *A Paradise Lost: The Imperial Garden Yuanming Yuan*, 59.

[37] See George Loehr, "The Sinicization of Missionary Artists and Their Works at the Manchu Court during the Eighteenth Century", *Cahiers D'histoire Mondiale* 8 (1963): 795–803. Compare Phillipe Foret, *Mapping Chengde*, 15–23. See also Pamela Kyle Crossley, *A Translucent Mirror: History and Identity in Qing Imperial Ideology*.

telescopes that had previously been introduced by the Jesuits. He also thought that such gifts would convince the Qianlong emperor of Britain's dominance in science and technology.[38]

Furthermore, when members of the Macartney mission visited the imperial glass-making workshop in 1793–1794, one of them noted that the site was now neglected. After Brossard's death in 1758, production at the Jesuit workshop had declined. Subsequently in 1827 all missionary property was confiscated, and in 1860 the Lofty Pavilion gardens were destroyed by Lord Elgin's troops, who had taken Beijing during the Second Opium War (1856–1860) and forced the imperial court to flee to their summer retreat in Chengde. The planetarium Macartney delivered had attracted some interest, but his observation of Chinese "ignorance" misleadingly convinced him that diplomatic success would naturally follow once British cultural superiority and scientific expertise were transmitted to the emperor.[39]

The Macartney mission may have been a harbinger of things to come, but neither Lord Macartney nor the Qianlong emperor could foresee what the industrial revolution in England would yield in terms of British military superiority and Protestant evangelism in 19th-century China. We should not read the events of the first and second Opium Wars back into the 18th century. Qianlong's arrogance encoded in the emperor's infamous 1793 edict to King George III and British imperialism were inseparable. Qing China did not yet require "ingenious objects" or "manufactures" from England anymore than India did before Manchester mills dislodged the Indian cottage industry. Indeed, the Jesuits had already provided the court and literati with ingenious clocks, fountains, palaces, and glass works.

Moreover, the Chinese were masters of porcelain while in the late 18th century Europeans avidly sought the manufacturing secret for making it but were unsuccessful until the late Qing. Englishmen such as Francis Bacon (1561–1626) had at one time thought that porcelain developed

---

[38] John Barrow, *Travels in China*, 110.

[39] George Macartney, *An Embassy to China*, 299. See also James Hevia, "Looting Beijing: 1860, 1900", in Lydia Liu, ed., *Tokens of Exchange: The Problem of Translation in Global Circulations*, 192–199.

from an "artificial cement" buried in the earth for a long time, not unlike Chinese views of the formation of amber (*hupo* 琥珀). John Donne (1572–1631) considered buried clay the source for porcelain. Song Yingxing's 宋应星 (1587–1666?) 1637 *Heavenly Crafts for Opening Things up to Practical Use* (*Tiangong kaiwu*《天工开物》) had included a section on pottery-making, but because the work described procedures for government monopolies the work was not available publicly, except for portions reproduced in the early 18th-century *Synthesis of Books and Illustrations Past and Present* (*Gujing tushu jicheng*《古今图书集成》) encyclopedia.[40]

The international trade in porcelain from the Jingdezhen pottery kilns, for example, was directed by the tens of thousands of pieces through Guangzhou to Southeast Asia and Europe. In early modern Europe, Chinese porcelain was called "china". Queen Mary II of Britain, for instance, had a porcelain room designed by the Dutch decorator Daniel Marot (1661–1752). Another such room was restored in 1688 in the Oranienburg Palace by Frederick III, elector of Brandenburg and Prussia, which consisted of a salon decorated with cabinets and pyramids of Chinese porcelain. The king of Saxony, Augustus the Strong, had over 12,000 blanc de Chine wares made in Dehua, Fujian. White-glazed porcelain was more fragile due to the crispy clay used in Fujian. Until their popularity waned in the 18th century, blanc de Chine wares were also collected at the English royal palace at Hampton Court and showed up in the 1688 catalog of holdings in the Cecil family's Burghley House. In the 18th century, British aristocratic wives adored porcelain for their teatime in the gardens to the chagrin of their profit-seeking husbands.[41]

Aeneas Anderson, who accompanied Lord Macartney on his 1793 mission, wrote: "There are no porcelain shops in the entire world which can compare in size, richness or delivery with those in Canton." As imperial patronage lessened in the 19th century, however, the porcelain

---

[40] See Lydia Liu, "Robinson Crusoe's Earthenware Pot", *Critical Inquiry* 25 (Summer 1999): 749–750. See also Arthur Hummel, ed., *Eminent Chinese of the Ch'ing Period*, 691.

[41] See John Ayers and Rose Kerr, *Blanc De Chine: Porcelain from Dehua*, passim. Compare David Porter, "Sinicizing Early Modernity: The Imperatives of Historical Cosmopolitanism", *Eighteenth-Century Studies* 43, 3 (Spring 2010): 299–306.

industry declined significantly, and Wedgwood in England and Japanese potters replaced Jingdezhen as the best-known pottery kilns in the world. The post-industrial material needs of China would follow on the heels of an illegal opium trade that artificially created a low cost commodity grown in a British colony, to which millions of Chinese became addicted.[42]

[42] Michel Beurdeley and Guy Raindre, *Qing Porcelain*, 35–37, 143–144, 192–210. See also Ssu-yü Teng and John Fairbank, eds., *China's Response to the West: A Documentary Survey, 1839–1923*, 19–20. Compare James Hevia, *Cherishing Men From Afar: Qing Guest Ritual and the Macartney Embassy of 1793*, 225–248, and Lydia Liu, "Robinson Crusoe's Earthenware Pot", 749–750.

Chapter 5

Western Learning and Evidential Research in the 18th Century

In the 18th century, classical scholars shared a simultaneous passion for antiquity and new forms of scholarship. Scholars re-appropriated the mathematical classics and early astronomy in the millennial quest for ancient wisdom. In a post-Jesuit world, the Qing court during the Qianlong era was fortuitously buffered from contemporary European wars and the revolutionary changes then preoccupying Britain and France. In this geopolitical vacuum, Qing literati sought to compare what they knew of European learning, brought principally by the Jesuits, with native learning. Though the priority was on the latter, the restoration of ancient learning allowed Manchus and Chinese to bring under control early modern European contributions in mathematics and astronomy.[1]

The Jesuits in China had devised a unique accommodation approach to gain the trust of the Qing court and its gentry elites, which they rarely employed in Persia, India, Japan, or Southeast Asia, not to mention the "New World". Matteo Ricci (1552–1610) and his immediate followers prioritized natural studies and mathematical astronomy during the late Ming and early Qing because they recognized that Chinese literati and Ming and Qing emperors were interested in such fields. Such literati interests in natural studies and "Western learning" continued in the 18th century despite the impact of the Rites Controversy that peaked between 1715 and 1721. Hence, the account here challenges the usual image of Chinese lack of curiosity concerning early modern European science (see also the discussion in Chapter 3).

[1] Compare Pingyi Chu, "Remembering Our Grand Tradition: The Historical Memory of the Scientific Exchanges between China and Europe", *Historia Scientiarum* 41 (2003): 193–215, and Minghui Hu, "Provenance in Contest: Searching for the Origins of Jesuit Astronomy in Early Qing China, 1664–1705", *The International History Review* 24, 1 (March 2002): 1–36.

The reverse of current claims about Chinese disinterest in European science is the parallel assertion that Christianity and science had only marginal influence on Chinese literati before the 19th century. Many still emphasize the requirement to understand, first and foremost, the key, internal issues inscribed in the classical debates of Ming–Qing scholars. In the round, this claim has many merits. This approach overlooks, however, parallel events in European and Chinese intellectual and social history that suggest that literati interests in European science were cut short not by Chinese disinterest but instead by the failure of the Jesuit mission to act as a reliable conduit of scientific and mathematical knowledge during and after the Kangxi reign. The Jesuits did not transmit "modern science" to China.[2]

The Chinese "lack of knowledge" about 18th-century scientific developments in Europe, notably Newtonian mechanics and continental calculus, represented a failure of scientific transmission — not a failure of curiosity — that can be tied directly to the demise of the Jesuits and their schools in Europe during the 18th century, which vicariously affected Chinese information about new trends there. Michel Benoist (1715–1774), for example, finally introduced an accurate account of Copernican cosmology in China only after Church's ban on Copernican astronomy ended in 1757. Anti-Jesuit polemics generated first by the Jansenists and later by the Enlightenment *philosophes*, however, led to suppression of the order, first in Portugal in 1759 and then by France, Spain, Naples, and Parma, before the Pope dissolved the order worldwide in 1773. China's "window on Europe" was shattered by forces internal to both European and Chinese history. It was not due to their lack of curiosity!

The Academy of Mathematics in Beijing

When the French Jesuits arrived in China after 1689, they successfully created a legitimate place for themselves in the direct service of the Qing ruler. In fact, they equated the Kangxi emperor with their own "Sun King", Louis XIV. In addition to their missionary work, they hoped to

[2]See Benjamin A. Elman, "Jesuit Scientia and Natural Studies in Late Imperial China", *Early Modern History: Contacts, Comparisons, Contrasts* 6, 3 (Fall 2002): 1–24.

introduce contemporary French science in China. For instance, the French mission's first superior, Jean de Fontaney (1643–1710), held the chair of mathematics at the Paris Jesuit college Collège Louis le Grand between 1676 and 1685 before leaving for China. Joachim Bouvet (1656–1730) hoped that the Kangxi emperor would establish his own Academy of Science that would emulate the Academy of Sciences in Paris.[3]

The "Studio for the Cultivation of Youth" (*Mengyang zhai* 蒙养斋) was established in the suburban Lofty Pavilion (*Yuanming yuan* 圆明园) imperial garden in 1712–1713 for astronomical and mathematical work inside the court. The Kangxi emperor recognized the need to continue to employ French Jesuits on the calendar despite his dissatisfaction with Rome's papal policies toward China after the Rites Controversy. He invited the French Jesuits to work for him as they worked for the French Academy while abroad.[4]

The Kangxi emperor also molded his own court's Academy of Mathematics (*Suanxue guan* 算学馆) on the model of the Parisian Academy of Sciences, but it was strategically named after the Tang dynasty school for mathematics. The Academy was established in 1713 in the "Studio for the Cultivation of Youth" for calendrical work, but only Qing literati and bannermen were appointed. No Jesuits were allowed in this inner coterie of imperial scholars, which included the third prince, Yinzhi 胤祉 (1677–1732). This post-Rites Controversy policy ensured that the Jesuits would not be unduly influential in court mathematics.[5]

The Kangxi court sought to escape the dynasty's reliance on the Jesuits in calendrical matters. Li Guangdi's group included Wang Lansheng 王兰生, who was granted the highest civil service degree in 1721 by the emperor because of his mathematical abilities and called a "palace graduate in mathematical astronomy" (*chouren jinshi* 畴人进士).

---

[3]See Han Qi, "Bai Jin de Yijing yanjiu he Kangxi shidai de xixue zhongyuanshuo" 〈白晋的易经研究和康熙时代的西学中源说〉, *Hanxue yanjiu*《汉学研究》(Taiwan) 16, 1 (1998): 185–200.

[4]See Horng Wann-sheng, "Li Shan-lan: The Impact of Western Mathematics in China during the Late 19th Century", 16–17.

[5]See Catherine Jami, "From Louis XIV's Court to K'ang-hsi's Court: An Institutional Analysis of the French Jesuit Mission to China (1688–1722)", in Hashimoto Keizō *et al.*, eds., *East Asian Science: Tradition and Beyond*, 493–499.

Wang then entered the "Studio for the Cultivation of Youth" where the French Jesuits helped him to translate the works included in the *Sources of Musical Harmonics and Mathematical Astronomy* (*Lüli yuanyuan* 《律历渊源》) collectanea, which Mei Juecheng 梅瑴成 (1681–1763) and Chen Houyao 陈厚耀 (1648–1722) helped on.

In 1712, Chen had proposed a new compendium of European mathematics to replace the late Ming *Calendrical Studies of the Chongzhen Reign* (*Chongzhen lishu* 《崇祯历书》) inspired by the Jesuits. The result was the *Sources of Musical Harmonics and Mathematical Astronomy*, which included the *Collected Basic Principles of Mathematics* (*Shuli jingyun* 《数理精蕴》). In 1713, the Kangxi emperor charged Mei Juecheng and Chen Houyao with supervising He Guozong 何国宗 (1712 *jinshi*), Minggatu (Ming Antu 明安图, 1692–1763), and others to complete the project. The *Sources* was printed in 1723. This special group of mathematical and calendrical specialists included Wei Tingzhen 魏廷珍 (1669–1756) and others whom Mei Wending 梅文鼎 (1633–1721) had trained before he died.[6]

The emperor recruited more than 100 promising scholars to join the Academy of Mathematics regardless of their civil examination status. Mei Juecheng was made chief and Minggatu assistant editor for preparation of the *Collected Basic Principles of Mathematics*. In addition to those in the Academy of Mathematics who studied mathematics, astronomy, and music, a large number of instrument makers were also hired for the technical needs of the new academy. A team of 15 calculators verified the computations based on the theoretical notions, mathematical techniques and applications, and numerical tables in the first part of the *Collected Basic Principles*.

Patterned after mathematical textbooks used in Jesuit colleges, the *Collected Basic Principles* introduced European algebra, while the last part had a section on logarithms to the base ten, which drew on European methods to compute decimal logarithms. The Chinese mathematics that informed the *Collected Basic Principles* included traditional equation

[6]See Li Yan and Du Shiran, *Chinese Mathematics: A Concise History*, translated by John Crossley and Anthony Lun, 218, Jean-Claude Martzloff, *A History of Chinese Mathematics*, translated by Stephen Wilson, 218–219, and Catherine Jami, "Learning Mathematical Sciences during the Early and Mid-Ch'ing", in Benjamin A. Elman and Alexander Woodside, eds., *Education and Society in Late Imperial China, 1600–1900*, 231, 238–240.

methods (*fangcheng* 方程) and techniques for computing the sides of a right-angled triangle (*gougu* 勾股), which were based on Mei Wending's reinterpretation of traditional techniques to solve simultaneous linear equations.

A successor to the late Ming *Calendrical Studies of the Chongzhen Reign*, this new and influential collectanea included: (1) the *Compendium of Observational and Computational Astronomy* (*Lixiang kaocheng* 《历象考成》); (2) the *Collected Basic Principles of Mathematics*; and (3) the *Exact Meaning of the Pitch-pipes* (*Lülü zhengyi* 《律吕正义》), which were all compiled in the "Studio for the Cultivation of Youth" starting in 1712. The collection was intended as a series of textbooks for the "Studio" and for students in the Imperial College's (*Guozi jian* 国子监) own Academy of Mathematics.

After the *Collected Basic Principles of Mathematics* was printed in 1723, no other European mathematical works were introduced into China until after the Opium War (1839–1842). Notably missing in China was the European discovery of the more dynamic differential and integral calculus by both Leibniz (1646–1716) and Newton (1643–1727), which had exceeded the static limits of Greek geometry and Islamic algebra. Moreover, the version of Euclid's (ca. 325–265 B.C.) *Elements of Geometry* in the *Collected Basic Principles* remained the official version until 1865.[7]

The Kangxi Era Compendium of Observational and Computational Astronomy

Kangxi era efforts at reform culminated in the 1724 promulgation of the *Compendium of Observational and Computational Astronomy* and a sequel. The European astronomy in the *Compendium* was mostly a century old, but the sequel of 1742 (*Lixiang kaocheng houbian* 《历象考成后编》) adapted more recent European discoveries, such as Kepler's elliptic orbits, to the ends of traditional calendar reform.[8]

---

[7]Catherine Jami, "Western Influence and Chinese Tradition in an Eighteenth-Century Chinese Mathematical Work", *Historia Mathematica* 15 (1988): 311–331.

[8]Nathan Sivin, "Copernicus in China", in *Colloquia Copernica II. Etudes sur l'audience de la theorie heliocentrique*, 63–75, 89–92.

Overall, the Qing experts appointed by the Kangxi emperor followed Mei Wending's lead in rejecting Jesuit efforts to insinuate Christianity into their astronomy. Mei's mathematical work fit in with the court's efforts to have a calendar that would fuse European and Chinese techniques into a greater system. When the *Compendium of Observational and Computational Astronomy* was drafted in 1722 and promulgated in 1724, for instance, it followed European models, but it was prepared by Chinese in the court with only indirect Jesuit input.[9]

The early Qing calendars produced by the Jesuits were based exclusively on European models, but the new system cobbled together in the *Compendium* fused European with "Chinese methods". Mei Juecheng and his group of native specialists affirmed Mei Wending's efforts to reach a higher synthesis, which would supersede European and Chinese systems. During the 1720s, the court specialists mastered Jesuit astronomical methods and made them part of the imperial repertoire for computational astronomy.

However, Qing specialists had no domestic incentive to go beyond the immediate needs of the Qing calendar, now successfully reformed. Nor were they intellectually pressed by the Jesuits to do so. By 1725, the latter were themselves no longer on the cutting edge of the early modern sciences, and their mathematics went no further than simple algebra, trigonometry, and logarithms, which had been domesticated by a small group of late Ming and early Qing specialists. In the 18th century, a larger community of Qing classical scholars associated with evidential studies (*kaozhengxue* 考证学) would restore traditional Chinese mathematics to a level of classical prestige.[10]

Although, the "Studio for the Cultivation of Youth" did not continue in the Yongzheng era (1723–1735), nevertheless, the development of official studies in mathematics by selected bannermen, initiated by the Kangxi emperor in 1670, was expanded in scope by the Yongzheng emperor in 1734. In 1739, the Qianlong emperor placed mathematics as a field of

[9]Wang Ping, "Qingchu lisuanjia Mei Wending"〈清初历算家梅文鼎〉, *Jindaishi yanjiusuo jikan*《近代史研究所集刊》2 (1971): 369.

[10]Hashimoto Keizō, "Rekisho Kōsei no seiritsu", in Yabuuchi Kiyoshi and Yoshida Mitsukuni, eds., *Min Shin jidai no kagaku gijutsu shi*, 49–92.

study under the purview of the Dynastic School system. Han Chinese students outside the Astro-calendric Bureau (*Qintian jian* 钦天监) could now study mathematics officially. Despite the Yongzheng rejection of Jesuit learning in court, the key Kangxi era advisors under the leadership of Mei Juecheng still played an influential role in the succeeding Qianlong court.[11]

When compared to eighteenth century developments in Europe, however, the fate of the Qing dynasty Academy of Mathematics is instructive. In France, the Paris Academy of Sciences became a building block for an increase in science professionals and the institutions that supported them. Such institutional changes encouraged the eclipse of the more general learned societies and favored the rise of more specialized institutions. The establishment of professional standards for scientific disciplines by the late eighteenth century was accompanied by the expansion of universities and research institutes where professionalized science slowly incubated in institutions of higher learning, and specialized laboratories eventually replaced gentlemanly academies. Not until the late nineteenth century, would such developments commence in China.[12]

Revival of Ancient Chinese Mathematics

Mei Jucheng lamented the destruction of Yuan–Ming astronomical instruments that had been in the Astro-calendric Bureau until 1672 when they were replaced by Ferdinand Verbiest's (1623–1688) new instruments. Mei had seen them in storage in 1713–1714, but in 1715 Bernard-Kilian Stumpf (1655–1720), then in charge of the Bureau, had several melted down to build a bronze quadrant. By 1744, only the armillary sphere, simplified sphere, and a celestial globe were left of the older instruments. Such material losses of the traditional calendrical heritage influenced Mei Juecheng's efforts to recover Song–Yuan "single unknown" (*tianyuan shu* 天元术; also translated as "heavenly element notation") algebraic techniques for manipulating several unknowns, an enterprise that became a major mathematical feature of evidential research.

[11] See also Minghui Hu, "Cosmopolitan Confucianism: China's Different Road to Modern Science (1664–1830)", Chapter 3.

[12] Roger Hahn, *The Anatomy of a Scientific Institution: The Paris Academy of Sciences, 1666–1803*, 275–285.

To this end, Mei focused on the Yuan minor official Li Ye's 李冶 (1192–1279; originally Li Zhi 李治) *Sea Mirror of Circular Measurement* (*Ceyuan haijing* 《测圆海镜》) of 1248, which was the oldest extant work on the "single unknown" technique. Under the Ming, the tradition of mathematical calculations associated with the *Computational Methods in Nine Chapters* (*Jiuzhang suanshu* 《九章算术》) had been continued. However, the pioneering algebraic methods for solving polynomial equations developed by Qin Jiushao 秦九韶 (1202–1261), Li Ye, and Zhu Shijie 朱世杰 (1249–1314) were not studied.[13]

Recovery and Collation of Ancient Chinese Mathematical Works

During the mid-Qing revival of interest in mathematics, Mei Juecheng and others also realized that they no longer had access to many of the works originally included in the medieval Ten Computational Classics (*Shibu suanjing* 十部算经). Moreover, in addition to Li Ye's *Sea Mirror*, the seminal works of Qin Jiushao on polynomial algebra had been unavailable in Mei Wending's time. In the midst of the "closed door" policies of the Yongzheng emperor and his successors, a large-scale effort to recover and collate the treasures of ancient Chinese mathematics became a major aspect of the late 18th- and early 19th-century internalist turn in evidential studies.[14]

In addition to the more famous evidential scholars such as Dai Zhen 戴震 (1724–1777), Qian Daxin 钱大昕 (1728–1804), Ruan Yuan 阮元 (1764–1849), and Jiao Xun 焦循 (1763–1820), who stressed mathematics in their research, the editing of ancient mathematical texts and the continued digesting of European mathematical knowledge was carried out by a series of literati mathematicians who were also active in evidential studies:

Chen Shiren 陈世仁 (1676–1722)
Minggatu
Li Huang 李潢 (1746?–1811)
Shen Qinpei 沈钦裴 (n.d.)
Luo Shilin 罗士琳 (1789–1853)
Dong Youcheng 董佑诚 (1791–1823)

---

[13] Arthur Hummel, ed., *Eminent Chinese of the Ch'ing Period* (hereafter ECCP), 569, and Jean-Claude Martzloff, *A History of Chinese Mathematics*, 20.

[14] See Benjamin A. Elman, "Geographical Research in the Ming-Ch'ing Period", *Monumenta Serica*, 35 (1981–1983): 1–18.

Wang Lai 汪莱 (1768–1813)
Li Rui 李锐 (1773–1817)
Xiang Mingda 项名达 (1789–1850)
Dai Xu 戴煦 (1805–1860)
Li Shanlan 李善兰 (1811–1882)[15]

Many of the collations of mathematical texts were carried out under imperial auspices during the last years of the Kangxi reign when the *Synthesis of Books and Illustrations Past and Present* encyclopedia (*Gujin tushu jicheng* 《古今图书集成》) was completed. When published under Yongzheng in 1726, it included some European calendrical and mathematical texts from the late Ming *Calendrical Studies of the Chongzhen Reign* in its Qing version known as the *Calendrical Studies According to New Western Methods* (*Xiyang xinfa lishu*《西洋新法历书》). Also included in the calendar section of the encyclopedia were five works of ancient and traditional Chinese mathematics:

1. *Zhou Dynasty Classic of Gnomonic Computations* (*Zhoubi suanjing*《周髀算经》)
2. *Notes on Bequeathed Mathematical Arts* (*Shushu jiyi*《数术记遗》)
3. *Mathematical Manual of Xie Chawei* (*Xie Chawei suanjing*《谢察微算经》)
4. Mathematics in the *Brush Talks from the Dream Brook* (*Mengxi bitan*《梦溪笔谈》)
5. *Systematic Treatise on Computational Methods* (*Suanfa tongzong*《算法统宗》)

When the first set of the Qianlong Imperial Library collection was completed between 1773 and 1781, its compilers also included several classical collators as well versed in mathematics as Dai Zhen: Kong Jihan 孔继涵 (1739–1784), Chen Jixin 陈际新, Guo Changfa 郭长发, and Ni Tingmei 倪廷梅. The "astronomy and mathematics" (*Tianwen suanfa* 天文算法) category incorporated 58 works into the collection (see below). Several older, lost mathematical texts were recopied from the early Ming *Great Compendium of the Yongle Era* (*Yongle dadian*《永乐大典》), which had survived in the imperial court relatively intact. The general

[15] Li Yan and Du Shiran, *Chinese Mathematics: A Concise History*, 223–224.

catalog of the Imperial Library, for example, included 25 notices on mathematics. Of these, nine were on the Tang Computational Classics, three were for Song–Yuan works, four on works from the Ming period, including the Ricci and Li Zhizao 李之藻 (1565–1630) partial translation of Euclid's *Elements*, and nine on works from the Qing, most importantly the *Collected Basic Principles of Mathematics* (*Shuli jingyun* 《数理精蕴》) and several works by Mei Wending.[16]

Eighteenth-century efforts to recover ancient mathematical works extended beyond the borders of the Qing dynasty. The role of Korea and Japan in preserving lost Chinese works is generally well known. Worthy of special mention in this regard, however, was Ruan Yuan's recovery of the lost *Primer of Mathematical Calculations* (*Suanxue qimeng*《算学启蒙》) by Zhu Shijie from a 1660 Korean edition. It had been used as textbook in Korea during the 15th century after it was reprinted there in 1433.[17]

Published in 1299 in Yangzhou, Zhu Shijie's work described the rudiments of polynomial algebra. Several Korean emissaries stayed in Beijing in the early nineteenth century when Ruan Yuan's scholarship was influential there, particularly Ruan's work on ancient technology entitled *Explications Using Diagrams of the Design of Wheeled Carriages in the "Artificer's Record"* (*Kaogong ji chezhi tujie* 《考工记车制图解》). Later Kim Chŏng-hui 金正喜 (1786–1856) visited Beijing in 1809 and met Ruan Yuan in 1810. After their meeting, Kim sent Ruan the Korean edition of the *Primer of Mathematical Calculations*, and Ruan reciprocated by sending a number of his works to Kim. Ruan and others were interested in Zhu Shijie's role in the formation of "single unknown" methods.[18]

Moreover, as the Ten Computational Classics were reconstituted, the Song–Yuan works of Qin Jiushao, Zhu Shijie, and Li Ye, among others,

---

[16] *Siku quanshu zongmu*《四库全书总目》, compiled by Ji Yun 纪昀 (1724–1805) *et al.*, hereafter SKQSZM, Chapters (*juan* 卷) 106–107. See also Jean-Claude Martzloff, *A History of Chinese Mathematics*, 32–33, and ECCP, 637.

[17] Sato Ken'ichi, "Re-evaluation of *Tengenjutsu* or *Tianyuanshu*: in the context of comparison between China and Japan", *Historia Scientiarum* 5, 1 (1995): 57–67.

[18] Lam Lay-Yong, "Chu Shih-chieh's *Suan-hsueh ch'i-meng* (Introduction to Mathematical Studies)", *Archive for History of Exact Sciences*, 21, 1 (1979): 1–31. See also Fujitsuka Chikashi (Rin), *Nichi Sen Shin no bunka kōryū*, 77.

also reappeared. A special, rare edition of seven of the Ten Computational Classics was reprinted by the Imperial Printing Office, including the *Zhou Dynasty Classic of Gnomonic Computations*, and *Computational Methods in Nine Chapters*, as well as 100 chapters (*juan* 卷) from the *Sources of Musical Harmonics and Mathematical Astronomy* of the Kangxi era. Traditional mathematical works were also reprinted in several collectanea such as the *Ripple Pavilion Collectanea* (*Weibo xie congshu* 《微波榭丛书》), the *Collectanea from the Can't Know Enough Pavilion* (*Zhibuzu zhai congshu* 《知不足斋丛书》), and the *Collectanea of the Yijia Hall* (*Yijiatang congshu* 《宜稼堂丛书》).[19]

Reconstruction of the Ten Computational Classics

In the late Ming, Xu Guangqi 徐光启 (1562–1633) had claimed in his preface to Ricci's *Translations of Guidelines for Practical Arithmetic* (*Tongwen suanzhi* 《同文算指》) that the Ten Mathematical Classics were inferior to Jesuit mathematics. As a result of the recovery and collation of ancient mathematical texts, Xu's claims about the superiority of Jesuit mathematics were increasingly disparaged by evidential scholars who appealed to the "Chinese origins of Western learning" (*Xixue Zhongyuan* 西学中源) as a historical reality and not just a political tactic to justify calendrical reform, as had been the case for Xu in the last years of the Ming.[20]

Indeed, only the *Zhou Dynasty Canon of Gnomonic Computations* was still printed and available widely in Ming times. The rest of the mathematics canon was either lost or not available to scholars. The sources left by the late Ming were derived from Southern Song editions or from the early Ming *Great Compendium of the Yongle Era*. Fortunately in the late Ming, Mao Jin 毛晋 (1599–1659) and his son collated seven of the mathematical classics from Southern Song editions for their Suzhou printing house known as the "Pavilion Reaching to the Ancients" (*Jigu ge* 汲古阁):

1. *Zhou Dynasty Canon of Gnomonic Computations*
2. *Sunzi's Computational Canon* (*Sunzi suanjing* 《孙子算经》)

[19] Li Yan and Du Shiran, *Chinese Mathematics: A Concise History*, 225–226.

[20] See Roger Hart, *Imagined Civilizations: China, the West, and Their First Encounter*.

3. *Computational Canon of the Five Administrative Departments* (*Wucao suanjing*《五曹算经》)
4. *Zhang Qiujian's Computational Canon* (*Zhang Qiujian suanjing*《张邱建算经》)
5. *Computational Canon of the Continuation of Ancient Techniques* (*Jigu suanjing* 《缉古算经》)
6. *The Marquis of Xia, Yang's Computational Canon* (*Xiahou Yang suanjing* 《夏侯阳算经》)
7. an incomplete version of the *Computational Methods in Nine Chapters*

Although criticized for their numerous errors due to some slipshod xylography, the "Pavilion Reaching to the Ancients" versions of the Classics and the Dynastic Histories were highly prized. Among Mao's specialties was a process for making facsimiles of Song editions by tracing every feature of the rare books he borrowed from other collectors. The *Zhou Dynasty Canon* and *Notes on Bequeathed Mathematical Arts* (*Shushu jiyi* 《数术记遗》) were also published in a Wanli era collection.[21]

Later, however, the "Pavilion Reaching to the Ancients" editions were dispersed and fell into book collectors' hands. In the process, only five of the ten mathematical classics were intact in the early Qing. A manuscript copy of the *Computational Methods in Nine Chapters* was sent to the Kangxi court and kept in an imperial pavilion. Subsequently, the collating of the Ten Computational Classics was accelerated upon the 1728 publication of the *Zhou Dynasty Canon of Gnomonic Computations* and *Notes on Bequeathed Mathematical Arts*, both in the *Synthesis of Books and Illustrations Past and Present* encyclopedia. The notoriety that Mei Wending had achieved as a mathematician, coupled with the publication of several new European mathematical works during the late Kangxi reign, brought mathematical astronomy into the mainstream of classical studies.

While serving on the Imperial Library commission in the 1770s, Dai Zhen collated seven of the ten mathematics classics from the *Great Compendium of the Yongle Era*. In addition, he recovered two more from manuscript copies originally held by the Mao family, which were

[21] Jean-Claude Martzloff, *A History of Chinese Mathematics*, 125.

published in the Imperial Printing Office Collectanea of Rare Editions (*Wuyingdian juzhenban congshu* 武英殿聚珍版丛书). Dai's colleague Kong Jihan had them reprinted in the *Ripple Pavilion Collectanea* in 1773 under the title Ten Mathematical Classics. Subsequent editions were based on these late Qianlong versions.[22]

The rediscovery and reconstruction of the mathematical classics stimulated interest in them, and they were increasingly studied by evidential scholars such as Dai Zhen, Li Huang, Shen Qinpei, and Gu Guanguang 顾观光 (1799–1862). They produced several important works on mathematical texts in the follow-up style of scholarship known as "additions and corrections" (*buzheng* 补正) in Qing dynasty classical book titles. Li Huang, for example, prepared works entitled *Careful Examination with Diagrams of the Computational Methods in Nine Chapters* (*Jiuzhang suanshu xicao tushuo* 《九章算术细草图说》), *Careful Examination with Diagrams of the Sea Island Computational Canon* (*Haidao suanjing xicao tushuo* 《海岛算经细草图说》), and *Examination of Annotations of the Ancient Mathematical Classics* (*Jigu suanjing kaoju* 《缉古算经考据》). Gu Guanguang, for instance, published his *Collation Notes for the Zhou Dynasty Canon of Gnomonic Computations* (*Zhoubi suanjing jiaokanji* 《周髀算经校勘记》).[23]

Recovery of Song–Yuan Mathematical Works

Scholarly efforts in the late Qianlong era to reconstruct the "single unknown" and "four unknowns" (*siyuan shu* 四元术) techniques for solving complex equations to several powers, led to increased collation and research on the mathematical texts produced by a long-neglected group of Song–Yuan minor officials and commoner mathematicians. Qin Jiushao's *Computational Techniques in Nine Chapters* (*Shushu jiuzhang* 《数书九章》, 1247), for example, provided general algorithms for solving simultaneous equations and remainder problems. His method was similar to the Horner–Ruffini rule devised in the early 19th century, which

[22] Li Yan and Du Shiran, *Chinese Mathematics: A Concise History*, 226–227.

[23] Ibib., 227–230. See Benjamin A. Elman, *From Philosophy to Philology: Social and Intellectual Aspects of Change in Late Imperial China*, 242–244.

invented a technique for the numerical calculation of the roots of polynomial equations. In the *Computational Techniques*, Qin used a parallel technique of alternating additions or subtractions, a procedure also used by Yang Hui 杨辉 (ca. 1238–1298), Zhu Shijie, and others.[24]

Although Qin's work followed somewhat the overall structure of the *Computational Methods in Nine Chapters*, his algorithms were much more sophisticated. In addition, because Qin had studied in the Song Astro-calendric Bureau in Hangzhou as a youth, his work also dealt with calendrical chronology as a problem in remainder theory. Finally, the *Computational Techniques* calculated the area of an arbitrary triangle as a function of the lengths of its three sides, which was similar to the proto-trigonometric relational features for computing the sides of a right-angled triangle (*gougu* 勾股). Later, this approach drew the attention of Dai Zhen and others interested in equating such proto-trigonometric relational features with Jesuit trigonometry.[25]

Qin's *Computational Techniques* was copied from the *Great Compendium of the Yongle Era* into the Qianlong Imperial Library by Dai Zhen. Jiao Xun and Li Rui each studied and wrote on Qin's findings. Later Shen Qinpei discovered a Ming manuscript of the *Computational Techniques*, which he compared to the version in the *Great Compendium of the Yongle Era*. A definitive edition of the *Computational Techniques* was then included in the 1842 *Collectanea of the Yijia Hall* and became the basis for the modern versions. Hua Hengfang's 华蘅芳 (1833–1902) *Mathematical Notes* (*Xuesuan bitan* 《学算笔谈》) for 1882–1888 presented a list of must-read books, which included both Chinese and Western works. Although he was well versed in modern mathematics, Hua still recommended the *Computational Techniques* for solving remainder problems.[26]

---

[24] Alexander Wylie, *Notes on Chinese Literature*, 116, and Jean-Claude Martzloff, *A History of Chinese Mathematics*, 149–152. See also Roger Hart, *The Chinese Roots of Liner Algebra*.

[25] See Ulrich Libbrecht, *Chinese Mathematics in the Thirteenth Century: The Shu-shu Chiu-chang of Ch'in Chiu-shao*, passim, and Jean-Claude Martzloff, *A History of Chinese Mathematics*, 2–12, 231–247.

[26] Alexander Wylie, *Notes on Chinese Literature*, 116, and Hu Mingjie, "Merging Chinese and Western Mathematics: The Introduction of Algebra and the Calculus in China, 1859–1903" (Princeton: Princeton University Ph.D. dissertation in History, 1998), 252–253.

After the Mongols conquered north China in 1232, Li Ye prepared two works: the *Sea Mirror of Circle Measurement* and the *Adding to Ancient Techniques for Computing Geometric Figures* (*Yigu yanduan*《益古演段》, 1259), which were copied into the Imperial Library. Although Li lived in reclusion in Shanxi after the Mongol triumph, he was called to the Mongol court to consult on governance and earthquakes. The *Sea Mirror* survived from a book in Li Huang's private library. The *Adding to Ancient Techniques* was copied into the *Compendium of the Yongle Era*. Li Rui later collated both, and they were also printed in the *Collectanea from the Can't Know Enough Pavilion*. Li Rui's edition of the *Sea Mirror* also included a series of explanatory notes.[27]

Although Yang Hui's mathematical works were all included in the *Compendium of the Yongle Era*, they were not copied by Dai Zhen into the Imperial Library. Part of Yang's *Continuation of Ancient Mathematical Methods for Elucidating Strange Numbers* (*Xugu zhaiqi suanfa*《续古摘奇算法》) was, however, edited for the *Collectanea from the Can't Know Enough Pavilion* and printed by the Hangzhou bookman Bao Tingbo 鲍廷博 (1728–1814). In 1840, Yang's commentary and supplement for the *Computational Methods in Nine Chapters* and the *Yang Hui's Calculation Methods* (*Yang Hui suanfa*《杨辉算法》) were included in the *Collectanea of the Yijia Hall*. Yang's complete work was lost in China but was rediscovered in the Qing by Li Rui. In the early 20th century, a 1433 Korean edition of the *Yang Hui's Calculation Methods*, based on a 1378 Ming edition, was found in Japan.[28]

On the other hand, Zhu Shijie's 1303 *Jade Mirror of the Four Unknowns* (*Siyuan yujian*《四元玉鉴》) and his 1299 *Primer of Mathematical Calculations* (see above) were not recovered in time to be included in the Imperial Library. Although the focus in Zhu's *Jade Mirror* was on practical issues dealing with architecture, finance, military logistics, etc., it energized late Qing evidential scholars who found in it a

---

[27] See Alexander Wylie, *Notes on Chinese Literature*, 116–117.

[28] Li Yan and Du Shiran, *Chinese Mathematics: A Concise History*, 230–231, and Jean-Claude Martzloff, *A History of Chinese Mathematics*, 149–152, 157–159. See also Lay-Yong Lam, *A Critical Study of the Yang Hui Suan Fa, a Thirteenth-Century Mathematical Treatise*.

Chinese algebra to extract roots (*kaifang* 开方) predating the Jesuits. Zhu's polynomial equations went beyond the second and third degrees up to the fourteenth.

During the Jiaqing reign, Ruan Yuan obtained a version of the *Primer of Mathematical Calculations* from Korea (see above) while governor of Zhejiang, which he used to reconstitute the *Jade Mirror of the Four Unknowns*. He then sent it to Bejing to be included as supplement for the Imperial Library. Ruan also gave Li Rui a copy to collate, which Li left uncompleted. Others such as Xu Youren 徐友仁 (1800–1860) and Shen Qinpei worked on it, although Shen's commentary was never published.[29]

Luo Shilin obtained a copy of the *Jade Mirror* in 1822, and after ten years of collation he presented a definitive reconstruction of "four unknowns" techniques, which he entitled the *Jade Mirror of the Four Unknowns with Detailed Calculations* (*Siyuan yujian xicao* 《四元玉鉴细草》) and had it printed in 1843 in Yangzhou. After the rediscovery of Zhu Shijie's *Primer of Mathematical Calculations*, Luo Shilin also had it reprinted in 1839 based on the Korean edition of 1660 recovered by Ruan Yuan. As noted above, the *Primer* was also an important clue to the fundamentals of the "single unknown" and "four unknowns" polynomial algebra developed by Qin Jiushao and Li Ye.[30]

Zhang Dunren 张敦仁 (1754–1834) in 1801 worked out every problem in the *Computational Canon of the Continuation of Ancient Techniques* based on "single unknown" procedures. Zhang also prepared an 1831 work on indeterminate analysis based on the expansion method for solving simultaneous congruencies. Li Rui, who was Zhang's personal secretary, helped complete this harvest of Song–Yuan mathematics and bring it into the mainstream of evidential studies in the first half of the 19th century.

The "internal turn" to native mathematics had enabled those Chinese classical scholars at the cutting edge of evidential studies to grasp the

[29] See Jock Hoe, "Zhu Shijie and His *Jade Mirror of the Four Unknowns*", in *First Australian Conference on the History of Mathematics: Proceedings of a Conference at Monash University* (Clayton, Australia) 6 & 7 (November 1980): 105.

[30] Luo Shilin 罗士琳, *Siyuan yujian xicao* 《四元玉鉴细草》, *juan* 1. See also Li Yan and Du Shiran, *Chinese Mathematics: A Concise History*, 115–117, 231–232, and Jean-Claude Martzloff, *A History of Chinese Mathematics*, 153–157.

importance of advanced algebraic techniques for solving complicated equations based on sophisticated mathematical problems. When the differential (*weifen* 微分) and integral (*jifen* 积分) calculus was finally introduced by Alexander Wylie (1815–1887) and John Fryer (1839–1928) in the middle of the 19th century, its sophistication was readily appreciated by Li Shanlan and others who had already mastered "single unknown" and "four unknowns" problem-solving skills.[31]

The Critique of Western Learning in the Qianlong Imperial Library

Biased in favor of Ancient Learning (*guxue* 古学) and evidential research, the editors of the ambitious Qianlong Imperial Library project (*Siku quanshu* 《四库全书》) initiated a well-publicized empire-wide search in the 1770s for every written work in the empire. They then engaged in a critical review of every book available to them, selected books worthy of inclusion in the collection, and carefully collated the final versions chosen for inclusion. Leading classical scholars in various fields were appointed to evaluate and collate books in their respective specialties. More than 360 scholars officially worked at the apex of a staff of several thousand in Beijing.

The editors, many of them partisans of Han Learning (*hanxue* 汉学), separated the "investigation of things and extension of knowledge" (*gewu zhizhi* 格物致知) from its longstanding association with Cheng–Zhu 程朱 Song Learning and connected it instead with their emphasis on verifiable knowledge derived from empirically based research. They contended, for instance, that Chen Yuanlong's 陈元龙 (1652–1736) early 18th-century encyclopedic entries in the *Mirror of Origins Based on the Investigation of Things and Extending Knowledge* (*Gezhi jingyuan* 《格致镜源》) "were all [examples of] broad learning and thus the work was entitled using the phrase for 'investigating things and extending knowledge.'"[32]

---

[31] See Alexander Wylie, *Notes on Chinese Literature*, 115–116, and Li Yan and Du Shiran, *Chinese Mathematics: A Concise History*, 242, 251.

[32] SKQSZM, 136.25a–26a. See also R. Kent Guy, *The Emperor's Four Treasuries: Scholars and the State in the Late Ch'ien-lung Era*.

In many cases, the editors elided any mention of European learning when they could. For example, their summary of Fang Yizhi's 方以智 (1611–1671) late Ming *Notes on the Principles of Things* (*Wuli xiaozhi*《物理小识》), which generally accepted Jesuit explanations of natural phenomena, presented Fang's "material investigations" in light of encyclopedias compiled since medieval times in China. The final version of the Imperial Library account made no effort to mention the notions of a spherical earth, limited heliocentrism, or human physiology that Fang had culled from Jesuit translations.[33]

The editors could not ignore the translations and other works that had been compiled in Chinese since the late Ming by the Jesuits and their collaborators. The Jesuit role in the Kangxi era *Compendium of Observational and Computational Astronomy* and its supplement were addressed in the *General Catalog of Complete Collection in the Imperial Four Treasures* (*Siku quanshu zongmu* 《四库全书总目》). The prestigious catalog mentioned 36 European works, 20 of which were copied into the Library. All were on natural studies. On the other hand, the ten works from the "applications" (*qi* 器, "implements") section of Li Zhizao's *First Collection of Celestial Studies* (*Tianxue chuhan*《天学初函》) were included, but not those from the "principles" (*li* 理) section, except for a geographical work by Giulio Aleni (1582–1649). Li Zhizao's 1628 *First Collection* placed the natural theology of the Jesuits under "principles" and material studies under "applications". Similarly, only two of Verbiest's works were given consideration in the Imperial Library. The Library catalogers included mention of Aleni's *A Summary of Western Learning* (*Xixue fan* 《西学凡》), but they downplayed it as a heterodox study and did not copy it into the Imperial Library.[34]

The editorial summary of Aleni's work presented the six-fold classifications of the sciences that corresponded to 16th-century European standards of learning. Seeing some parallels with native traditions, the editors analogized the Jesuit field of rhetoric (*wenke* 文科), which included grammar, history, poetry, and the art of writing, to the native category of "lesser

[33] SKQSZM, 122.29a–b.

[34] SKQSZM, 106.28a–36a. See also Ji Wende, *Cong Siku quanshu tanjiu Ming Qing jian shuru zhi Xixue* 〈从四库全书探究明清间输入之西学〉, 404–436.

learning" (*xiaoxue* 小学). They likened philosophy (*like* 理科), which for Jesuits included logic, physics, metaphysics, mathematics, and ethics, to classical teachings in the *Great Learning* (*Daxue*《大学》). In addition, the editors praised the system of knowledge in Aleni's *Summary* and interpreted it in light of the "investigation of things and fathoming principles" (*gewu qiongli* 格物穷理). For Aleni the investigation of things to fathom principles was used as the Chinese term for contemporary European philosophy.[35]

Despite such correspondences, however, the Qianlong editors linked the teachings of the *Great Learning* to an emphasis on verifiable knowledge derived from empirical studies. Moreover, the editors concluded that Aleni had forced equivalences between European and native learning in order to substantiate Christianity as an ancient teaching in China, which late Ming scholars under the influence of Wang Yangming's 王阳明 (1472–1528) "studies of the mind" (*xinxue* 心学) had failed to reveal:

> "They also investigate things to exhaust principles, and seek to understand substance for practical use; in that they are roughly similar to our literati. However, the things they investigate are mostly petty, and the principles they seek to exhaust are mostly esoteric and untestable. That is why this book is considered heterodox."

On this basis, the editors also ridiculed Alfonso Vagnoni's (1566–1640) 1633 *Treatise on the Composition of the Universe* (*Kongji gezhi*《空际格致》) because it had tried to introduce a new reading for *qi* 气 (= Aristotelian "air") that favored the ancient Greek notion of the four elements.[36]

---

[35] See SKQSZM, 106.1a–51a, 107.23a–24a, 125.27b–35b, 134.10a–11b, and Nicolas Standaert, S.J., "The Investigation of Things and the Fathoming of Principles (*Ko-wu ch'iung-li*) in the Seventeenth-Century Contact Between Jesuits and Chinese Scholars", in John W. Witek, S.J., ed., *Ferdinand Verbiest (1622–1688): Jesuit Missionary, Scientist, Engineer and Diplomat*, 412–417.

[36] SKQSZM, 125.31b–34a, 125.34b–35a. Compare Erik Zürcher, "Renaissance Rhetoric in Late Ming China: Alfonso Vagnoni's Introduction to his *Science of Comparison*", in Federico Masini, ed., *Western Humanistic Culture Presented to China by Jesuit Missionaries (XVII–XVIII centuries)*, 331–359.

Astronomy and Mathematics in the Qing Imperial Library

The *General Catalog of Complete Collection in the Imperial Four Treasures*, completed in 1782, attempted to present the gist of a book in outline form so that readers could get a general idea of its contents. Except for entries related to the Jesuits, such accounts usually gave the reader an accurate and concise idea of the nature and importance of the work. Even though the editors chose to elide or ridicule Jesuit works that were included in the *General Catalog*, they did innovate when they presented surviving colophons on works dealing with astronomy and mathematics (*tianwen suanfa* 天文算法).

Dai Zhen worked for the Imperial Library commission on the mathematical, astronomical, and calendrical texts included in the "Astronomy and Mathematics" section, which we have seen above enabled him to recover several of the Ten Computational Classics. Of the 10,000 titles reviewed by the editors, about one-third finally were copied into the imperial manuscript collection, including 58 works on "Astronomy and Mathematics". Forty of the latter were recovered by Dai Zhen and his colleagues from the *Great Compendium of the Yongle Era*.[37]

The late 18th-century classification of knowledge reveals the manner in which the variety of learning was perceived at that time. We also see the nature and structure of the concepts used to order that variety within the emerging disciplines in Qing classical scholarship. The bibliographic clustering of subjects in the Qianlong Imperial Library in particular presents the culturally conditioned biases in Qing dynasty Han and Song Learning. Moreover, the 18th-century structure of knowledge shaped evidential studies and influenced how new research on mathematical astronomy would be understood.

Representing the classical scheme of disciplines in the late 18th century, the Imperial Library was based on the "four classifications system" (*sibu* 四部), which incorporated mathematical astronomy and calendrical studies, as well as medicine, as subcategories under the pre-Han "Masters" (*zibu* 子部) main category (see Table 1 in Chapter 2). Similarly the

[37] Ji Wende, *Cong Siku quanshu tanjiu Ming Qing jian shuru zhi Xixue*, 410–426.

mathematical aspects of music were subsumed under the Classics, while chronography and geography were listed under the main category of History.[38]

The editors rejected placing mathematics in the "lesser learning" section, which the late 17th-century compilers of the *Ming History* had done, in favor of what they regarded as a more commonsense linking of mathematics to astronomy. Because of their association with esoteric and exotic fields such as astrology, chronomancy, the five phases, milfoil divination, prognostication, and geomancy, the editors separated the "numerological arts" (*shushu* 数术) from their traditional association with mathematics and granted them an independent status.[39]

Although they gave virtually no credit to the Jesuits for their innovation, the Qianlong compilers broke new ground by placing mathematics and astronomy under the same framework. The editorial overview for the "Astronomy and Mathematics" section of the *General Catalog* represented a native response to the Jesuit impact on "new methods" (*xinfa* 新法) that had successfully enabled more mathematically precise calculations of the calendar. Moreover the pride the editors took in acknowledging how Qing scholars had balanced and unified "Western and Chinese" mathematics in the process of recovering the "single unknown" techniques made it clear that they regarded the "Chinese origins of Western learning" as the key ingredient for overcoming the less informed mathematics of their Ming predecessors.[40]

Ruan Yuan and the Biographies of Mathematical Astronomers

We have seen in chapter 2 that Ruan Yuan's assemblage of a staff to compile the *Biographies of Astronomers and Mathematicians* (*Chouren zhuan* 《畴人传》) while serving in Hangzhou from 1797 to 1799 marked

---

[38] SKQSZM, *juan* 106–107. Alexander Wylie's 1867 *Notes on Chinese Literature*, 106–130, presents a catalog of Chinese works in the category of *Tianwen suanfa*. See also, Benjamin A. Elman, *From Philosophy to Philology: Social and Intellectual Aspects of Change in Late Imperial China*, 202–204.

[39] See the editorial introduction to the "Astronomy and Mathematics" classification in the SKQSZM, 106.1a–2a.

[40] SKQSZM, 106.1b.

the climax of the celebration of natural studies within the Yangzi delta literati world of the late 18th century. Ruan Yuan's technical interests were influential because of his status as a patron of Yangzi delta scholarship, particularly evidential scholars from Yangzhou. He had become famous among his peers in 1791 when his prose-poem for the Hanlin Academy (*Hanlin yuan* 翰林院) special examination topic on early Yuan astronomy was singled out by the Qianlong emperor for special praise.[41]

Ruan was aided in his Hangzhou project by many of the leading evidential scholars of the late Qianlong period: Li Rui, Qian Daxin, Jiao Xun, Ling Tingkan 凌廷堪 (1757–1809), and Tan Tai 谈泰, among others. Their efforts in astronomy have been described by Nathan Sivin as "a programmatic synthesis of traditional and Western astronomy designed to encourage the study of the latter in order to improve the former. Ruan and his co-editor Li Rui emphasized the old idea that the roots of modern astronomy are to be found in ancient China."[42] Their efforts reaffirmed the value of mathematics and astronomy as part of a classical education.

Ruan did not include in the collection those associated with fortune telling and numerology, and he opposed connecting mathematical astronomy with musical theory or studies of the *Change Classic*. As a conservative, however, he was critical of three new findings introduced by Michel Benoist: (1) heaven and earth are round; (2) planets follow elliptical paths; and (3) the sun is stationary. Although Benoist had finally presented the "heliocentric" Copernican system to China, Ruan Yuan found such views unacceptable, in part because they contradicted earlier Jesuit presentations of Copernicus (1473–1543), which had reduced the latter's views to the "geoheliocentric" Tychonic system. Ruan sought a fusion of European and Chinese mathematics based on shared conceptions. For astronomy, he sought an accurate, predictive computational system that would be based on improved techniques, not Copernican cosmology. Ruan's views were influential empire-wide because in 1799 Ruan also served as director of the mathematics section of the Dynastic School in Beijing.[43]

---

[41] Wang Ping, "Ruan Yuan yu *Chouren zhuan*" 〈阮元与畴人传〉, *Jindai shi yanjiusuo jikan*《近代史研究所集刊》 (1973): 601–611.

[42] See Nathan Sivin, "Copernicus in China", 45–50.

[43] Ibid. See also Jean-Claude Martzloff, *A History of Chinese Mathematics*, 166–172.

Containing summaries of the works of 280 mathematicians and astronomers, including 37 Europeans, the *Biographies* was followed by four supplements in the 19th century. The collection was reissued in 1829 with only Qing biographies, and it was later enlarged and reprinted in 1849. In 1840, for example, Luo Shilin added 43 sections on Song–Qing mathematical astronomers based on new sources for the "single unknown" and "four unknowns" techniques recovered from the Song and Yuan, such as *Yang Hui's Calculation Methods* and the *Jade Mirror of the Four Unknowns*. In 1857, Alexander Wylie worked with Wang Tao 王韬 (1828–1897) to improve on the views presented in the collection, particularly critiquing the "Chinese origins" narrative.[44]

Interest in mathematical astronomy among literati, which had developed steadily since Mei Wending, had grown in importance outside the imperial court by the late 18th century. This growth was tied to the popularity of evidential studies outside the patronage networks of the Manchu court, which had sponsored Manchu and Mongol bannermen to try to control such knowledge. By connecting mathematics and astronomy to classical studies, Ruan Yuan successfully integrated mathematical astronomy with evidential studies. Because mathematics and natural studies remained dependent on classical studies, Ruan Yuan and his colleagues revived the ancient category of calendricists (*chouren* 畴人), whom he now considered "mathematical astronomers".

In the mid-18th century, the official *Ming History* had already described the ancient dispersion of classical calendricists to the Western region (*Xiyu* 西域), specifically to the Islamic world. Moreover, the term for classical "calendricist" had been used in the canonical *Artificer's Record* (*Kaogong ji* 《考工记》) and Sima Qian's 司马迁 *Records of the Official Historian* (*Shiji* 《史记》) during the Han dynasty. We have seen above that earlier in the 18th century, the Kangxi emperor had singled out "palace graduates in mathematical astronomy" such as Wang Lansheng, Minggatu, and Mei Juecheng for special honors.

[44] See Li Yan and Du Shiran, *Chinese Mathematics: A Concise History*, 232–233, and Paul Cohen, *Between Tradition and Modernity: Wang T'ao and Reform in Late Ch'ing China*, 176–177.

Such usage was now reworked by Ruan Yuan and Tan Tai in their lead accounts of the meaning and scope of "mathematical astronomers" in the *Biographies*. They employed the term as a classical sanction for a new intellectual and social category of contemporary scholar-literati such as Mei Wending, Dai Zhen, and Qian Daxin, which also referred to a genealogy of professionalized skills in mathematics and astronomy going back to antiquity. This orthodox term was the first of several used in the 18th and 19th centuries to describe European "scientists" in classical Chinese.[45]

The category for mathematical astronomer also had its conceptual roots in mathematics as one of "six arts" (*liuyi* 六艺; see Chapter 2) of ancient, aristocratic scholars. Beginning in the late Ming, as literati increasingly engaged in the study of mathematics and astronomy, two types of specialists emerged: (1) specialists in calendar making; and (2) literati with an academic interest in mathematics. These two categories were most evident during the Kangxi era when mathematical study, which was a required tool for calendrical reform, was upgraded from a trivial skill. The academic climate among evidential scholars, along with imperial patronage, helped make mathematics and astronomy a collateral branch of classical learning.[46]

Despite their distance from the court, literati who favored mathematical astronomy, such as Ruan Yuan, never thought it might eventually gain an independent position from other fields of classical learning. As calendrical difficulties declined in importance during the 18th century due to the success of the Ming–Qing astronomical reforms inspired by the Jesuits and their Chinese counterparts, mathematics emerged as an independent field of inquiry in evidential research and among Han Learning scholars, particularly in Yangzhou. As cultural and political disputes over the calendar declined, literati debates shifted in the early 19th century to acclaim the achievements of native mathematics.[47]

---

[45] Ruan Yuan, "Chouren zhuan fanli" 〈畴人传凡例〉, in Ruan Yuan, *Chouren zhuan* 《畴人传》, 1–5, and Tan Tai, "Chouren jie" 〈畴人解〉, in *Chouren zhuan*, 1–4.

[46] Bai Limin, "Mathematical Study and Intellectual Transition in the Early and Mid-Ch'ing", *Late Imperial China* 1, 2 (October 1995): 23–61.

[47] See Pingyi Chu, "Western Astronomy and Evidential Study: Tai Chen on Astronomy", in Yung Sik Kim and Francesca Bray, eds., *Current Perspectives in the History of Science in East Asia*, 144.

Literati Natural Studies: Classics and Mathematics

Literati before the Opium War were not doomed by the premises of Qing classicism to overlook natural studies and mathematics. Lord Macartney had recognized the Chinese interest in astronomy, but he had erred when he reduced it to "astrological trifling, the goal of which is the calculation of auspicious times". Moreover, because he was unaware of Chinese expertise in mathematical astronomy, Macartney indicated that the Chinese had no notion of algebra and possessed only a limited understanding of geometry and plain trigonometry. Englishmen such as Macartney were continuing a tradition of denigrating Chinese natural studies by Europeans that had begun with the Jesuits.[48]

Despite their textual focus, 18th-century evidential scholars successfully restored a place for mathematical studies within the framework of the "Chinese origins of Western learning". Even the massive *Qing Exegesis of the Classics* (*Huang Qing jingjie* 《皇清经解》), a scholarly collection published in the early 19th century and devoted exclusively to evidential scholarship, included a significant number of works on natural studies and mathematical astronomy. Because it was the first comprehensive collection of Qing contributions to classical scholarship, the 1829 publication of the *Qing Exegesis* in Guangzhou, after four years of compiling and editing at the Sea of Learning Hall (*Xuehaitang* 学海堂) under the auspices of Ruan Yuan, then governor-general there, was greeted with acclaim in China, Korea, and Japan.[49]

An imposing anthology of some 180 diverse works by 75 17th- and 18th-century authors in more than 360 volumes totaling some 1,400 chapters (*juan*), the *Qing Exegesis of the Classics* represented a major tribute to the research carried out by evidential scholars. It served as a collection of exemplary works from the Yangzi delta community in the 17th and 18th centuries. Although a continuation to earlier commentaries and annotations of the Five Classics and Four Books, the *Qing Exegesis* was also a

[48]George Macartney, *An Embassy to China: being the journal kept by Lord Macartney during his embassy to the Emperor Ch'ien-lung, 1793–1794*, edited with an introduction and notes by J. L. Cranmer-Byng, 264.

[49]Fujitsuka Chikashi (Rin), *Nichi Sen Shin no bunka kōryū*, 108.

response to early Qing collectanea and encyclopedias that had stressed Cheng-Zhu learning.[50]

What was noteworthy about Ruan Yuan's summation of Qing classical scholarship was not only its spotlight on the Han Learning and evidential studies produced by literati scholars from intellectual centers in the Yangzi delta. His partisan focus also allowed Ruan to expand the scope for imperial classical studies beyond the domain of Song Learning (*Songxue* 宋学). Via the accolades for Han Learning, Ruan and his staff incorporated the works of many of the Qing authors he had included in the *Biographies of Mathematical Astronomers*.

For example, Ruan Yuan included verbatim major sections from the *Biographies* dealing with Qing mathematical astronomers, beginning with two chapters on the mathematical astronomer Wang Xichan 王锡阐 (1628–1682), followed by a section on Mei Wending and an extensive chapter on Dai Zhen. Remarkably, this was followed by four chapters from the *Biographies* on "Westerners", with mention of Greek scholars, including Aristotle (384–322 B.C.) and Ptolemy (ca. 90–168), and Europeans such as Copernicus, Tycho Brahe (1546–1601), Matteo Ricci, Adam Schall (1592–1666), Ferdinand Verbiest, and Nicolas Smogolenski (1611–1656). Isaac Newton was also briefly mentioned for his revision of Tycho Brahe's length of the solar year, which Ruan et al. copied from the 1742 *Supplement to the Compendium of Observational and Computational Astronomy*. Inclusion of foreigners to this degree was unprecedented in a repository of Chinese classical learning. The era of eliding European astronomical scholarship, dominant since the Yongzheng reign, was effectively redressed 40 years before the onset of the Opium War.[51]

In addition, three works on the *Artificer's Record* chapter of the *Rituals of Zhou* (*Zhouli* 《周礼》) were incorporated in the *Qing Exegesis*, including Dai Zhen's and Cheng Yaotian's 程瑶田 (1725–1814) illustrated studies of ancient ceremonial bronze bells. Ruan also included his own *Explications Using Diagrams of the Design of Wheeled Carriages in*

[50] See Benjamin A. Elman, *From Philosophy to Philology: Social and Intellectual Aspects of Change in Late Imperial China*, 126–133.

[51] See Ruan Yuan, ed., *Huang Qing jingjie* 《皇清经解》, Chapters 1059–1068, especially 1067.1a–b (Vol. 15, 11,324) on Newton.

the "Artificer's Record" (*Kaogong ji chezhi tujie*《考工记车制图解》), his first published work. Similarly, Ruan printed two important studies of the historical geography in the "Tributes of Yu" (*Yu gong* 禹贡) chapter of the *Documents Classic* (*Shangshu*《尚书》). The lead work was Hu Wei's 胡渭 (1633–1714) highly praised *A Modest Approach to the Tributes of Yu* (*Yugong zhuizhi*《禹贡锥指》), which corrected many mistakes in earlier commentaries. In addition, numerous geographical accounts and the chronology of events in the classics were published in Ruan's collectanea.[52]

Sheng Baier's 盛百二 (fl. ca. 1756) study of the astronomy in the *Documents Classic* was copied in the *Qing Exegesis* even though Sheng was not known for any other significant works, though he had taught in academies for over a decade. Sheng's expertise in astronomy and trigonometry had influenced many of his students. His work was filled with illustrations based on the Tychonic geoheliocentric system from the late Kangxi *Compendium of Observational and Computational Astronomy*. It also had many references to Jesuit scholarship on star maps, eclipses, and the motion of the planets.[53]

Ruan Yuan added Chen Maoling's 陈懋龄 1797 *Examination of Mathematics and Astronomy in Classical Works* (*Jingshu suanxue tianwen kao*《经书算学天文考》to the collection. Chen had become interested in Mei Wending's work in 1793 and subsequently inquired about Western learning to a European in Guangzhou who was on his way to Beijing to serve in the Astro-calendric Bureau. Chen's own inquiry stressed that the *Computational Methods in Nine Chapters* classic represented the origins of remainder problems, "single unknown" procedures, and other calculation techniques. Based on Ruan Yuan's findings, for example, Chen affirmed the theory of a round earth. He then claimed that the geoheliocentric position, which he argued was first enunciated in the *Rituals of Zhou*, proved that the earth was round, a position that Ruan

---

[52]See the table of contents to Ruan Yuan, ed., *Huang Qing jingjie*, Vol. 1, 9–32. See also Benjamin A. Elman, *From Philosophy to Philology: Social and Intellectual Aspects of Change in Late Imperial China*, 243.

[53]Ruan Yuan, ed., *Huang Qing jingjie*, Chapters 485–490 (Vol. 7, 5305–5394), especially 488.2b–4b (5348–5349).

Yuan had rejected. In effect, Chen's *Examination* served as a repository of native mathematical astronomy that had been reinvigorated by the impact of Jesuit studies.[54]

Many other straightforward classical works that also incorporated natural studies were included, but their technical content was hard to locate. To make the technical subjects in the *Qing Exegesis* more accessible, Cai Qisheng 蔡启盛, a student of the Hanlin academician Yu Yue 俞樾 (1821–1907), later prepared a topical index to Ruan Yuan's collection as part of late Qing efforts to document the Chinese priority in scientific knowledge at a time when modern science was being introduced by Protestant missionaries. Yu Yue and Cai Qisheng deemed this index important for students at the "Refined Study for the Explication of the Classics" (*Gujing jingshe* 诂经精舍) Academy in Hangzhou where Yu Yue taught for 30 years beginning in 1867. While governor of Zhejiang in 1801, Ruan Yuan had established the "Refined Study" in Hangzhou to honor Later Han classicists and to link a classical education with a commitment to "concrete studies" (*shixue* 实学). Ruan had seen to it that students there would be examined in astronomy, mathematics, and geography, in addition to their literary and textual studies.[55]

Cai Qisheng's index of topics in the *Qing Exegesis* at first sight looked very much like the table of contents for a Ming–Qing encyclopedia. It opened with the category of astronomy (*tian* 天), which was tied to the use of mathematics (*suanxue* 算学) to measure the heavens (*tian* 天). The index included 42 other categories ranging from human relations, morality, and political matters (*zhengshi* 政事) to rituals, food and drink, and things (*wu* 物). "Things" were delimited to animals (with feathers or hair) and plants from grasses and vegetables to crops (*nongye chanwu* 农业产物), trees and bamboo. Despite the similarity to earlier encyclopedias, however, the terminology Cai employed was drawn from a new era. What he called "astronomy" (*tianwen* 天文) was the earlier term for astrology.

[54]Ibid., 1328.1a–2a (Vol. 19, 24459), especially 1328.22a–33a (14471–14472). See also *Chouren zhuan*, Vol. 2, 48.634–637.

[55] See Benjamin A. Elman, "The Hsueh-hai T'ang and the Rise of New Text Scholarship in Canton", *Ch'ing-shih wen-t'i* (now *Late Imperial China*), 4, 2 (December 1979): 51–82.

Instead of pharmacopoeia (*bencao* 本草), the index used the new term for plants (*zhiwu* 植物) derived from the study of horticulture, which had been introduced through Protestant translations after the Opium War (see Chapter 6).[56]

Mathematics among Literati in an Age of Evidential Research

In addition to the classics, we can also discern a more general interest in mathematics and natural studies in late Qing novels. Casual references to some aspects of Chinese mathematics had been included in the late 18th-century novel *A Country Codger's Words of Exposure* (*Yesou puyan* 《野叟曝言》, ca. 1780) by Xia Jingqu 夏敬渠 (1705–1787). The topics of mathematics and natural studies were presented in far more detail, however, in the early 19th-century novel *Flowers in the Mirror* (*Jinghua yuan* 《镜花缘》, ca. 1821–1828) by Li Ruzhen 李汝珍 (ca. 1763–1830). There, mention of such technical subjects occurred in the form of chats and quizzes by women at parties held to celebrate their success in the civil examinations. Seven of the female scholars had some knowledge of mathematics.[57]

The fantasy of a realm of women knowledgeable in the classics and mathematics was prescient in the early 19th century, although Li Ruzhen's ending for the novel reaffirmed the literati world of his day. That women in the novel understood mathematics parodied an academic world in which most literati were still not well informed about natural studies. Those we have focused on above, like their counterparts in Europe, were exceptional. The larger questions *Flowers in the Mirror* raised about conventional concepts and practices, such as women's subordination to men

---

[56] See Yu Yue's "Huang Qing jingjie jianmu xu" 〈皇清经解检目序〉 to Cai Qisheng's *Huang Qing jingjie jianmu* 《皇清经解检目》 (1886 edition), 1a–2a. My thanks to Minghui Hu for providing me with a copy of the index. See also ECCP, 944–945, and Benjamin A. Elman, *From Philosophy to Philology: Social and Intellectual Aspects of Change in Late Imperial China*, 162.

[57] Li Ruzhen, *Jinghua yuan* 《镜花缘》, 415–418, 484–492, 527–534. See also *Flowers in the Mirror, by Li Ju-chen*, abridged translation by Lin Tai-yi, 133–141, 229–235, 242–244. Compare Yu Wang Luen, "Knowledge of Mathematics and Science in Ching-Hua-Yuan", *Oriens Extremus* 21, 2 (1974): 217–236.

and foot-binding, made Li Ruzhen's inclusion of mathematical puzzles in the novel both entertaining and enigmatic.[58]

Li Ruzhen can be tied to a circle of Yangzhou scholars who embodied the overlap between evidential studies and mathematics that he ascribed to women in the novel. Li's teacher was Ling Tingkan, a Yangzhou scholar who was an acknowledged expert in mathematics and astronomy and had helped Ruan Yuan compile the *Biographies of Astronomers and Mathematicians*. Ling had studied under Dai Zhen and was influenced by Dai's works on mathematics such as the *Calculations Using Counting Rods* (*Cesuan* 《测算》), the *Record of Measuring Segments of a Circle by Computing the Sides of a Right-angled Triangle* (*Gougu geyuan ji* 《勾股割圆记》), as well as the mathematical classics Dai had retrieved from the Imperial Library.

Another Yangzhou native, Cheng Yaotian, was linked to both Ling and Dai and had written two works on the *Zhou Dynasty Classic of Gnomonic Computations* dealing with properties of the square. Jiao Xun, another of Dai Zhen's Yangzhou partisans, prepared several mathematical works on squares and explanations of mathematical processes. He had also completed a textbook entitled *General Explanations of Extracting Roots* (*Kaifang tongshi* 《开方通释》) and the more technical *An Explanation of Single Unknown Procedures* (*Tianyuan yishi* 《天元一释》). In addition, Li Ruzhen's brother-in-law, Xu Guilin 许桂林 (1778–1821), had prepared instructional manuals for "single unknown" procedures and arithmetic.[59]

Measurements of the circumference of a circle in Li Ruzhen's novel were based on the formula that the circumference equals the diameter times π. The value of π that Li chose is interesting because some evidential scholars preferred a less accurate value from antiquity over a more accurate one from the Song dynasty. The *Zhou Dynasty Classic of Gnomonic Computations*, for example, gave the value of π (*pi*) = "3", but since the Han dynasty there had been many efforts to obtain a more precise value. In his notes for the *Computational Methods in Nine Chapters*

[58]Li Ruzhen, *Jinghua yuan*, 229–235. See also Maram Epstein, "Engendering Order: Structure, Gender, and Meaning in the Qing Novel Jinghua yuan", *Chinese Literature: Essays, Articles, and Reviews* 18 (December 1996): 105–131.

[59]Lin Taiyi, "Introduction", *Flowers in the Mirror, by Li Ju-chen*, 5–9, and Yu Wang Luen, "Knowledge of Mathematics and Science in Ching-Hua-Yuan", 235–236.

completed circa 263 A.D., Liu Hui 刘徽 (fl. third century) inscribed an equilateral hexagon (with one side equal to the radius) in a circle to show that π had to be greater than "3". Liu arrived at a lower value of "3.14" plus 64/625 by calculating the perimeter of a cyclic polygon of 96 sides inscribed in the circle. He also calculated the upper value of π as "3.14" plus 169/625 by inscribing a circle inside a touching polygon. The real value was between these two limits.

In medieval times, Zu Chongzhi 祖冲之 (429–500) had arrived at a more accurate calculation by obtaining a range for π between "3.1415927" and "3.1415926". Subsequently the intermediate figure of 3.14159265 was calculated. Both the Sui and Jin (金) dynasty treatises on the medieval calendar had used the calculation of "3.14159265" that Zu had made. In the late Kangxi era, the *Collected Basic Principles of Mathematics* gave "3.141592653" as the value of π, while Zhu Hong 朱鸿 gave the value to 39 decimal points later in the 18th century. Unlike evidential scholars such as Qian Daxin and Li Rui who settled for a value of π = "3.16", Li Ruzhen used the figure of "3.14" for the novel, which was based on Zu's calculations.[60]

Aspects of the natural world were also discussed by women in *Flowers in the Mirror*. To determine the weight of matter and the velocity of sound, for instance, Li Ruzhen presented the problem in light of the velocity of sound through the medium of lightning and thunder. The speed the novel gave was 1,285.7 feet (*chi* 尺) per second. Newton had given a theoretical value of 979 feet per second and experimental value of 1,142 feet per second for the speed of sound. It is likely that Li Ruzhen's figure was derived from the *Collected Basic Principles of Mathematics*, where the solution to the velocity of sound of a canon blast was given as 1,285.7 feet (*chi*) per second. Converted to meters, the velocity was about 397.3 meters per second, about 20% higher than the modern value.[61]

The level of knowledge of mathematics and natural studies in Li Ruzhen's novel reflected the academic climate among evidential scholars,

---

[60]Li Ruzhen, *Jinghua yuan*, 531–532. See also *Flowers in the Mirror, by Li Ju-chen*, 243, and Yu Wang Luen, "Knowledge of Mathematics and Science in Ching-Hua-Yuan", 221–224.

[61]Li Ruzhen, *Jinghua yuan*, 533–534. See also Yu Wang Luen, "Knowledge of Mathematics and Science in Ching-Hua-Yuan", 233–234.

which had helped make mathematics and astronomy an important part of classical learning. Through the Kangxi emperor's promotion of the *Collected Basic Principles of Mathematics* in the early 18th century, the Jesuit introduction of European mathematics and astronomy was taken seriously by leading classical scholars. Despite the Yongzheng emperor's closed door policy from 1723, which helped impede the transmission of 18th-century science in Europe, the revival of native mathematics in the 18th century was noteworthy.

Dai Zhen, for example, insisted on deferring to the native tradition when using foreign knowledge. Dai maintained that the essential elements of astronomy and mathematics could be located in ancient classical texts. If studied properly, according to Dai, the Classics would prove themselves repositories of mathematical and astronomical knowledge that had been lost due to neglect and lack of understanding. He set out, for instance, to show — mistakenly as it turned out — that a cryptic passage in the *Documents Classic* revealed that the ancients had been aware of the complicated path of the sun on the celestial sphere. In the process, Dai concluded that this and other examples "are clear proof that Western methods were derived from the *Zhou dynasty Canon of Gnomonic Computations* (*Zhoubi suanjing*)", which he had recovered from the Ming archives.[62]

Similarly, other evidential scholars also valued the recovery of ancient mathematical texts. Li Rui's biography of Qian Tang 钱塘 (1735–1790) in the *Biographies of Astronomers and Mathematicians*, for instance, cited Qian's confirmation that the ancient value of π had been "3.16", and that this was more acceptable than Liu Hui's and Zu Chongzhi's more accurate — but later — finding of "3.14". That Li Rui and Qian Daxin both upheld the more ancient finding indicates that many evidential scholars skewed their "search for the truth from facts" (*shishi qiushi* 实事求是) in light of ancient learning.

In this way, the revival of ancient mathematics precluded the development of mathematics itself. On the other hand, Wang Lai preferred to use the European mathematical notation from the Imperial Observatory, for which he was condemned, and Dong Youcheng was critical of Qian Tang's

[62] Benjamin A. Elman, *From Philosophy to Philology: Social and Intellectual Aspects of Change in Late Imperial China*, 118–119.

affirmation of π = "3.16" and affirmed Liu Hui's value of "3.14" by referring to the Jesuit inspired *Collected Basic Principles of Mathematics.* Wang Lai ironically noted: "Contemporary philologists always focus on what their predecessors did, and do nothing but copy what has already been written. They are never able to discover what their predecessors had not yet discovered."[63]

The literary focus of Qing scholarly interests in mathematics and astronomy did not prevent scholars such as Dai Zhen and Qian Daxin from concentrating on mathematics and astronomy, but they connected such studies to what they considered was their more central objective: the reconstruction of antiquity. Their lesser concern for new discoveries in mathematics prevented them from realizing the full potential of natural studies as an independent field of inquiry before the late 19th century. On the other hand, there were little contacts with Europeans after 1750 until the 1793 Macartney mission, and even after that no Europeans during the French Revolution and the following Napoleonic era transmitted the new sciences and calculus in England and France to China.

Their concern with documents restricted 18th-century evidential scholars to a textual focus, even if they occasionally made use of archaeological findings or carried out some astronomical investigation. Once keeping the calendar accurate was no longer an insurmountable technical problem, research on ancient astronomy was valued mainly for its application to classical studies. The exact fields of astronomy and mathematics were rarely perceived as anything more than fields ancillary to more important classical and historical concerns. Inquiries into natural phenomena, for the most part, were still dependent on textual evidence and not experimentation.

The Usefulness of Recovering Ancient Mathematics

Evidential scholars in the 18th century were not doomed to a lack of curiosity about the natural world or mathematics, but the philological biases

---

[63] Ruan Yuan, *Chouren zhuan*, 42.545. See also Horng Wann-sheng, "Chinese Mathematics at the Turn of the 19th Century", in Cheng-hung Lin and Daiwie Fu, eds., *Philosophy and Conceptual History of Science in Taiwan*, 183–190, especially 186.

that dominated their scholarship did not independently support the nascent research and experimentation required in the step-by-step quantification of the natural world. In light of the important place mathematics and astronomy occupied in evidential research, however, we cannot assume that because a scientific revolution did not take place in China it could not have taken place. It is remarkable how quickly — not overnight to be sure — the Chinese people adapted to the needs of science and technology in the late 19th century. Politics and economics, not science or technology, made it seem as if little progress had been made before the Qing defeat during the Sino-Japanese War of 1894–1895 (*Jiawu zhanzheng* 甲午战争).

The triumph of modern science in China, as in Europe, required an intellectual transformation that would exceed the boundaries of textual scholarship, and social, economic, and political changes that would challenge the social pedigree of classical studies. Such intellectual and technological transformations began during the Newtonian century in Europe, a century before they were fully understood in China. European science in 18th-century China was not built upon and developed partly because of faulty Jesuit transmission that failed to challenge native classicism or provide an academic alternative. The preeminent position of classical studies along with its historical focus, remained intact. That preeminence would be shaken in the aftermath of the Taiping Rebellion (1850–1864).[64]

Nevertheless, important mathematical research did take root, and the educational institutions required for precise scholarship were already in place when Protestant missionaries made their way to the China coast in the post-Napoleonic era. Moreover, the successful reconstruction of Song–Yuan mathematical texts made it easier for a significant number of Chinese literati in the 19th century to recognize the precise significance of and the need to master the new developments in advanced algebra, analytic geometry, and the differential and integral calculus that would be introduced via translation by Protestant

[64] Nathan Sivin, "Why the Scientific Revolution did not take place in China — or didn't it?" reprinted in Nathan Sivin, *Science in Ancient China: Researches and Reflections*, VII, 45–66.

missionaries. These translations were possible because of the cooperation of "traditional" Chinese mathematicians who fully understood the "single unknown" and "four unknowns" techniques that had been successfully restored.

In the early 19th century, for example, Luo Shilin commented on the relative strengths of traditional and European mathematics before the impact of the calculus in China. Luo was versed in European mathematics after he studied in the Astro-calendric Bureau as a stipend student for seven years. His early work followed the Bureau's European tradition, but he changed his mind when he went to capital in 1822 for the provincial examination, which he never passed. While in Beijing he had been finally able to read Zhu Shijie's *Jade Mirror of the Four Unknowns* on "single unknown" methods. In Zhu's work he discovered a method powerful enough to solve sophisticated mathematical problems.

Luo, who perished in the Taiping assault on Yangzhou in 1853, pointed out that European mathematics — in terms of trigonometry, logarithms, and its method of borrowing roots and powers — was not as powerful as Song–Yuan "single unknown" and "four unknowns" techniques, by then completely reconstructed, which could solve problems that Jesuit algebra could not. Luo successfully explored the geometrical properties of the cone using "single unknown" operations, and his 1840 *Supplement to the Calculations of Segments of Circles* (*Hushi suanshu bu* 《弧矢算术补》) extended Li Rui's work on arcs by adding many problems Luo solved using "single unknown" procedures. Not yet aware of the calculus, he urged Chinese scholars not to follow European mathematics too closely for their practical measurements.[65]

With the introduction of the differential and integral calculus in the mid-19th century, for which the Chinese tried but could not find a native precedent, Li Shanlan and other Chinese mathematicians such as Hua Hengfang admitted that although the "four unknowns" notation was perhaps superior to Jesuit algebra, which Alexander Wylie acknowledged, the Chinese had not developed the calculus. Moreover, after the Opium War

[65] Horng Wann-sheng, "Chinese Mathematics at the Turn of the 19th Century", 187, 199. See also Hu Mingjie, "Merging Chinese and Western Mathematics", 214–223, ECCP, 538–539, and Alexander Wylie, *Notes on Chinese Literature*, 125.

the most influential Chinese mathematicians no longer were devoted exclusively to the revival of ancient Chinese mathematics. They merged European and Chinese mathematics into a new synthesis, which drew extensively on evidential studies and the renovation of Song–Yuan mathematics during the Qianlong era.[66]

---

[66]Andrea Bréard, *Re-Kreation eines mathematischen Konzeptes im chinesischen Diskurs: "Reihen" vom 1. bis zum 19. Jahrhundert.*

Chapter 6

The China Prize Essay Contest and the Late Qing Promotion of Modern Science

The second half of the 19th century was the "seeding time" for the modernization of China. During the period from 1850 to 1870, many works on astronomy, mathematics, medicine and related fields of botany, geography, geology, mechanics, and navigation were translated by a core group of Protestant missionaries and Chinese co-workers in Guangzhou, Beijing and Shanghai. Parallel to the arsenals and official schools, private initiatives were also needed to popularize "modern science" (*gezhixue* 格致学) in the treaty ports and among Qing officials and literati. To attract the interest of the literati mainstream, John Fryer (1839–1928) and Wang Tao 王韬 (1828–1897) devised the "China Prize Essay Contest" (*Gezhi shuyuan keyi* 格致书院课艺) in 1886. The essay writing contest was conceived by Fryer as a means to attract the many Chinese literati proficient in civil examination essay writing to write about foreign subjects, including science and technology. What is interesting about the China Prize Essay Contest, which lasted through 1893, is that it used the prestige of the imperial civil service examinations to aid missionary efforts to promote modern science during the late Qing, which in turn created a working partnership between Western translators and Qing high officials, particularly those in provinces with treaty ports.

After the civil examinations were abrogated in 1904, however, the Qing dynasty quickly lost control of its science policies, which increasingly fell under the control of local Chinese intelligentsia and overseas students in Japan and Europe. The political consequences of this loss of control were inevitable when the Qing bureaucracy gave up one of its major institutions that had for centuries successfully induced literati acceptance of the imperial system. The essay contest helped to popularize the new knowledge contained in the science translations from 1865 to 1885 by employing a traditional vehicle to valorize that knowledge. The Sino-Japanese War of 1894–1895 (*Jiawu zhanzheng* 甲午战争) aborted such efforts, but had the Chinese won that war, which was possible, such efforts to spread the new sciences through traditional institutional forms might have been legitimated and expanded.

Translations at the Jiangnan Arsenal

In 1867, a Translation Department was initiated at the Jiangnan Arsenal (*Jiangnan zhizaoju* 江南制造局) in Shanghai. The initiative was led by Xu Shou 徐寿 (1818–1882), Hua Hengfang 华蘅芳 (1833–1902), and Xu Jianyin 徐建寅 (1845–1902), classical scholars with scientific interests. In addition to emphasizing foreign manufacture, Zeng Guofan 曾国藩 (1811–1872) and Li Hongzhang 李鸿章 (1823–1901) regarded translation as the foundation for learning the techniques of modern manufacture and the mathematics on which they were based. Their precedent was the late Ming and early Qing translation projects that had enabled the imperial calendar to be successfully reformed based on new techniques and models introduced by the Jesuits and used in the Astronomy Bureau.[1]

Invited to join the Department in 1867, John Fryer wrote in 1880, for instance, that the Jiangnan Arsenal had published translations of Western works since 1871. By June 30, 1879, some 98 works were published in 235 volumes (*juan* 卷). Of these, 22 dealt with mathematics, 15 were on naval and military science, and 13 covered the arts and manufactures. Fryer reported that another 45 works in 142 volumes were translated but not yet published, and 13 other works were in process with 34 volumes already completed.

Altogether, the Translation Office had sold 31,111 copies representing 83,454 volumes, and this had been accomplished without advertisements or postal arrangements. A work on the German Krupp guns translated in 1872 sold 904 copies in eight years. Another work on coastal defense published in 1871 sold 1,114 copies in nine years. *A Treatise on*

[1] Knight Biggerstaff, "Shanghai Polytechnic Institution and Reading Room: An Attempt to Introduce Western Science and Technology to the Chinese", *Pacific Historical Review* 25 (May 1956): 127–134.

Practical Geometry (1871) sold 1,000 copies in eight years; and *A Treatise on Algebra* (1873) sold 781 copies in seven years. Fryer's work on coal mining published in 1871 sold 840 copies in nine years. Publicizing these works beyond Shanghai, Beijing, and the treaty ports was difficult, and even for the latter venues such numbers were disappointing but not insignificant.[2]

For example, the controversial reformer and classicist Kang Youwei 康有为 (1858–1927) purchased all the Arsenal works when he was in Shanghai in 1882. Between 1890 and 1892, his disciple Liang Qichao 梁启超 (1873–1929) also purchased many of the Arsenal's translations as well as Fryer's science journal entitled *The Chinese Scientific and Industrial Magazine* (*Gezhi huibian* 《格致汇编》). Liang developed an influential reading list based on these materials known as the *Bibliography of Western Learning* (*Xixue shumu biao* 《西学书目表》), which was revised and published in 1896. Of these 329 published works, 119 (36%) were translated by Fryer. Tan Sitong 谭嗣同 (1865–1898) began writing on scientific topics in the 1890s and mentioned *The Chinese Scientific and Industrial Magazine* as one of his sources of scientific learning. Tan had visited Fryer in Shanghai in 1893 and bought many of the Arsenal's works.[3]

Shanghai Polytechnic and The Chinese Scientific and Industrial Magazine

The Shanghai Polytechnic Institution (*Gezhi shuyuan* 格致书院) and Reading Room (*Shushi* 书室), where *The Chinese Scientific and Industrial Magazine* was published, had been founded in 1874–1875. Its programs promoted the sciences, arts, and manufactures of the West through exhibitions, lectures and classes, and a Chinese Library and Reading Room. Because the Polytechnic did not draw the expected interest hoped for, Fryer and Xu Shou also created the scientific journal in China to reach out

[2] John Fryer, "An Account of the Department for the Translation of Foreign Books at the Kiangnan Arsenal, Shanghai", *North-China Herald*, January 29, 1880: 77–81.

[3] Adrian Bennett, *John Fryer: The Introduction of Western Science and Technology into Nineteenth-Century China*, 42–44.

to the Chinese in the treaty ports. In 1885, a project for classes and public lectures was finally implemented, and the scientific essay contest was initiated and became popular. Instructional plans included an appointment for a foreign professorship of science, which did not materialize.

Fryer's idea for the new journal immediately drew support from the Society for Diffusion of Useful Knowledge (*Guangxue hui* 广学会) in Beijing, which was closing down in 1875 and ending its illustrated monthly known as *The Peking Magazine* (*Zhongxi wenjian lu* 《中西闻见录》). The Society's members in Beijing transferred their subscriptions to the Polytechnic. Although the Shanghai journal was published and sold through the Polytechnic, it was a separate enterprise that Fryer and his Chinese assistant took responsibility for. Because Fryer's written classical Chinese was not good enough to produce the journal by himself, he employed Luan Xueqian 栾学谦 as his private secretary to help translate the unattributed Chinese articles in the journal. In the past it was assumed these unauthored pieces were all composed by Fryer. Luan was trained at Calvin Mateer's (1836–1908) Hall of the Culture Society (*Wenhui guan* 文会馆) school in Shandong, where science and advanced mathematics courses were taught in Chinese.[4]

Fryer had worked with Luan since at least 1877 when *The Chinese Scientific and Industrial Magazine* commenced. Luan Xueqian, for example, prepared an account of a chemistry class at Shanghai Polytechnic before he collaborated with Fryer. Moreover, Luan was probably also involved in the *Science Outline Series* (*Gezhi xuzhi* 《格致须知》) and *Science Handbook Series* (*Gezhi tushuo* 《格致图说》) that were published from 1882 to 1898. In addition, Luan managed the Chinese Scientific Book Depot (*Gezhi shushi* 格致书室) for Fryer from 1885, which Fryer eventually turned over to Luan in 1911.[5]

Copies of the Polytechnic's journal were available at first at 24 and then 27 of the most important treaty ports and trading centers in China and

[4] See Knight Biggerstaff, "Shanghai Polytechnic Institution and Reading Room: An Attempt to Introduce Western Science and Technology to the Chinese", 144.

[5] Wang Yangzong, "Gezhi huibian zhi Zhongguo bianjizhe kao" 〈格致汇编之中国编辑者考〉, *Wenxian* 《文献》 63 (January 1995): 237–243. See also David Wright, *Translating Science: The Transmission of Western Chemistry into Late Imperial China, 1840–1900*, 319–325.

Japan. There were 30 agents in early 1880, which increased to seventy by the end of the year. Although *The Chinese Scientific and Industrial Magazine* continued from *The Peking Magazine*, it went beyond the latter by focusing on the natural sciences and technology in Europe and the United States. With more Chinese participants in the production process, the translations in the Shanghai journal were much better.

The journal initially printed 3,000 copies and usually sold out in several months. Nine months later, the first nine issues were reprinted in a second edition to meet the demand. With 4,000 copies printed per issue at its peak in the 1880s and 1890s, it reached some 2,000 readers in the treaty ports. Fryer hoped that in this way, popular-oriented presentations of mathematics and the industrial sciences would become more acceptable among literati and merchants. He expected that *The Chinese Scientific and Industrial Magazine* would also compensate for the limited scope of the Jiangnan Arsenal's translations, which were usually printed in runs of only a few hundred or more copies. Later in 1891, reprints of previous issues were sold.[6]

Altogether, 60 issues were published intermittently over seven years. After 1880, the magazine shifted its emphasis from introductory essays on science to accounts of the basic fields of science. Moreover, Fryer increasingly paid attention to mathematics as the foundation of scientific knowledge. The Polytechnic's teaching program was most effective after 1885 when it pioneered the teaching of mathematics and science in late Qing China. According to a General Affairs Office (*Zongli yamen* 总理衙门) memorial of May 18, 1887, which advocated modifying the civil examinations to allow candidates to be examined in mathematics, the Shanghai Polytechnic was training half of the talented students of mathematics in the empire.[7]

[6]The venues to buy issues were advertised in the journal. Compare San-pao Li, "Letters to the Editor in John Fryer's *Chinese Scientific Magazine*, 1876–1892: An Analysis", 743. See also Ferdinand Dagenais, *John Fryer's Calendar: Correspondence, Publications, and Miscellaneous Papers with Excerpts and Commentary*, Version 3, 1891:1, Adrian Bennett, *John Fryer: The Introduction of Western Science and Technology into Nineteenth-Century China*, 50–55, and David Wright, "Careers in Western Science in Nineteenth-Century China: Xu Shou and Xu Jianyin", *Journal of the Royal Asiatic Society* 5 (1995)： 49–90.

[7]*Guangxu chao donghua lu*《光绪朝东华路》(Records from within the Eastern Gate during the Guangxu reign), 82.11a. See also Knight Biggerstaff, "Shanghai Polytechnic Institution and Reading Room: An Attempt to Introduce Western Science and Technology to the Chinese", 148–49.

In addition *The Chinese Scientific and Industrial Magazine* published some 317 inquiries sent in to the journal as "Letters to the editor". The letters and their context stressed the practical value of technology and showed less interest in pure science. About 123 letters (38.2%) showed some interest in or knowledge of scientific theories or abstract scientific models, which is a relatively high percentage for a journal more oriented to popular science or popular mechanics. The queries paralleled the technological interests of those involved in the Self-Strengthening Movement (*Ziqiang yundong* 自强运动). The large volume of letters to Fryer anticipated a more widespread awakening of interest in science after 1895 and paved the way for the general acceptance of Western science in China.[8]

Besides their use in the increasing number of missionary schools, such translations were also institutionalized as texts within a regional matrix of arsenals, factories, and technical schools that formed the 19th-century roots of the 20th-century industrial revolution in China. Hence, we should also acknowledge the scope and scale of scientific translation and military arsenals elsewhere in China after 1860.[9]

In his "Report of the Chinese Scientific Book Depot, Shanghai, 1887", John Fryer summarized the results of the first three years of its operations, which had aimed at "facilitating the spread of all useful knowledge literature in the native language throughout China, and especially of books, maps and other publications of a scientific or technical character". Branch depots were established in Tianjin, Hangzhou, and Shantou in 1886, and four more in Beijing, Hankou, Fuzhou, and Xiamen were added in 1887. Fryer noted that 17,000 dollars worth of books and maps had been sold since 1885. That translated into approximately 150,000 volumes of scientific and educational literature that had "found their way to the most distant parts of China as well as to Japan and Corea".[10]

The 1886 catalog of the Scientific Book Depot, for instance, included mention of 371 works for sale. Fifty-nine titles were on science, 49 on

---

[8] San-pao Li, "Letters to the Editor in John Fryer's *Chinese Scientific Magazine*, 1876–1892: An Analysis", *Jindaishi yanjiu suo jikan* 《近代史研究所集刊》 4 (1974): 730–731, 762.

[9] Ting-yee Kuo and Kwang-Ching Liu, "Self-Strengthening: the pursuit of Western technology", in Denis Twitchett and John Fairbank, eds., *The Cambridge History of China*, Volume 10, *Late Ch'ing, 1800–1911*, Part 1, 519–537.

[10] For the 1887 "Report", see the *North-China Herald*, December 28, 1887: 702–703.

Chinese studies, and 35 on mathematics. The catalog also included 44 works that Fryer had translated, 28 of them for the Jiangnan Arsenal. By 1888, there were 650 titles on Western topics at the Scientific Book Depot. Of these, 228 were original Chinese works. Books on the Sino-Japanese War were popular after 1895 when the Depot became a mecca for young Chinese students of the sciences and mathematics in China.[11]

Another hopeful sign that Fryer included in his report was that knowledge of mathematics and science was now required in the Qing civil service examinations. W. T. A. Barber from the Wuchang High School later reported in July 1888, for example: "At the recent examination for the Siu Ts'ai [*xiucai* 秀才] degree held in Han-yang-fu, the district [prefecture] containing the mart of Hankow [Hankou, Hubei], three candidates presented themselves in mathematics." Two passed by answering questions dealing with four questions on right-angled triangles to find the sides and two questions on solid contents to find the dimensions. Barber regarded the answers as equivalent to the "Pass B.A." at Cambridge or Oxford. He added that three years earlier three of the 66 graduates had received their provincial degrees for mathematics in Hunan. Candidates for the local degree, he thought, were learning sufficient mathematical knowledge to pass. He was apparently unaware that these problems had long been solvable using traditional Chinese "four unknowns" notations for algebraic equations.[12]

Despite the considerable efforts Fryer, Xu Shou, and Wang Tao expended on maintaining the viability of the Polytechnic and its affiliated if irregular *The Chinese Scientific and Industrial Magazine*, as well as the Chinese Scientific Book Depot, they were aware of the limits they faced in reaching educated Chinese outside the exclusive audience of literati in Shanghai and the other treaty ports. The vast mainstream of Chinese literati was still drawn to what the Protestants belittled as the useless focus in the prestigious civil examinations on Chinese literature and poetry.[13]

---

[11] Adrian Bennett, *John Fryer: The Introduction of Western Science and Technology into Nineteenth-Century China*, 64–66.

[12] Mathematics questions had been allowed in civil examinations by an 1887 imperial edict. See W. T. A. Barber, "A Chinese examination paper", *North-China Herald*, July 7, 1888: 15.

[13] See Fryer's "Report of the Chinese Scientific Book Depot, Shanghai, 1887", *North-China Herald*, December 28, 1887: 702–703. See also Benjamin A. Elman, *A Cultural History of Civil Examinations in Late Imperial China*, 584, 728, on the 23% increase of gentry with regular degrees to 920,000 after 1850.

Contest Procedures and Official Patronage

To attract the interest of the literati mainstream, Fryer and Wang Tao devised the "China Prize Essay Contest" in 1886. The essay writing contest was conceived by Fryer as a means to attract the many Chinese literati proficient in civil examination essay writing to write about foreign subjects, including science and technology:

> To popularise Western knowledge among the *literati* it is necessary to take advantage of all such existing national characteristics; and hence it was conceived that in essay writing there existed a most powerful means for inducing the better classes of Chinese to read, think, and write on foreign subjects of practical utility, and thus carry out one of the main objects for which the Polytechnic Institution was founded.

Fryer failed to add, however, that numerous prize essays solicited by the Royal academies in London, Paris, and Berlin had similarly been of strategic importance to focus interest on specific issues in the development of the sciences and mathematics during the 17th and 18th centuries in Europe. Moreover, school examinations in the sciences in Britain had been instrumental in promoting such fields of learning.[14]

In time this experiment became one of the most successful undertakings of the Shanghai Polytechnic to spread Western learning beyond the treaty ports. Fryer described the selection process as follows:

> A high official is asked to give a subject on which prize essays are invited, and to promise not only to look over the essays himself but to bestow certain

[14] See John Fryer, "Chinese Prize Essays: Report of the Chinese Prize Essay Scheme in connection with the Chinese Polytechnic Institution and Reading Rooms, Shanghai, for 1886 and 1887", *North-China Herald*, January 25, 1888: 100–101. Compare Knight Biggerstaff, "Shanghai Polytechnic Institution and Reading Room: An Attempt to Introduce Western Science and Technology to the Chinese", 141–143. On prize essay questions in 18th-century Europe, see Roger Hahn, "Laplace and the Mechanistic Universe", in David C. Lindberg and Ronald Numbers, eds., *God & Nature: Historical Essays on the Encounter between Christianity and Science,* 266. See also Keith Hoskin, "Examinations and the Schooling of Science", in Roy MacLeod, ed., *Days of Judgement: Science, Examinations, and the Organization of Knowledge in Late Victorian England.*

> sums of money upon some of the more successful essayists in addition to the regular quarterly amount of Tls. 25 [34.75 silver dollars] voted from the funds of the Institution for this special purpose. Every quarter a fresh subject is advertised in the native newspapers, and a date is fixed after which no essay will be received. The bundle of essays is then forwarded to the co-operating official who reads them carefully over and adjudicates their order of merit, affixing a criticism of greater or less length in his own handwriting, pointing out the features of excellence or defect in each.... Three receive the highest awards and ten or more receive smaller sums.... At the end of each year the three highest names for each quarter are honoured by having their essays, with the criticisms, printed in the form of a book, complimentary copies of which are sent to co-operating high officials and successful essay writers.[15]

Mimicking the palace examination (*dianshi dengke lu* 殿试登科录), three major and ten minor prizes were usually given annually for the China Prize Essay Contest, and the winners were announced in the Chinese and Western press. The best essays were released to newspapers such as the influential *Shanghai Journal* (*Shenbao* 《申报》). From 1886 to 1893, the three major prize winners had their essays printed together in a book that was placed on public sale for others to emulate. Wang Tao was also editor for the special prize science essay volumes published by the Polytechnic, which paralleled collections of 8-legged essays for the civil examinations that were widely in print in Ming and Qing China. This tactic also mimicked the official publication of the policy essays of the top three finishers on the regular palace examination. When Fryer prepared his "Second Report of the Chinese Prize Essay Scheme" in 1889, he proudly announced:

> By its means the existence of the Polytechnic Institution has become known far and wide; the cooperation of some of the highest officials in the Empire secured; and an interest in western ideas has been created in some of the most influential quarters. By the annual expenditure of only a hundred Taels (139 silver dollars) or thereabouts, and by working in harmony with

[15] John Fryer, "Chinese Prize Essays: Report of the Chinese Prize Essay Scheme in connection with the Chinese Polytechnic Institution and Reading Rooms, Shanghai, for 1886 and 1887", 100.

> the Chinese methods of thought, and time-honoured systems of literary competition, a result has been obtained which the use of large sums of money in other ways would have failed to produce.[16]

For Fryer, one of the most encouraging features of the essay competition was the support it received from Qing officials, who quickly saw the efficacy of applying the civil examination ethos, which was so well entrenched among the literati, to Western learning. The Polytechnic's essay contest closely paralleled the civil examination process of reward and fame for prized essays. Li Hongzhang and Liu Kunyi 刘坤一 (1830–1902) as the Northern and Southern Superintendents of Trade respectively, for example, each consented to give an extra theme every year during the spring and fall, in addition to the quarterly arrangement already in place. Fryer noted that when Li Hongzhang's extra theme for the first half of 1889 was issued, 30 essays were forwarded to him for ranking. Li's list of the names of the 27 successful writers, and the extra awards of $204 given them, together with his personal criticisms, were published in the local Chinese papers.[17]

Literati Participation

Literati who sent in essays for the competition between 1886 and 1893 were getting their information for the topics mainly from Jiangnan Arsenal translations, materials prepared at the Beijing School of Foreign Languages (*Tongwen guan* 同文馆) and articles on the sciences that had appeared in *The Chinese Scientific and Industrial Magazine*, *The Peking Magazine*, as well as the *Shanghae Serial* (*Liuhe congtan*《六合丛谈》). The popularity of the "Answers to readers' queries" in *The Peking*

---

[16] John Fryer, "Second Report of the Chinese Prize Essay Scheme in connection with the Chinese Polytechnic Institution and Reading Rooms, Shanghai, From July 1887 to July 1889", *North-China Herald*, July 20, 1889: 85–86. On civil examination essay collections, see Benjamin A. Elman, *A Cultural History of Civil Examinations in Late Imperial China*, 400–420.

[17] The essays were collected by Wang Tao and published in Shanghai in 1897 under the title of *Gezhi keyi huibian*《格致课艺汇编》. Wang Ermin, *Shanghai Gezhi shuyuan zhilue*《格致书院志略》, 54–55, presents a list of officials involved.

Magazine and "Letters to the editors" in *The Chinese Scientific and Industrial Magazine* also indicates the popularity these journals had elicited among literati in and outside the treaty ports, although the real boom in reprinting such publications came after the Sino-Japanese War in 1895.

Following these pioneering compilations, other compendia on the sciences were also widely available. Compiled from 1877 to 1903, they reprinted many science works from the Jiangnan Arsenal Translation Department and from Fryer's "Science Outline Series". Literati found out about these new compilations via book advertisements in the emerging Western and modern Chinese press and the catalogs of Western books mentioned earlier. Advertisements in the 1897 issues of the *Shanghai Journal*, for instance, included mention of new works on chemistry and public affairs that aimed at the civil examination market. Candidates were told in one advertisement that a particular book on Western history was absolutely essential for success on the policy essays required in the civil examinations.[18]

The essay competition met with an enthusiastic response, and many of the essays were not only printed in newspapers but also included in reformist encyclopedias such as the *Collected Writings on Political Economy from the August* [*Qing*] *Dynasty* (*Huangchao jingji wenbian* 《皇朝经济文编》). In a "Table Showing the Results of the Chinese Prize Essay Scheme", Fryer presented a breakdown of the competition. An average of 48 essays were received for each of the 15 contests with 19 essayists (39.5 %) receiving prizes. Fryer's table presented figures through spring 1889, and I have updated these to include contests through 1893. Based on the figures for 42 contests held between 1886 and 1893, in which

[18]The lead article "The Progress of Foreign Studies" in the *North-China Herald*, April 14, 1893: 513–514, described some sources for answering the questions. See also Shang Zhicong, "1886–1894 nianjian jindai kexue zai wan Qing zhishi fenzi zhong de yingxiang — Shanghai Gezhi shuyuan Gezhi lei keyi fenxi" 〈1886–1894 年间近代科学在晚清知识份子中的影响 — 上海格致书院格致类科艺分析〉, *Qingshi yanjiu* 《清史研究》 (August 2001): 73, 82n6, and David Wright, *Translating Science: The Transmission of Western Chemistry into Late Imperial China, 1840–1900*, 163–173.

2,236 candidates presented essays, an average of 46.1% were awarded prizes. Overall, about 53 candidates sent in essays for each contest.[19]

Seventeen Chinese officials presented a total of 86 questions, with several presenting questions a number of different times:

Gong Zhaoyuan 龚照瑗	4 times 4 questions	(Shanghai Circuit)
Li Hongzhang	5 times 15 questions	
Liu Kunyi	3 times 7 questions	
Sheng Xuanhuai 盛宣怀 (1844–1916)	6 times 6 questions	
Wu Yinsun 吴引孙	6 times 12 questions	(Ningbo Circuit)
Xue Fucheng 薛福成 (1838–1894)	3 times 3 questions[20]	

Altogether 1,878 Chinese submitted essays over the nine years of the competition. Among the 92 essayists who were honored (4.9%), several produced five or more essays that were awarded prizes:

Yang Minhui 杨敏辉	Licentiate	14 essays
Li Dingyi 李鼎颐	2nd class provincial	7 essays
Xu Keqin 徐克勤	Supplementary student	7 essays
Wang Zuocai 王佐才	Senior tribute student	6 essays
Yin Zhilu 殷之辂	Polytechnic graduate	6 essays
Li Jingbang 李经邦	2nd class provincial	6 essays
Zhong Tianwei 钟天纬	Expectant official	5 essays
Ye Han 叶瀚	County purchase	5 essays[21]

---

[19] John Fryer, "Second Report of the Chinese Prize Essay Scheme in connection with the Chinese Polytechnic Institution and Reading Rooms, Shanghai, From July 1887 to July 1889": 86. Complemented by figures from *Gezhi shuyuan keyi* (hereafter, GZSYKY) from the Fryer Private Library, University of California, Berkeley, Vols. 1–2, 1889–1893. See also Knight Biggerstaff, "Shanghai Polytechnic Institution and Reading Room: An Attempt to Introduce Western Science and Technology to the Chinese", 149.

[20] Xiong Yuezhi, *Xixue dongjian yu wan Qing shehui* 《西学东渐与晚清社会》, 385–386.

[21] Xiong Yuezhi, *Xixue dongjian yu wan Qing shehui*, 387–391. Compare Shang Zhicong, "1886–1894 nianjian jindai kexue zai wan Qing zhishi fenzi zhong de yingxiang — Shanghai Gezhi shuyuan Gezhi lei keyi fenxi", 81, and Wang Ermin, *Shanghai gezhi shuyuan zhilue*, 69–72.

Despite such success and patronage, Fryer was still troubled by the small number of essays Chinese sent in for the extra spring 1889 competition: "The fact that only thirty essayists dared to tackle all three subjects is an evidence of the general ignorance of the literati on everything outside the ordinary curriculum of Chinese study; while at the same time it shows how effectively this prize essay scheme is doing its work." Fryer also saw some limits in the sorts of questions that other officials such as the Shanghai circuit intendent or governor-general in Nanjing prepared: "[A]lthough their questions relate, perhaps, more to political economy and commerce than to the severer branches of science, it is still gratifying to see how patriotic they are, and how they regard knowledge from the practical, utilitarian point of view, rather than from the theoretical alone."[22]

In time, the Shanghai Polytechnic prize essays themselves, discussed below, became sources of information, as indicated by the publication of the *Compendium of Prize Essays on Science* (*Gezhi keyi huibian* 《格致课艺汇编》) in 1897 and the *Shanghai Polytechnic Prize Essay Competition* (*Gezhi shuyuan keyi* 《格致书院课艺》) again in 1898. Key bibliographies of Western learning were also compiled in the 1890s such as *Books on Eastern and Western Learning* (*Dong Xixue shulu* 《东西学书录》), when published in 1899, praised *The Chinese Scientific and Industrial Magazine* published by the Polytechnic's for introducing the sciences to a generation of literati since the 1870s.

These essay competitions had a great deal of impact as model essays on the reformed 1901–1904 civil examinations, which were promulgated after the Boxer Rebellion (*Yi He Tuan* 义和团) and required all candidates to be knowledgeable of the new fields in the sciences and world affairs. Often, candidates who had prepared essays for the Polytechnic Prize Essay Competition, such as Zhong Tianwei (1840–1900), went on to take the reformed civil examinations. In addition, many of the policy questions used in the reformed civil examinations were derived from the topics chosen by officials such as Li Hongzhang for the essay competition.

[22] John Fryer, "Second Report of the Chinese Prize Essay Scheme in connection with the Chinese Polytechnic Institution and Reading Rooms, Shanghai, From July 1887 to July 1889", 85.

Although the civil examinations were abrogated in 1904, these scientific texts and science topics remained important educationally in the new schools (*xuetang* 学堂) formed after 1905.[23]

Hence, from the 1880s to the end of the Qing dynasty, we can document a significant increase in numbers of literati, such as Du Yaquan 杜亚泉 (1873–1933), who were educated in the modern sciences in this period. In 1911, Du became the editor of *Eastern Miscellany* (*Dongfang zazhi* 《东方杂志》), a scholarly journal published by Commercial Press (*Shangwu yinshuguan* 商务印书馆) in Shanghai, but he had gained his knowledge of science at the end of the 19th century and first served as a science educator in Zhejiang province before taking charge of the science publications section of the Press in 1904. The growth of a significant community of scholar-officials conversant with science, which accompanied the growth of thousands of technicians, engineers, and artisans in the empire-wide arsenals, began in the 1880s and 1890s — before the Sino-Japanese War.[24]

Prize Essay Topics and Their Scientific Content

Between 1886 and 1894, 46 prize essay contests were held on Western learning. A total of 92 essay topics were selected by the Polytechnic and Qing officials. Of these, 33 (36%) dealt with political economy and industry (*fuqiang* 富强). The sciences were next with 24 topics (26%). Fryer noted the importance of the latter in his first report for 1886 and 1887. The questions that Li Hongzhang and his Qing colleagues prepared also influenced the reformed civil examinations initiated after 1900 as part

[23] Xiong Yuezhi, "Gezhi huibian yu xixue chuanbo" 〈《格致汇编》与西学传播〉, *Shanghai yanjiu luncong* 《上海研究论丛》 1 (1989): 65–66, and Wang Ermin, *Shanghai gezhi shuyuan zhilue*, 69–72.

[24] *Zhongwai shiwu cewen leibian dacheng* 《中外时务策问类编》(1903 edition), 16.1a–b (essay prepared by Zhong Tianwei), 18.2b–3a (civil examination question taken from Xu Xingtai's 徐兴台 spring 1887 Polytechnic essay competition topic). See also Benjamin A. Elman, "Naval Warfare and the Refraction of Qing Nineteenth Century Industrial Reforms into Failure", *Modern Asian Studies*, 38, 2 (Fall 2003): 283–326.

of the "New Governance" (*Xinzheng* 新政) policies initiated in the last decade of the Qing dynasty.[25]

The revamping of the civil examinations held empire-wide for some 150,000 provincial candidates meant that science questions were now regularly introduced under the category of science. For example, five of the eight post-1900 essay topics on the natural sciences were phrased as follows:

1. Much of European science originates from China; we need to stress what became a lost learning as the basis for wealth and power.
2. In the sciences, China and the West are different; use Chinese learning to critique Western learning.
3. Substantiate in detail the theory that Western methods all originate from China.
4. Prove in detail that Western science studies mainly were based on the theories of China's pre-Han masters.
5. Explain why Western science studies are progressively refined and precise.
6. Itemize and demonstrate using scholia that theories from the Mohist Canon preceded Western theories of calendrical studies, optics, and mechanics.[26]

Such questions and their answers revealed that in official terms, the wedding between the traditional Chinese sciences and Western science was still in effect among imperial examiners.

---

[25] See GZSYKY, Vol. 1, 1887–1890, Vol. 2, 1891–1893. Compare Xiong Yuezhi, *Xixue dongjian yu wan Qing shehui*, 374n3 and Shang Zhicong, "1886–1894 nianjian jindai kexue zai wan Qing zhishi fenzi zhong de yingxiang — Shanghai Gezhi shuyuan Gezhi lei keyi fenxi", 79–80, with Wang Ermin, *Shanghai gezhi shuyuan zhilue*, 56–68. Shang presents all 24 essay topics on science from 1886 through 1894. Fryer's collection at the University of California, Berkeley, library ends with the 1893 essays.

[26] See *Zhongwai shiwu cewen leibian dacheng*, *mulu* 目录, 13a–b.

Beginning with the first civil examinations since the enactment of the post-Boxer reforms in 1902, many of the essays in the Shanghai Polytechnic Prize Essay Contest became model essays for the required science policy questions. Xu Xingtai's 徐兴台 spring 1887 theme comparing the sciences of China and the West, and Nie Jigui's 聂缉槼 (1855–1911) spring 1894 question on locating the Western principles of calendrical studies, optics, and mechanics in the ancient text of *Master Mo's Teachings* (*Mozi*《墨子》) became a model policy question that Qing examiners might use for civil examination policy questions between 1902 and 1903.

The Polytechnic's prize essays thus give us a unique vantage point from which to evaluate the more common understanding of modern science among Qing literati. Normally we focus on the perspectives of a few leading classical scholars such as Kang Youwei, Liang Qichao, or Tan Sitong when we present the impact of science in the late Qing, which is overdetermined in favor of a few prominent reformers who were more interested in the political importance of science rather than the study of science itself. The 92 questions prepared by 18 Qing officials for the Polytechnic's 46 regular and special essay competitions from 1886 to 1894 provide a better overall frame of reference.[27]

For example, Zhong Tianwei (1840–1900), an expectant appointee to be a magistrate's aide in Guangdong province from Huating, Jiangsu, prepared an essay for Li Hongzhang's spring 1889 "Extra Theme" on native and Western science. Parts of his answer, presented below, were written following the exact parallelism of an 8-legged civil examination essay (*baguwen* 八股文):

> With regard to the sciences, China and the West are different.
>
> Speaking of it from the angle of what is above form,
>
> then earlier literati scholars clarified everything leaving nothing out.

[27] See *Zhongwai shiwu cewen leibian dacheng*, *mulu*, 13a–b. For discussion of the civil examination reforms of 1901–1902, see Benjamin A. Elman, *A Cultural History of Civil Examinations in Late Imperial China*, 594–602. See also Xiong Yuezhi, *Xixue dongjian yu wan Qing shehui*, 362–363.

Speaking of it from the angle of what is below form,

then the new principles of the West daily emerge without end.

It is likely that,

China has emphasized the Way while undervaluing the arts.

Therefore, Chinese science solely prioritized meaning and principles as important.

The West has emphasized the arts and undervalued the Way.

Therefore, Western science has focused more on the principles of things

This is why China and the West have diverged.[28]

In another spring 1889 essay prepared for Li Hongzhang's "Extra Theme" on the development of Western science since Aristotle (384–322 B.C.), Wang Zuocai, a student at Shanghai Polytechnic, argued:

> Therefore, Master Zhu [Xi] 朱熹 [1130–1200] appended a chapter to the [*Great Learning*] commentary . . . , but what was elucidated was a form of science [*gezhi*] that stressed meanings and principles and not the *gezhi* that emphasized the principles of things. China has stressed the Way and undervalued the arts. Anything to do with statutes and institutions, or rituals and music for moral edification and civilizing, were always stressed without leaving anything undone. If a sage were to reappear, he would have nothing to add. Only the makeup of the principles of things, which have examinable forms, have been discarded.[29]

This perspective stressed that since the Greeks the West had focused on things themselves as objects of analysis in contrast to the political, moral, and institutional focus of classical scholars in imperial China.

---

[28] See GZSYKY, Vol. 1, 1889, 20b. The essay is cited from another edition in Xiong Yuezhi, *Xixue dongjian yu wan Qing shehui*, 371–372. See also Benjamin A. Elman, *A Cultural History of Civil Examinations in Late Imperial China*, 389–399, for stylistic examples of the 8-legged essay.

[29] See GZSYKY, Vol. 1, 1889, 6b, which is also cited from another version in Xiong Yuezhi, *Xixue dongjian yu wan Qing shehui*, 371.

Medical Questions as Prize Essay Topics

Since Dr. Benjamin Hobson's (1816–1873) pioneering translations of 19th-century Western medicine in the 1850s, the anatomical and surgical strengths of Western medicine, when compared to the therapeutic efficacy of traditional Chinese medicine, were increasingly noted. The missionary vocation that informed Hobson's work and that of his successors had focused on medical education and missionary hospitals. For example, the English missionary and physician John Dudgeon (1837–1901) introduced courses in anatomy and physiology at the School of Foreign Languages in Beijing in 1872. Dudgeon also published a six-volume work in Chinese on anatomy, which the Qing government subsidized in 1887. A companion volume on physiology also came out.[30]

Between 1874 and 1905, the number of professional medical missionaries had risen from ten to some 300. In 1876, there were 40 missionary hospitals and dispensaries treating 41,281 patients. Three decades later, there were approximately two million treated annually in 250 mission hospitals and dispensaries. Dr. John G. Kerr (1824–1890), who was associated with the American Presbyterian Mission, took over Peter Parker's (1804–1888) Guangzhou hospital, and supervised the treatment of over a million patients during his half century of service to his Cantonese patients.

Other large missionary hospitals were established in Hangzhou and Tianjin. The latter was endowed by Li Hongzhang's wife to repay Dr. John K. Mackenzie (1850–1888) for saving her life. In addition, a generation of Chinese trained as modern physicians also emerged after 1870. Many, such as Sun Yat-sen 孙中山 (1866–1925), were trained in missionary hospitals in China or Hong Kong. By 1897, there were about 300 Chinese doctors who had graduated from missionary medical schools, with another 250–300 then in training. Many more fully trained native assistants made up the staffs of most such hospitals and dispensaries.[31]

---

[30] See Su Jing, *Qingji Tongwen guan ji qi shisheng* 《清季同文馆及其师生》, 252–255. See also Ting-yee Kuo and Kwang-Ching Liu, "Self-Strengthening: the pursuit of Western technology", 531–532, and Bridie Andrews, "Tailoring Tradition: The Impact of Modern Medicine on Traditional Chinese Medicine", in Viviane Alleton and Alexeï Volkov, eds., *Notions et Perceptions du Changement en Chine*, 152.

[31] See Paul Cohen, "Christian Missions and Their Impact to 1900", in Denis Twitchett and John Fairbank, eds., *The Cambridge History of China*, Volume 10, *Late Ch'ing, 1800–1911*, Part 1, 574–575.

Bridie Andrews has noted that several Chinese revolutionaries studied Western medicine early in their careers. Sun Yat-sen, for example, was a member of the first graduating class of the Hong Kong College of Medicine in 1892. Both Guo Moruo 郭沫若 (1892–1978) and Lu Xun 鲁迅 (1881–1936) had traveled to Japan to study modern medicine, and Guo completed his premedical and medical courses there. Lu Xun turned to literature after viewing depressing news reels of the Russo-Japanese War fought on Chinese soil, but he described his youthful zeal for Western medicine as a reaction against the "unwitting or deliberate charlatans" who posed as traditional Chinese physicians. Lu also gleaned from existing translations that the Japanese revival had been based in part on the "introduction of Western medical science to Japan".[32]

Not yet triumphant until the 20th century, Western medicine still faced a sizeable opposition among traditional Chinese physicians in the late Qing, although many Chinese had joined Xu Shou in his critique of the conceptual weaknesses in the etiology provided by Chinese medical theories. Moreover, the Western anatomy of blood vessels and the nervous system were gradually more integrated by traditional Chinese physicians. Accordingly, when the Polytechnic included medical topics for the Prize Essay Contest, the rout of traditional medicine was not yet on.[33] In spring 1891, for instance, the topic chosen dealt with the medieval traditions of *materia medica* that had linked particular foods to human health. This query was directed at "students who researched the principles of things" (*wuli* 物理). Moreover, it was framed in light of Ji Kang's 嵇康 (233–262) medieval advocacy of medical and spiritual techniques for "nourishing life" (*yangsheng* 养生), which informed many late Ming encyclopedias.

The prize essays appealed to the methods for investigating things that had informed medieval lexicographies such as the Jin scholar Yang Quan's 杨泉 *On the Principles of Things* (*Wuli lun*《物理论》) and Zhang Hua's

[32] Bridie Andrews, "Medical Lives and the Odyssey of Western Medicine in Early Twentieth-Century China", paper presented at the History of Science Society Annual Meeting, San Diego, CA, November 8, 1997. See also Lu Xun, *Selected Works of Lu Hsun*, translated by Yang Hsien-yi and Gladys Yang, 2, and Howard Boorman and Richard Howard, eds., *Biographical Dictionary of Republican China*, 170–189. Compare Benjamin A. Elman, "Wang Kuo-wei and Lu Hsun: The Early Years", *Monumenta Serica*, 34 (1979–80): 389–401.

[33] Bridie Andrews, "Tailoring Tradition: The Impact of Modern Medicine on Traditional Chinese Medicine", 152–153.

张华 (232–300) *A Treatise on Curiosities* (*Bowu zhi* 《博物志》). In addition, the essays engaged in a comparative pharmacopoeia that related Chinese and Western foods. Most importantly, however, the essays pointed to the strength of modern chemistry to elucidate the alchemical findings of medieval masters of *esoterica* (*fangshi* 方士). Citing Martin's *Elements of Natural Philosophy and Chemistry* (*Gewu rumen* 《格物入门》), one essay noted that native traditions for "nourishing life" could be complemented by Western chemistry to get a better sense of the efficacy of *materia medica* (*bencao* 本草) in China. When Wang Tao added his comments for the published version, he agreed that medieval *esoterica* had paralleled Western chemistry in important ways.[34]

Similarly, when Liu Kunyi prepared an essay topic on medicine for the special fall 1892 Prize Essay Contest, he asked authors to address which medical tradition, Chinese or Western, was superior theoretically. The Zhejiang literatus Xu Keqin, who submitted seven prize essays, stressed the achievements of ancient Chinese physicians, especially Zhang Zhongjing 张仲景 (150–219), whose *Treatise on Cold Damage Disorders* (*Shanghan lun* 《伤寒论》) had by early Ming times reached canonical status. Xu added that Western physicians emphasized the nervous system but they were unaware of the meridian system of twelve circulation tracts (*jingluo* 经络) needed to understand the human body and its susceptibility to illness. In particular, Xu and the other essays focused on the strength of "heat factor" (*rebing* 热病) and "cold factor" (*shanghan* 伤寒) therapies in Chinese medicine, while pointing to the dangers in the surgical techniques employed by Western physicians.[35]

In his prize essay for the summer 1893 contest, Xu Keqin again addressed a query by the Ningbo circuit intendent Wu Yinsun on the comparative strengths and weaknesses in Chinese and Western medicine. This time, however, the essayists were asked to demonstrate their knowledge of the history of Western medicine and its most famous physicians. Xu provided such a summary, but he still concluded that Chinese therapies

[34] Gezhi shuyuan keyi 《格致书院课艺》 (hereafter, GZSYKY), Vol. 2, 1891, 1b (topic), 1a–17a (prize essays).

[35] GZSYKY, Vol. 2, 1892, 3a–b (topic), 4b–6b, 33b–34b (prize essays). See also Shang Zhicong, "1886–1894 nianjian jindai kexue zai wan Qing zhishi fenzi zhong de yingxiang — Shanghai Gezhi shuyuan Gezhi lei keyi fenxi", 81.

were superior to the more invasive, i.e., surgical techniques, used in Western countries and now in China. He admitted, however, that each tradition had certain strengths that should be selected out and combined.

The 1893 prize essay on medicine by the Anhui literatus Li Jingbang 李经邦 (1852–1910) pointed to the stress on the brain (*nao* 脑) in Western medicine, but Li also noted the institutional importance of the Western hospital for the success of Western medicine. Li claimed that traditional medicine had also stressed the brain, but his apologetics were a transparent effort to apply the still prominent claim of the "Chinese origins of Western learning" (*Xixue Zhongyuan* 西学中源) to medical studies.

Li claimed, for example, that during the Roman empire Westerners had come to China and that they had taken back copies of the *Basic Questions* (*Suwen* 《素问》), part of the *Inner Canon of the Yellow Emperor* (*Huangdi neijing*《黄帝内经》), among other canonical medical texts. Over time, Western medicine had built on these Chinese texts to produce new findings drawn from chemistry. Such traditionalistic claims were increasingly suspect by 1900 as more Chinese were trained as modern physicians. Moreover, Li placed the blame for the decline of Chinese medicine on its recent practitioners, who had failed to live up to the comprehensive understanding that the ancient physicians had achieved.

Yang Minhui's prize essay for the summer 1893 medicine theme betrayed the fact that Western medicine was increasingly respected in official and popular circles and feared by native physicians. Yang tackled this change by appealing to the theoretical superiority of traditional medical principles while at the same time admitting the advantages of Western medical practices. Western knowledge of electricity and chemistry, according to Yang, were two areas that when applied therapeutically superseded native medicine. Altogether Yang presented ten areas in which Western medicine was superior but only four for native medicine. To salvage the strengths in Chinese medical principles, Yang proposed that a "great medium" (*dazhong* 大中) could be achieved if ancient principles were informed by the new medical procedures from the West.[36]

The Polytechnic's Prize Essay topics on medicine also influenced the science questions that official examiners used in the reformed civil

[36] GZSYKY, Vol. 2, 1893, 2a (topic), 4a–9b (prize essays).

examinations after 1901. Essays from both the 1891 Polytechnic query on the medieval *materia medica* and the 1892 theme of "which medical tradition was superior theoretically" were presented as model policy essays for the civil examinations. Moreover, both the Shandong licentiate Sun Weixin's 孙维新 1891 prize essay on the *materia medica* and Li Jingbang's 1893 essay on the history of Western medicine were included verbatim as model essays in a 1903 collection of "New Governance" era policy questions and answers. More significantly, medicine was now regarded officially as one of the modern natural sciences, whereas it had only been considered an unofficial part of the investigation of things and extension of knowledge before the 19th century.[37]

Natural Theology, Darwin, and Evolution

Li Hongzhang's spring 1889 "Extra Theme" had included a straightforward request for clarification of Darwin's (1809–1882) and Spencer's (1820–1903) writings:

> "With respect to the 'Science' referred to in the 'Great Learning,' from Ching-kang-ching [Zheng Xuan 郑玄] downwards, there have been several tens of scholars who have written on the subject. Do any of them happen to agree with Western scientists? Western science began with Aristotle in Greece; then came Bacon [1561–1626] in England who changed the previous system and made it more complete. In later years, Darwin's and Spencer's writings have made it still more comprehensible. Give a full sketch of the history and bearings of this whole subject."[38]

The essays sent in for this topic, however, reveal a remarkable range of ignorance and knowledge concerning Darwin's then increasingly famous claims about the biological evolution of all life forms.

---

[37] *Zhongwai shiwu cewen leibian dacheng, mulu*, 14a–b, 19. 15a–22b, especially 18b–20b.
[38] The translation for this question is taken from John Fryer, "Chinese Prize Essays: Report of the Chinese Prize Essay Scheme in connection with the Chinese Polytechnic Institution and Reading Rooms, Shanghai, for 1886 and 1887", 100. Fryer's "Ching-kang-ching" is the Later Han classicist, Zheng Kangcheng, i.e., Zheng Xuan (127–200).

For example, Jiang Tongyin's 蒋同寅 prize essay betrayed how some Han Chinese essayists simply bluffed their way through the complexity of Western learning. The essay judges, including Li Hongzhang, apparently did not know the difference. Jiang received the first prize for his essay on the development of science in the West, but when it came to identifying Darwin, he simply noted that Darwin was a famous geographer who also wrote on chemistry. Jiang described Spencer as an expert in mathematics. These were odd portrayals for someone who was a supplementary student in Baoshan county, just outside of Shanghai. The reason becomes clearer when we note that Jiang added: "Today the works by both gentlemen are widely prevalent abroad, … but unfortunately I regret not yet seeing translations anywhere."[39]

Wang Zuocai, because he studied at Shanghai Polytechnic, presented Darwin and Spencer in his second-place essay in somewhat more discerning terms. Wang noted that Darwin had argued that all organisms changed from the course to the spiritually elevated in the process of adapting to the world. Without such adaptation they could never survive eternally. Although Wang recognized that Darwin had discovered a principle that no one had ever understood before, his account missed the key role of the survival of the fittest. Moreover, when he came to Spencer, Wang bluffed his way through by linking Spencer to Darwin and simply stressed the popularity of the latter's writings.[40]

As the third-place finisher on the spring 1889 "Extra Theme", Zhu Dengxu 朱登叙, like the first-place essayist, revealed total ignorance of Darwin's and Spencer's views. A supplementary student in Shanghai county, Zhu described Darwin as a geographer who had become famous because of his travels. Darwin's research, Zhu added, was based on the study of science and chemistry, but most of his work had not yet been translated into Chinese. According to Zhu, Spencer was skilled in practical applications of mathematics for the study of science. Such myopic characterizations of Darwin and Spencer in 1889 by Chinese literati in the Shanghai area are surprising, perhaps, given John Fryer's ties to British science since his early 1870s home visit. After returning to Shanghai in 1873,

[39] GZSYKY, Vol. 1, 1889, 1a.

[40] GZSYKY, Vol. 1, 1889, 8b.

it is interesting that Fryer failed to deal with the Darwinian controversy. No works of Darwin or Spencer were ever translated at the Arsenal.[41]

This lack of attention to Darwin by Fryer and other missionary translators from England and the United States becomes a bit more worrying, however, when we examine more carefully the essay on the spring 1889 "Extra Theme", prepared by the fourth-place finisher, Zhong Tianwei, which we have briefly looked at. Zhong had received a local degree in the civil examinations at the age of 26 *sui*. Failing to advance through the more competitive provincial examinations in Nanjing, Zhong at age of 33, circa 1873, entered the Foreign Language School (*Guang fangyan guan* 广方言馆) located in the Jiangnan Arsenal, where he studied Western learning for three years.

Subsequently, Zhong traveled abroad in Europe for two years, and then he worked with Fryer and others in the Jiangnan Arsenal's Translation Department. Because of his ties to the Arsenal, both as a student and translator, and his travels abroad, he likely had access to accounts of Darwin and Spencer unavailable to others in Shanghai. Later, after 1901, Zhong's answer to a question on "comparing the vicissitudes in Chinese and Western mathematical astronomy" was chosen by examiners as a model essay in the reformed civil examinations. The question and Zhong's reply were drawn on information that Wylie (1815–1887) and Li Shanlan 李善兰 (1810–1882) had included in their article "Progress of Astronomical Discovery in the West" ("Xiguo tianxue yuanliu" 〈西国天学源流〉) that had appeared in the *Shanghae Serial* (*Liuhe congtan* 《六合丛谈》) in 1857.[42]

Among the works Fryer had brought back with him in 1873 after his leave in England had been Henry Roscoe's (1833–1915) series of *Science Primers* published in 1872, which had included the participation of Thomas Huxley (1825–1895), a champion of Darwin since 1860 when Huxley had debated the bishop of Oxford at the British Association for the Advancement of Science. Fryer must have heard of Darwin's *Origin of the Species* while on home-leave in England to find appropriate science textbooks to translate in the Jiangnan Arsenal and later for the missionaries' School and Text-Book Series Committee. Moreover, he chose to have

[41] GZSYKY, Vol. 1, 1889, 12b.

[42] *Zhongwai shiwu cewen leibian dacheng, mulu*, 12a.

the Arsenal's Translation Department translate Roscoe's series of *Science Primers* (*Gezhi qimeng*《格致启蒙》), despite the participation of Thomas Huxley. Hence, we find in Zhong's prize essay a remarkably accurate account of Darwin's theory of evolution. Why was it ranked fourth, then?

David Wright has explained that the essays submitted for the Chinese Prize Essay Contest contain some of the earliest references in Qing China to Charles Darwin and his theory of evolution. The first, brief documented reference to Darwin came in the 1873 Jiangnan Arsenal translation of Lyell's (1797–1875) *Elements of Geology* (6th edition; *Dixue qianshi*《地学浅释》) by Hua Hengfang and Dr. Daniel Jerome Macgowan (1814–1893), and again that year in an article in the *Shanghai Journal.*

A vague mention of human evolution also appeared in an anonymous September 1877 article in *The Chinese Scientific and Industrial Magazine* entitled "The Theory of Chaos" ("Hundun shuo"〈混沌说〉), in which humans' evolution from apes was discussed in an account of theories for how the world might end. This article, which argued that it was more important to consider the end of life forms rather than their origins, may have been written by Fryer and Luan Xueqian, suggesting perhaps that it was best for an unnamed Christian to present and critique Darwin anonymously to the Chinese. Jesuits had introduced Copernicus (1473–1543) in a similar, misleading manner and later were chided by 18th-century literati for the contradictions in their presentation of the Copernican system.[43]

Zhong Tianwei's 1889 spring essay, cited above, had opened with a well-prepared and well-informed account of Greek science. Then it described the evolution of science from Aristotle to Bacon, Darwin, and Spencer. Zhong, however, went well beyond the natural theology that the missionary essays on science in the *Shanghae Serial* and the *Scientific and Industrial Magazine* had encoded in their translations of botany and

---

[43] For the essay on the "The Theory of Chaos", see *Gezhi huibian*, Vol. 2, 13–15 (September 1877). The dating of articles in this reprint is defective, and some sections are disordered. Reference to the original editions, though limited in number, is recommended. See also *Gezhi keyi huibian*, collected by Wang Tao as editor, 4.1a, 4.6a, 4.9b, 4.16a, 5.42b, *Dixue qianshi*, 13.16a, and *Shenbao* (Tongzhi 12th year, intercalary 6th month, 29th day; August 21, 1873), 2. James R. Pusey, *China and Charles Darwin*, 4–5, stressed 1895 for when Darwin's views were first presented.

biology. After surveying the three-stage development of Greek science from natural philosophy (*gezhi lixue* 格致理学) to metaphysics (*xinglixue* 性理学) and dialectics (*bianlixue* 辯理学), Zhong summarized the modern contributions of Francis Bacon, Charles Darwin, and Herbert Spencer:

> Two thousand and three years later, the Englishman Bacon first appeared and transformed Aristotle's theories. … At the age of thirteen he entered the state school to study, but he dismissed old learning, which demonstrated his independent stance.… Then he focused on the study of science.… The main point of his study was that in all scientific matters it was necessary to provide substantiation through demonstrable proof. On this basis each principle can be exhaustively grasped without first enunciating a principle as is. In this manner, through an evidential analysis of the nature of things, its principle will become manifest.…
>
> Darwin was born in 1809 … the grandson of a physician and the son of a scientist.… Growing up he was selected to attend Edinburgh University in Scotland. Later he traveled around the globe accompanying an English naval vessel carrying out surveys and preparing drawings while investigating each plant and animal in its ecological setting.… In 1859 he prepared his magnum opus "on the origins of the species of all things" (*lun wanwu fenzhonglei zhi genyuan* 论万物分种类之根源). He also declared the "principle of the survival of the fittest" [*lun wanwu qiangcun ruomie zhi li* 论万物强存弱灭之理, "the principle that the strong survive and the weak perish"].
>
> All species of plants and animals undergo changes over time and never remain unchanged from time immemorial to the present. Those plants and animals that are not successful in adapting slowly perish. Those that successfully adapt survive for the long term. This is the natural principle of the heavenly way [*tiandao ziran zhi li* 天道自然之理, i.e., natural selection]. His theory, however, contradicted the teachings of Jesus, and thus scholars from each country together did not follow his words. At first he was greatly attacked, but today those who honor him have gradually increased. Hence, science underwent a great change, and Darwin can be called a superior man who arises once in a thousand autumns.
>
> As for Herbert Spencer, … he was with Darwin for eleven years in his youth. His life works mainly expanded on Darwin's theories, enabling people to grasp the principles of psychology [*linghun zhi li* 灵魂之理]. …

> What he claimed was knowable was only the external appearance of all things. The inner subtleties of all things were in fact unknowable.... Comparing it to what Christianity has called God [*shangdi* 上帝], what science calls a fundamental element [*yuanzhi* 原质], although it cannot be known or measured through the power of human intellect, yet the point is that without any doubt such elements actually exist. Moreover the changes that all things go through go back to an origin in one thing. This one thing is the root, and all other things are its branches.[44]

Although couched in the rhetoric of evidential studies (*kaozhengxue* 考证学) and the investigation of things (*gewu* 格物), Zhong Tianwei's remarkable essay for the Polytechnic's essay contest represented a succinct summary of Darwin's theories and introduced Spencer's methodology almost a decade before Yan Fu's 严复 (1853–1921) Chinese translation of Huxley's 1893 Romanes Lecture on "Evolution and Ethics" appeared in 1898 under the title of *On Evolution* (*Tianyan lun*《天演论》). Yan, for example, translated "natural selection" as "heavenly selection" (*tianze* 天择), which like the missionary introduction of evolution in botany avoided direct mention of the Darwinian struggle of the fittest to survive.

By contrast, Zhong avoided the natural theology that had informed Protestant accounts of evolution. Zhong noted in his essay that he had seen a recent translation of Spencer's first work, which was called *Essential Guides for Study* (*Siye yaolan*《肆业要览》). Zhong's student account of the survival of the fittest through natural selection in 1889, presented above, contrasted with the more classically domesticated notion of "heavenly selection" that Yan Fu later presented a decade later as a translation for natural selection.[45]

---

[44] See GZSYKY, Vol. 1, 1889, 21a–b. The essay is also cited from another version in Xiong Yuezhi, *Xixue dongjian yu wan Qing shehui*, 365–366.

[45] Yan Fu's translation of Spencer's *Study of Sociology*《群学》appeared after 1898. James R. Pusey, *China and Charles Darwin*, 59, traces the first introduction of Darwin to Yan Fu's 1895 essay "On the Origins of Strength" ("Yuan qiang"〈原强〉). See also David Wright, "Yan Fu and the Tasks of the Translator", in Michael Lackner, Iwo Amelung, and Joachim Kurtz, eds., *New Terms for New Ideas: Western Knowledge & Lexical Change in Late Imperial China*, 235–255, especially 253.

The lower rank that Zhong's essay received, despite its more expert account of Darwin and Spencer, becomes understandable, however, when we review the comments by Wang Tao, who was the overall supervisor of the essay contests. In the edition of the 1886–1893 essays that were published together as an annual group of essays, Wang Tao's comments were often included in the top margins of the page, as had been the comments of Ming and Qing examiners and outside teachers in the collections of model literary and policy essays for the civil examinations. In his comments on Zhong's explication of the "principle of the survival of the fittest", Wang wrote of Darwinism:

> This essay describes the flourishing of all living things whereby those most suitable survive the longest. What is referred to as "those most suitable" [*yizhe* 宜者] means "those most benefited" [*yi ye* 宜也]. The theory that "under heaven the strong survive and the weak perish" [*wu qiangcun ruomie zhi shuo* 无强存弱灭之说] has no basis in fact [*si qian kaoju* 似欠考据].[46]

Wang's comments indicate that the Darwinian view of evolution was unacceptable in the 1880s and 1890s for the Protestant missionaries in Qing China. Such opposition carried over to their longtime collaborators such as Wang Tao, who earlier had worked with the Scottish missionary James Legge (1815–1897) to translate the Chinese Classics into English. Hence, the lower ranking of Zhong's 1889 essay was an indication of the antagonism Darwin's views provoked among the contest judges, even though Zhong's essay was the only one to describe clearly the controversy of how Darwin's views "contradicted the teachings of Jesus".

Nevertheless, Zhong's essay was awarded fourth place and published in 1889 and reprinted in later collections. As a low-level translator and educator, Zhong supported educational reform in China within the balanced framework of adapting Western science and technology to Chinese learning, thus unifying the practical "arts" (*yi* 艺) and the Way. His precocious analysis of Darwin, however, revealed the potential for

[46] GZSYKY, Vol. 1, 1889, 21a. See also Benjamin A. Elman, *A Cultural History of Civil Examinations in Late Imperial China*, 407–420, on official collections of civil examination essays.

Chinese literati to supersede the Christian packaging of the modern sciences, a process that began among Chinese literati in earnest after the Sino-Japanese War, at the same time that Zhong continued earlier efforts to award Chinese learning place of eminence in theoretical matters.[47]

Just as the Chinese eventually learned about Copernicus despite the Jesuits' efforts to conceal his theory of a heliocentric cosmos, so too they learned about Darwin in spite of the Protestants' attempts to replace the theory of natural selection with a natural theology of "Christian Darwinism". The Chinese Prize Essay Contest reveals that the Protestant enterprise, once its religious agenda was exposed, was no more convincing to many Chinese in the late 19th century than the Jesuit translation agenda had been in the 17th and 18th. Hence, after the Sino-Japanese War, Chinese literati quickly turned to Meiji Japan as the source for the latest currents in modern science.

In the late 1880s, most Chinese scholar-officials such as Li Hongzhang, who posed the spring 1889 question, were still ignorant of Darwin's theory of evolution based on the survival of the fittest and followed instead the revised natural theology of Protestants, which informed the 1879–1880 and 1886 versions of the *Primers for Science Studies* and most of the articles on science in *The Chinese Scientific and Industrial Magazine*. Ironically, the essay contest exposed that some Chinese such as Zhong Tianwei were more in tune with the Darwinian turn in biology, zoology, and botany than their missionary teachers or their older Chinese predecessors in science studies such as Wang Tao.

After Wang Tao's death and Fryer's departure for Berkeley University, the essay contests were not as enthusiastically promoted, although they were still held regularly, sometimes monthly, sometimes quarterly in 1901, 1904, 1906, and 1907. The Sino-Japanese War heightened the disjunction between events before the war and those after. In particular, many of the events from 1865 to 1894 leading up to the establishment of modern science in late imperial China were replaced by the public attention that reformers, iconoclasts, and revolutionaries received after 1895. Their generation was designated as the decisive one, and not those

[47] Shang Zhicong, "1886–1894 nianjian jindai kexue — Shanghai Gezhi shuyuan Gezhi lei keyi fenxi", 81.

who had been part of the Protestant era or the arsenals and schools associated with the dispersed Foreign Affairs Movement.

One example of this displacement of pre-1894 literati efforts to master Western learning and modern science was the meteoric rise of Yan Fu as a public figure. His reputation vis-à-vis his predecessors as an iconoclast and the pioneer translator of Spencer and introducer of Darwin's theories replaced his earlier career as a graduate of the Fuzhou Navy Yard and naval school teacher and administrator. This spotlight on Yan Fu after 1895 has overlooked the historical events in the rise of modern science and Western learning in China before the Sino-Japanese War.

Similarly, Liang Qichao overlooked "the explosion of the newspaper market" after 1850 in his own self-serving accounts of the heroic emergence of a new and critical journalism in late 19th-century China. In the process, Yan and others such as Tan Sitong, Liang Qichao, and Kang Youwei, who all rose to prominence after the war, have received the credit for many of the contributions that Li Shanlan, Hua Hengfang, Wang Tao, and others such as Zhong Tianwei had already made in breaking new intellectual ground in the 1870s and 1880s. Both Tan Sitong and Kang Youwei as publicists appropriated science to legitimate their millennial visions but understood very little of science on its own terms.[48]

****

What is interesting about the China Prize Essay Contest is that it used the prestige of the imperial civil service examinations to aid missionary efforts to promote modern science during the late Qing, which created a working partnership between Western translators and Qing high officials, particularly those in provinces with treaty ports. At the same time, those who were drawn to scholarly and technical work in the new industrial arsenals after the Taiping Rebellion (1850–1864) in Fuzhou, Shanghai,

---

[48] See *North-China Herald*, January 29, 1902: 180, September 22, 1905: 697–698, January 25, 1907: 202, February 21, 1908: 418–419. See also Xiong Yuezhi, *Xixue dongjian yu wan Qing shehui*, 373. Compare Natascha Vittinghoff, "Unity Vs. Uniformity: Liang Qichao and the Invention of a 'New Journalism' for China", *Late Imperial China* 23, 1 (June 2002): 91–143.

and elsewhere, or to translation positions in the official Foreign Language Schools in Beijing, Shanghai, and Guangdong, tended to be relatively marginal literati. They often had failed the more prestigious civil examinations several times and saw Western learning and the sciences as an alternative route to fame and fortune.[49]

The missionaries generally remained sanguine about the civil examinations until the 1894–1895 Sino-Japanese War. Thereafter, Chinese naval defeats contributed to the transformation of official, elite, and popular perceptions of the Self-Strengthening era. New public opinions appeared in the Chinese and missionary press that shaped the emerging national identity and sense of crisis among Han Chinese, who increasingly opposed the Manchu regime in power. Disappointment with the military losses convinced many Chinese that the Foreign Affairs Movement (*Yangwu yundong* 洋务运动) had failed and that more radical political, educational, and cultural changes were required to follow Japan's lead in modernizing and coping with foreign imperialism.

The Sino-Japanese War provoked a striking switch in Protestant confidence about the future of Qing China. An account of the Chinese defeat prepared by one of the leading Protestant missionaries and translators in Beijing, Young J. Allen (1836–1907), when translated into Chinese, was frequently pirated, for example, and became required reading for the 1896 Hunan provincial examination in Changsha. Allen's account of the defeat outlined his views of needed reforms in China. In the essay, Allen traced China's backwardness to three root causes: (1) superstition (*mixin* 迷信); (2) opium; and (3) civil examinations. In this series, he also stressed the importance of science as a corrective for the causes of China's backwardness.

Native studies, according to Allen, had failed to grasp the universal lessons of modern science. In particular, China's assimilation of Western science was missing the importance of the "study of the principles of things" (*wuli zhi xue* 物理之学), or what in the late 1890s would increasingly be called "physics" based on Japanese translations of Western scientific texts. Moreover, Allen used "superstition" as a modern cultural category to pigeon

[49] David Wright, "The Great Desideratum: Chinese Chemical Nomenclature and the Transmission of Western Chemical Concepts", *Chinese Science* 14 (1997): 35–70.

hole the entire Chinese classical tradition, a reduction that would become de rigueur among many Chinese radicals in the 20th century.[50]

Similarly, before his departure for California to take up the chair of Oriental Languages at Berkeley University, which he had been offered in 1895, John Fryer publicly announced a competition for "new age novels" (*xin xiaoshuo* 新小说) in Chinese that would enhance the morals of China and eviscerate the triple evils of opium, stereotypical examination essays, and footbinding. This appeal for a new literature written in "easy and clear language with meaningful implications and graceful style" attracted the interest of Liang Qichao and other reformers who would provide the foundations for the call for a new culture in China, which was premised on the failure of traditional Chinese civilization symbolized by the bound feet of Chinese women.[51]

Like the Protestant missionaries, Chinese literati also attacked the civil examinations and footbinding after 1895. The race to establish Chinese institutions of higher learning that would stress modern science accelerated after the occupation of the capital by Western and Japanese troops in 1900. The Boxer popular rebellion in north China and the response of the Western powers and Japan to it unbalanced the power structure in the capital so much that foreigners were able to put considerable pressure on provincial and metropolitan leaders such as Li Hongzhang. Foreign support of reform and Western education thus strengthened the political fortunes of provincial reformers such as Yuan Shikai 袁世凯 (1859–1916) and Zhang Zhidong 张之洞 (1837–1909), who had opposed the Boxers.[52]

After the civil examinations were abrogated in 1904, the Qing dynasty quickly lost control of its science policies, which increasingly fell under the control of local Chinese intelligentsia and overseas students.

---

[50] The essay is abridged in *Wanguo gongbao wenxuan*《万国公报文选》, edited by Qian Zhongshu 钱钟书 and Zhu Weizheng 朱维铮, 179–201.

[51] Patrick Hanan, "The Missionary Novels of Nineteenth-Century China", *Harvard Journal of Asiatic Studies* 60, 2 (December 2000): 440–441, and Anonymous, "The New Novel Before the New Novel", in Judith Zeitlin and Lydia Liu, eds., *Writing and Materiality in China: Essays in Honor of Patrick Hanan*, 317–340.

[52] Benjamin A. Elman, *A Cultural History of Civil Examinations in Late Imperial China*, 608–618.

The political consequences of this loss of control were to be expected when the Qing bureaucracy meekly gave up one of its major institutions that had for centuries successfully induced literati acceptance of the imperial system. Some noted that the post-Boxer reforms were so radical that they helped precipitate the downfall of the dynasty. By spring 1906, for instance, Yan Fu admitted that many, including himself, were now having second thoughts about getting rid of the civil examinations so precipitously.[53]

What should be added, however, is that the radical reforms in favor of new schools allowed the sciences to develop independently of the political system for the first time. Via new school-based national examinations, the Manchu court tried in vain to maintain control over the delegitimated remnants of the older examination constituencies while at the same time gain control over education in the new schools. The Qing dynasty never reestablished control of the provincial and local educational systems, which it had irrevocably lost in 1905. Power had shifted to the new schools and more importantly to the Han Chinese gentry constituencies they served.

In this rapid parade of events, the 1886–1893 Chinese Prize Essay Contest was quickly discarded as a failed experiment and ultimately forgotten. I would suggest, however, that the second half of the 19th century was the "seeding time" for the modernization of China. During the period from 1850 to 1895, many works on astronomy, mathematics, medicine and related fields of botany, geography, geology, mechanics, and navigation were translated by a core group of Protestant missionaries and Chinese co-workers in Guangzhou, Beijing and Shanghai. The essay contest helped to popularize the new knowledge contained in those translations by employing a traditional vehicle to valorize that knowledge. The first Sino-Japanese War made such efforts appear stillborn, but had the Chinese won that war, which was not impossible, such efforts to spread the new sciences through traditional institutional forms might have been legitimated and expanded.[54]

---

[53]Yan Fu, "Jiuwang juelun"〈救亡决论〉, in *Wuxu bianfa ziliao*《戊戌变法资料》, 3/60–71.

[54]Knight Biggerstaff, "Shanghai Polytechnic Institution and Reading Room: An Attempt to Introduce Western Science and Technology to the Chinese", 127. See also Benjamin A. Elman, "Naval Warfare and the Refraction of Qing Nineteenth Century Industrial Reforms into Failure", 283–326.

Chapter 7

The Great Reversal: The "Rise of Japan" and the "Fall of China" after 1895

Chinese, Western, and Japanese scholarship has long debated the success or failure of the government schools and regional arsenals established between 1865 and 1895 to reform Qing China. For example, Quan Hansheng contended in 1954 that the Qing failure to industrialize after the Taiping Rebellion (1850–1864) was the major reason why China lacked modern weapons during the Sino-Japanese War.[1] This position has been built on in recent reassessments of the "Foreign Affairs Movement" (*Yangwu yundong* 洋务运动) and Sino-Japanese War of 1894–1895 (*Jiawu zhanzheng* 甲午战争) by Chinese scholars. They argue, with some dissent, that the inadequacies of the late Qing Chinese navy and army were due to poor armaments, insufficient training, lack of leadership, vested interests, lack of funding, and low morale. In aggregate, these factors are thought to demonstrate the inadequacies of the "Self-Strengthening era" and its industrial programs.[2]

Allen Fung reconsidered this "shopping-list of problems" in a review article that explored the "witch-hunt for the inadequacies of the Chinese army and navy" that ensued after 1895. Fung focuses on the defeat of the Chinese army in the Sino-Japanese War because Japanese land victories

---

[1] Quan Hansheng, "Jiawu zhanzheng yiqian de Zhongguo gongyehua yundong" 〈甲午战争以前的中国工业化运动〉, *Lishi yuyan yanjiusuo jikan*《历史语言研究所集刊》 25, 1 (1954): 77–78.

[2] Compare: Qi Qizhang, ed., *Jiawu zhanzheng jiushi zhounian jinian lunwen ji* 《甲午战争九十周年纪念论文集》; Qian Gang, *Haizang: Jiawu zhanzheng 100 nian* 《海葬：甲午战争 100 年》; Dai Yi *et al.*, *Jiawu zhanzheng yu Dong-Ya zhe*ngzhi 《甲午战争与东亚政治》; and Shi Quan, *Jiawu zhanzheng qianhou zhi wanQing zhengju*《甲午战争前后之晚清政局》. See also Kwang-Ching Liu and Richard Smith, "The Military Challenge: The North-West and the Coast", in John Fairbank and Kwang-Ching Liu, eds., *The Cambridge History of China*, Volume 11, *Late Qing, 1800–1911*, Part 2, 202–273.

gave them a clear path to march on Beijing. This threat to the capital forced the Qing court to seek an immediate settlement of the war. In contrast to accounts in China that still accuse the key Qing minister, Li Hongzhang 李鸿章 (1823–1901), of cowardice for his peace-at-any-cost policy, Fung maintains that Chinese armies were well-equipped during the early stage of the war with Japan and that the Chinese field commanders were not incompetent. He refutes earlier claims that China's land defeats in the Sino-Japanese War were due to the failure of the Chinese ordinance industry. Fung concludes that the primary explanations for China's losses in the land war are: (1) the better military training Japanese troops and officers received; and (2) the fact that Qing troops were outnumbered by the Japanese at the major battles.[3]

I will reassess the naval wars that the Qing dynasty lost in the late 19th century. Along with the Sino-Japanese War, the Sino-French naval battles of 1884–1885 have also been used as a litmus test for measuring the failure of the Self-Strengthening reforms initiated after the Taiping Rebellion. The rise of the new arsenals, shipyards, technical schools, and translation bureaus, which are usually undervalued in such "failure narratives", will also be reconsidered in light of the increased training in military technology and education in Western science available to Chinese after 1865. Longstanding claims made by contemporaries of the Sino-Japanese War that China's defeat demonstrated the failure of the Foreign Affairs Movement to introduce Western science and technology successfully will be scrutinized.[4]

In addition, my account will address how Chinese naval defeats contributed to the transformation of official, elite, and popular perceptions of

---

[3] Allen Fung, "Testing the Self-Strengthening: The Chinese Army in the Sino-Japanese War of 1894–95", *Modern Asian Studies* 30, 4 (1996): 1007–1031, which appears in a special issue on "War in Modern China". Compare Dai Yi, Yang Dongliang, and Hua Lizhu, *Jiawu zhanzheng yu Dong-Ya zhengzhi*, 104–105. Richard Smith, "Foreign Training and China's Self-Strengthening: The Case of Feng-huang-shan", *Modern Asian Studies* 10, 2 (1976): 195–223, also stressed the late Qing failure to train a modern officer corps.

[4] On the inadequacy of China's army and navy, see Ralph Powell, *The Rise of Chinese Military Power*, 36–50. Compare Richard Smith, "Reflections on the Comparative Study of Modernization in China and Japan: Military Aspects", *Journal of the Hong Kong Branch of the Royal Asiatic Society* 16 (1976): 11–23.

the Self-Strengthening era. New public opinions appeared in the Chinese and missionary press that shaped the emerging national identity and sense of crisis among Han Chinese who increasingly opposed the Manchu regime in power. Disappointment with the military losses convinced many Chinese that the Foreign Affairs Movement had failed and that more radical political, educational, and cultural changes were required to follow Japan's lead in modernizing and coping with foreign imperialism. The earlier adaptation of new technological and scientific learning before 1895 was quickly forgotten. Euro-American missionaries and experts who had aided in the Qing dynasty's scientific translation projects, which were used as textbooks in the arsenals and technical schools, now also thought that the Chinese nation, language, and culture were doomed.[5]

In military terms, the Chinese losses at sea in 1894–1895 meant that the Qing development of an ocean-going navy that might have rivaled the Chinese blue water fleets of the early Ming dynasty, when the navy sailed into the Indian Ocean, and the early Qing, when the navy stormed the Pescadores and Taiwan, was derailed in favor of a land-based army built around well-trained infantry forces. One of the ironies accompanying the revival of the Chinese navy after 1865 was that since the Southern Song dynasty China had at times supported a substantial navy, which the Mongols had expanded to try to invade Japan from Korea in 1274 and 1281 and to attack Java in 1293. Subsequently, the early Ming navy had carried out enormous excursions into Southeast Asia and the Indian Ocean, which ended when the navy was scrapped in the 1460s to prepare for possible land wars against the Oirat Mongols. Nevertheless, the Ming coastal navy successfully defended the South China coast from Japanese pirates in the 16th century.[6]

Chinese naval power revived when Ming loyalists under the command of Zheng Chenggong 郑成功 (Koxinga, 1624–1662) and his father resisted the Manchus in major naval and land battles along the Fujian

[5] See Hans J. van de Ven, "War in the Making of Modern China", *Modern Asian Studies* 30, 4 (1996): 737–756, which draws on Peter Paret, *Understanding War*, especially 209–226.

[6] Lo Jung-pang, "The Decline of the Early Ming Navy", *Oriens Extremus* 5, 2 (1958): 147–168. On the army, compare Stephan MacKinnon, *Power and Politics in Late Imperial China*, 90–136, and Edward McCord, *The Power of the Gun: The Emergence of Modern Chinese Warlordism*, 17–45.

coast in the 1640s and 1650s. Zheng's land and sea forces took heavy losses, however, when they moved up the Yangzi River to Nanjing in 1659, and he was forced to retreat to Xiamen (Amoy), where he repulsed Qing forces in 1660. Zheng's naval forces then challenged the Dutch garrison in northern Taiwan at Castle Zeelandia in April 1661 with a force of 600 ships and 25,000 sailors. The Dutch garrison capitulated after a bitter nine-month siege. Zheng subsequently and unsuccessfully demanded via a Dominican missionary that the Spanish in Manila recognize his suzerainty.[7]

For its part, the Qing government in 1662 ordered coastal inhabitants from Shandong in the north to Guangdong in the south to move inland to cut Zheng's supply lines and to negate the value of the coast as a battleground. In addition, the Manchus developed a naval fleet to defend the coastline. When Shi Lang 施琅 (1621–1696), one of the Southern Ming's most capable admirals, joined the Manchus in 1646 because of a dispute with Zheng, he took command of Qing naval forces in the 1650s and 1660s along the Fujian coast. In July 1683, Shi Lang commanded the Qing fleet of 300 warships and 20,000 sailors, which first subdued the Pescadores. Taiwan fell to the Qing navy in October, and for the first time the island became part of the "Chinese" empire.[8]

The Qing navy no longer remained on a war footing after Taiwan was annexed, and the Manchu emperors became increasingly preoccupied with the land-based expansion of the Russians from Siberia into the Manchu homelands and the renewed dangers posed by the Zunghars in Central Asia. In addition, the Qing sought expansion of its empire in Tibet and Turkestan. By the end of the 18th century, the Qing had doubled its size. When the Opium War broke out in 1839, therefore, the Qing fleet was again mainly a coastal navy used principally for defense against outside pirates and local marauders.[9]

---

[7] Christine Vertente *et al.*, *The Authentic Story of Taiwan: An Illustrated History, based on Ancient Maps, Manuscripts, and Prints*, 96–110. See also Arthur Hummel, ed., *Eminent Chinese of the Ch'ing Period* (hereafter, ECCP), 108–109.

[8] Christine Vertente *et al.*, *The Authentic Story of Taiwan: An Illustrated History, based on Ancient Maps, Manuscripts, and Prints*, 127–130, and ECCP, 653.

[9] Peter Perdue, "Boundaries, Maps, and Movement: Chinese, Russian, and Mongolian Empires in Early Modern Central Eurasia", *The International History Review* 20, 2 (June 1998): 263–286.

In many ways, the revival of the Qing navy after the Opium wars might have been heralded as a return to the brighter days of the imperial navy during the 15th and 17th centuries. Instead, however, after the Sino-Japanese War the late Qing navy was ridiculed at a time when anti-Manchu patriots appealed to the Ming fleets as a sign of China's past greatness. Moreover, the alleged superiority of Japan in modern science and technology was generally accepted after 1895 because of the success of its navy.

The Scope and Scale of the Foreign Affairs Movement

In the late 1950s American scholars such as Mary Wright contended that the imperial system and its classical ideology, which she labeled Confucian, were incompatible. The Taiwan scholar Wang Ermin subsequently challenged Wright's claim that classical learning and modernization were incompatible when he traced the beginnings of China's military industrial complex to the Self-Strengthening era. More recently Francis Moulder has maintained that China's failure to modernize should be understood in light of China's higher level of incorporation into the world economy than Japan, which enabled the more deleterious impact of imperialism in China. All three views have been presented in the aftermath of the crucial Japanese victory in 1895, particularly at sea, which decisively refracted Chinese and Western perceptions of the Self-Strengthening era period after 1865. Renewed attention to the "decisiveness of battle" to evaluate such broad conclusions is required before we can reach a new consensus that corrects the misrepresentations that become dominant after the 1895 "witch-hunt".[10]

In his influential study of modern Chinese naval development, for example, John Rawlinson contended that traditional institutions based on

---

[10] Mary Wright, *The Last Stand of Chinese Conservatism: The T'ung-chih Restoration, 1862–1874*, See also Wang Ermin, *Qingji bing gongye de xingqi*《清季兵工业的兴起》. Compare Francis Moulder, *Japan, China, and the Modern World Economy: Toward a Reinterpretation of East Asian Development, ca. 1600 to ca. 1918*. Stephen Thomas has argued that the adverse affects of imperialism were not felt in China until the 1890s, after the war. See Stephen Thomas, *Foreign Intervention and China's Industrial Development, 1870–1911*, and Hans J. van de Ven, "War in the Making of Modern China", 739–742.

classical ideology gave the Tongzhi Restoration (*Tongzhi Zhongxing* 同治中兴, 1862–1874) and the Foreign Affairs Movement its essential character and limited its achievements. The Qing failure to develop a national navy yielded a number of competing regional squadrons primarily because of weak imperial institutions and strong regional loyalties. Others such as Thomas Kennedy have assessed both the external and internal forces that influenced the efficacy of the Foreign Affairs Movement and its programs.

According to Kennedy, China's modern ordnance industry was an institutional innovation that ushered in a new era of mass production. Moreover, the Qing state managed the arsenals as bureaus within the traditional government, which lead to corruption and inefficiency. Poor imperial leadership and lack of coordination among provincial officials limited the success of the modernization programs. Financial troubles at the arsenals and generally poor Euro-American technicians resulted from China's semi-colonial status at the time. In this view, military firepower at sea and on land was not the key to the outcome of the Sino-Japanese War.[11]

David Pong has contended that the Beijing court failed to create a unified imperial navy because of its inability to change the system of public financing and because of insufficient funds. On the other hand, Albert Feuerwerker has noted that the Qing government compensated somewhat for the depressed rural economy after the Taiping Rebellion by instituting two new taxes to finance the arsenals and shipyards successfully: (1) customs duties on foreign trade; and (2) the *lijin* 厘金 (likin) tax on inter-provincial domestic trade. Feuerwerker has added that the Qing government could not tap into local economic resources or manage economic life and that the fundamental problem of revenues for reform revealed more internal weakness than outside imperialism. Pong's study uses the Fuzhou Navy Yard as an example of both the successes and failures of Foreign Affairs Movement. He has stressed the potential for

[11] See John Rawlinson, *China's Struggle for Naval Development, 1839–1895*, 198–204. Compare Thomas Kennedy, *The Arms of Kiangnan: Modernization in the Chinese Ordnance Industry, 1860–1895*, 146, 150–160.

change in this era and has avoided characterizing Self-Strengthening as a failure.[12]

Japanese scholars of late Qing reform such as Hatano Yoshihiro have singled out the Qing bureaucratic system as the principal factor that ensured the maintenance of the old order in China, not economic backwardness, imperialism, or the inherited ideology and culture. According to this point of view, the Qing bureaucracy and its financial system rewarded imperial officials inordinately while the land market and economic corruption encouraged the status quo. Neither the Qing peasantry nor merchants were protected from official abuses and commercial exploitation.

According to Hatano, Qing officials and literati never fully understood the external crises they faced because they were schooled in the traditional civil examination system and thus were ignorant of the outside world. To make his point, Hatano cites the Jiangnan Arsenal (*Jiangnan zhizaoju* 江南制造局) and Fuzhou Navy Yard. Because each was under the control of regional governors or governor-generals, they were inefficient, wasteful, and lacked centralized coordination. Non-military enterprises such as the China Merchants' Steam Navigation Company, the Kaiping Coal Mine, and the Imperial Telegraph Administration were all defense-oriented, but unlike the arsenals they were organized on a profit-making basis and competed successfully against foreign companies.[13]

---

[12] David Pong, "Keeping the Foochow Navy Yard Afloat: Government Finance and China's Early Modern Defense Industry, 1866–75", *Modern Asian Studies* 21, 1 (February 1987): 121–152, and Albert Feuerwerker, "Economic Trends in the Late Ch'ing Empire, 1870–1911", in John Fairbank and Kwang-Ching Liu, eds., *The Cambridge History of China*, Volume 11, Part 2, 59–68. See also David Pong, *Shen Pao-chen and China's Modernization in the Nineteenth Century*, 11–20. Kwang-Ching Liu, "Nineteenth-Century China", in Ping-ti Ho and Tang Tsou, eds., *China in Crisis*, Vol. 1, 93–178, had earlier described the *Yangwu yundong* as an early stage of China's institutional reform and not a failure.

[13] Hatano Yoshihiro, *Chūgoku kindai kōgyō shi no kenkyū*. Compare Ellsworth Carlson, *The Kaiping Mines, 1877–1912*, and Hatano Yoshihiro, "The Response of the Chinese Bureaucracy to Modern Machinery", *Acta Asiatica* 12 (1968): 13–28. Compare K. H. Kim, *Japanese Perspectives on China's Early Modernization: A Bibliographical Survey*.

Similarly, Itō Shūichi has contended that science and technology in late Qing China required the dethronement of the official orthodoxy and civil examination reform before they could be advanced. In Itō's view, although they served as a catalyst of the modern intellectual revolution in China — the publication of translated Western scientific works and technical books and the creation of new technical schools after 1865 contributed to the spread of Western social and political ideas among Chinese intellectuals — the translations per se produced no scientists or engineers.[14]

To challenge negative views of the imperial Chinese state in Japanese scholarship, the eminent sinologist Miyazaki Ichisada refuted claims that Qing bureaucratic control had ruined most of China's early industrial enterprises. According to Miyazaki, Li Hongzhang's desire to check foreign domination of shipping in China had motivated him to sponsor the China Merchants' Steam Navigation Company in 1872. As an "officially supervised and merchant-operated" (*guandu shangban* 官督商办) enterprise, the China Merchants' Company actually was a government venture, according to Miyazaki. Supported by a total of 2.15 million taels (3 million silver dollars) in long-term, interest-free government loans, the company became the largest shipping firm operating in China by 1876. Miyazaki ironically has noted that the company declined only after 1909 when it was privatized under the industrialist Sheng Xuanhuai's 盛宣怀 (1849–1916) personal control.[15]

In his study of the Foreign Affairs Movement, Onogawa Hidemi has described it as the first phase of a broader late Qing reform movement. The first phase focused on technical innovation, while the second phase shifted to institutional innovation after the 1894–1895 Sino-Japanese War. In Onogawa's view, the key figures of the first phase of technical and industrial reforms during the 1870s and 1880s, such as Xue Fucheng

[14] Itō Shūichi, "Kindai Chūgoku ni okeru kagaku gijutsu no chii", *Tōyō gakujutsu kenkyū* 5, 5–6 (1967): 65–77.

[15] Chi-kong Lai, "Li Hung-chang and Modern Enterprise: The China Merchants' Company, 1872–1885", in Samuel Chu and Kwang-Ching Liu, eds., *Li Hung-chang and China's Early Modernization*, 216–247, and Wellington Chan, "Government, Merchants, and Industry to 1911", in John Fairbank and Kwang-Ching Liu, eds., *The Cambridge History of China*, Volume 11, Part 2, 422–429.

薛福成 (1838–1894), Ma Jianzhong 马健中 (1844–1900), Guo Songtao 郭嵩焘 (1818–1891), and Zeng Jize 曾纪泽 (1839–1890), who served as administrative experts and advisors to many of the chief ministers of the late Qing, advanced mercantilist proposals for developing mining, railroads, and foreign trade to create the material wealth needed for military Self-Strengthening.

In late 19th century, however, Wang Tao 王韬 (1828–1897) proposed sweeping changes in the civil examination, military, and educational systems. Others were critical of Li Hongzhang's policies in mid-1880s because of his focus on the navy rather than basic reforms in internal administration, which they regarded as more pressing. The doctrine of "Self-Strengthening" eventually evolved into a doctrine of reform. Onogawa's stress on the shift from science and technology to the institutional changes needed in China in the 1880s suggests that the technical achievements before 1895 were recognized but deemed insufficient not in terms of a failure in science and technology but in light of institutional systems that needed reform. Similarly, some accounts from mainland China concluded that the Self-Strengthening reforms were in some ways successful despite China's naval defeats.[16]

The Role of Regional Arsenals in the Self-Strengthening Movement (*Ziqiang yundong* 自强运动)

In the summer of 1865, Li Hongzhang as Jiangsu governor and Ding Richang 丁日昌 (1823–1882) as the Shanghai customs intendant rented a machine shop in the Hongkew section of Shanghai from Thomas Hunt and Company, an American firm in the Shanghai Foreign Settlement that was the largest foreign machine shop in China. Li also approved the purchase of the machine shop and the shipyard of Hunt and Company for use by the Suzhou "Foreign Arms Office". Additional machinery was imported, and subsequently the Jiangnan Machine Manufacturing General Bureau (*Jiangnan jiqi zhizao zongju* 江南机器制造总局), usually called the

[16] Onogawa Hidemi, *Shimmatsu seiji shisōkenkyū*, 8–85. Compare Qi Qizhang, ed., *Jiawu zhanzheng jiushi zhounian ji'nian lunwen ji*, 443–459.

Jiangnan Arsenal, was established to administer the industrial works and educational offices.

Initially, the Jiangnan Arsenal used 250,000 taels (348,000 silver dollars) for production facilities, drawn mainly from maritime customs funds collected at Shanghai. Ding Richang was appointed director in 1865, and Ying Baoshi 应宝时 (1821–1890) was appointed 1866–1868. The Arsenal was moved just outside the Chinese city of Shanghai in the summer of 1867. According to Mary Wright, by 1870 the Arsenal had become the greatest manufacturing center of modern arms in East Asia and "one of the great arsenals of the world".[17]

In her revisionist account of the Jiangnan Arsenal, Meng Yue has described how for Zeng Guofan 曾国藩 (1811–1872), Li Hongzhang, and their advisors the manufacture of machines represented the fundamental building block for industry. In their view, the three basic ingredients for constructing new industry were: (1) manufacturing machines; (2) creating a new institutional category of engineers (制器之人, *zhiqi zhiren*, i.e., "machine workers"); and (3) the translation of scientific and technical texts. Via armaments manufacture, the Qing state would break the Euro-American monopoly of warships and cannons and master contemporary useful knowledge.[18]

Technical work at the Arsenal was left in hands of foreigners such as the American T. F. Falls, Hunt's chief engineer, who was the superintendent. Eight of Hunt's machinists were retained, and 600 workers from Hunt and Company were transferred directly to the Jiangnan Arsenal. Many others were later added. They produced serviceable muskets and small howitzers after initial failures in rifle production. By mid-1867 the arsenal was producing 12 muskets and 100 12-pound shrapnel daily. Twelve-pound howitzers were produced at a rate of 18 per month and used as munitions in the Nian Rebellion 捻军起义 of the 1860s. By 1871 the Arsenal produced Remington breech-loading rifles. At the end of 1873,

[17] Mary Wright, *The Last Stand of Chinese Conservatism: The T'ung-chih Restoration, 1862–1874*, 211–212.

[18] Meng Yue, "Hybrid Science *versus* Modernity: The Practice of the Jiangnan Arsenal", *East Asian Science, Technology, and Medicine* 16 (1999): 13–52.

4,200 were produced, but they were more costly and proved inferior to imported Remingtons. In 1874–1875, Li Hongzhang advised establishing a branch to produce powder and cartridges instead.[19]

Technical Learning in the Jiangnan Arsenal

Before the English missionary John Fryer (1839–1928) joined it, the translation project for the Jiangnan Arsenal was very modest. The Chinese and their collaborators planned to produce an encyclopedia of knowledge and information that would resemble the *Encyclopedia Britannica*, but this goal was quickly deemed too elementary and perhaps too traditional, i.e., a mimicking of Ming–Qing encyclopedia (*leishu* 类书) traditions. Instead, the Translation Bureau began producing a series of industrial treatises focusing on technology and machinery, rather than mathematics and the natural sciences, after hiring a core group of Chinese and Western translators.[20]

From 1863, when the Imperial Court approved creation of the Shanghai School of Foreign Languages (*Tongwen guan* 同文馆, "School of Translated Learning"), it had remained an independent school of translation. In 1869, however, the Shanghai *Tongwen guan* was moved into the Jiangnan Arsenal and renamed the "Foreign Language School" (*Guang fangyan guan* 广方言馆). Its new buildings were paid for by the Shanghai Maritime Customs. Fryer's work now turned to translating Western books on manufacturing as Chinese textbooks for the new school,

[19] Thomas Kennedy, "The Establishment and Development of the Kiangnan Arsenal, 1860–95" (Columbia University Ph.D. dissertation in History, 1968), Chapter 2, and ECCP, 721–722. See also Knight Biggerstaff, *The Earliest Modern Government Schools in China*, 165–166, and Ting-yee Kuo and Kwang-Ching Liu, "Self-Strengthening: the pursuit of Western technology", in Denis Twitchett and John Fairbank, eds., *The Cambridge History of China*, Volume 10, *Late Ch'ing, 1800–1911*, Part 1, 519–521.

[20] John Fryer, "An Account of the Department for the Translation of Foreign Books at the Kiangnan Arsenal, Shanghai", *North-China Herald*, January 29, 1880: 78–79. Adrian Bennett, *John Fryer: The Introduction of Western Science and Technology into Nineteenth-Century China*, 34–35, notes that after 1880, Fryer placed a greater concentration on natural sciences than technology.

which would include the fields of engineering, navigation, military technology, and naval affairs.[21]

Classical learning was continued in the Jiangnan Arsenal after the Shanghai school moved into the Arsenal. It remained separate from the Translation Department in the hope that its graduates would go on to pass the more prestigious civil examinations. Hence, the school attracted the sons of Shanghai merchants and Christian converts in a more foreign environment. Arsenal students were also drilled in the 8-legged essay (*baguwen* 八股文) at the same time that mathematics was given high priority. For the latter, the "Ten Mathematical Classics" (*Suanjing shishu* 算经十书), several of which had been reconstituted by Qing scholars in the 18th century (see Chapter 5), were used to teach traditional Chinese mathematics.[22]

Students studied Western algebra, geometry, trigonometry, astronomy, and mechanics in the lower division curriculum. They were also trained in international law, geography, and mechanical drawing. The upper-division curriculum for students emphasized seven fields:

1. Mineralogy and metallurgy;
2. Metal casting and forging;
3. Wood and iron manufacturing;
4. Machinery design and operation;
5. Navigation;
6. Naval and land warfare; and
7. Foreign languages, customs, institutions.

It took three years to complete the two divisions. Outstanding graduates, it was hoped, would then take special provincial exams in Beijing.[23]

---

[21] Meng Yue, "Hybrid Science *versus* Modernity: The Practice of the Jiangnan Arsenal", 32–33. See also Knight Biggerstaff, *The Earliest Modern Government Schools*, 166–167, 173, and Adrian Bennett, *John Fryer: The Introduction of Western Science and Technology into Nineteenth-Century China*, 18–25. Curiously, the *Guang fangyan guan* staff avoided using the texts that Fryer and the Translation Bureau produced. See John Fryer, "An Account of the Department for the Translation of Foreign Books at the Kiangnan Arsenal, Shanghai", *North-China Herald*, January 29, 1880: 81.

[22] Jean-Claude Martzloff, *A History of Chinese Mathematics*, translated by Stephen Wilson, 225–232.

[23] Knight Biggerstaff, *The Earliest Modern Government Schools*, 166–171.

At its highest stage of development, the Jiangnan Arsenal contained four institutions:

1. A translation department;
2. A school for training translators and linguists;
3. A school for training skilled workmen; and
4. A machine shop.

Meng Yue notes that the Jiangnan Arsenal had 13 branch factories. By 1892, it occupied 73 acres of land, with 1,974 workshops and a total of 2,982 workers. The Arsenal possessed 1,037 sets of machines and produced 47 kinds of machinery under the watch of foreign technicians who supervised production.[24]

Shipbuilding in the Jiangnan Arsenal

From 1868 to 1876, according to Meng Yue, shipbuilding in the Jiangnan Arsenal was highly productive, when 11ships were built in eight years. Ten were warships. Five of these had wooden hulls; the other five were provided iron hulls. All parts of each ship, including the engine, were built at the Arsenal. The Arsenal also experimented with different designs, from single to double-screw, wooden and iron hulls, and simple warships to turreted vessels. When compared to the warships made in the Yokosuka Dockyard in Japan in the 1870s, the level of shipbuilding technology at the Jiangnan Arsenal was actually higher than in the leading Japanese dockyard.[25]

---

[24] *Jiangnan zhizaoju ji* 《江南制造局记》, compiled by Wei Yun'gong 魏允恭 (1867–1914), 151–168. See Meng Yue, "Hybrid Science *versus* Modernity: The Practice of the Jiangnan Arsenal", 29–30, and Adrian Bennett, *John Fryer: The Introduction of Western Science and Technology into Nineteenth-Century China*, 18. See also Knight Biggerstaff, *The Earliest Modern Government Schools*, 172.

[25] Meng Yue, "Hybrid Science *versus* Modernity: The Practice of the Jiangnan Arsenal", 16–17. Han van de Wen notes that most scholars are "in agreement that China's navy was superior to the Japanese", although the final verdict remains open. See his "War in the Making of Modern China", 740. Others would argue that such claims are valid only if one uses aggregate measures such as tonnage and weight of shell. The Japanese navy by 1894 was newer, faster, and equipped, as we will see below, with faster firing guns.

Meng Yue notes that the Yokosuka Dockyard did not produce its largest wooden warships until 1887–1888. Two were armed with 12 guns and boasted 1,622 horsepower. Neither was the match for the largest warship built at the Jiangnan Arsenal in 1872, which had 1,800 horsepower and was armed with 26 guns. Five iron-hulled warships were produced at the Jiangnan Arsenal before 1875, while the first iron Japanese gunboats were not completed until after 1887. In terms of armaments, those manufactured at the Jiangnan Arsenal were by and large superior to Japan's.[26]

Overall, however, the Chinese fleet of iron and wooden ships quickly fell behind the new ironclad ships of Europe. Moreover, the compound engine in Europe, which China did not begin to build until 1877 because of the lack of funds, superseded the outmoded single or double-screw engines in Chinese vessels. Hence, China's ships were still behind Europe's in the 1870s. Moreover, because Chinese shipyards could not produce enough ships, more warships were built in Europe for the Chinese navy. Although foreign technicians were again employed for building large modern warships, Chinese ships were still outmoded by the 1890s because Chinese training could not keep pace with Western technological progress. Japanese officers and sailors, in contrast, were better trained to manage their ships and guns by 1894.[27]

Shipbuilding in the Jiangnan Arsenal dramatically slowed after 1876. In 1885, when the Arsenal completed its first steel gunboat, it ceased to be a military shipyard. The technological switch toward steel and armored warships in Europe highlighted the difficulty of transporting iron and coal from inland provinces to make steel in coastal China.[28] At the same time

---

[26] Meng Yue, "Hybrid Science *versus* Modernity: The Practice of the Jiangnan Arsenal", 16–24, especially Tables 1 and 2. Compare Takehiko Hashimoto, "Introducing a French Technological System: The Origin and Early History of the Yokosuka Dockyard", *East Asian Science, Technology, and Medicine* 16 (1999): 53–65.

[27] Meng Yue, "Hybrid Science *versus* Modernity: The Practice of the Jiangnan Arsenal", 17. See also David Pong, *Shen Pao-chen and China's Modernization in the Nineteenth Century*, 224, and Knight Biggerstaff, *The Earliest Modern Government Schools in China*, 246–247.

[28] Kenneth Pomeranz, *The Great Divergence: Europe, China, and the Making of the Modern World Economy*, notes that China's fossil fuels were relatively inaccessible when compared to Europe's.

imported steel remained prohibitively expensive to make the ships domestically. Nevertheless, shipbuilding technology in Jiangnan Arsenal and the Fuzhou Navy Yard probably remained slightly better than in Japanese arsenals until 1889, when a French engineer came to Japan and designed new steel and iron warships for the Yokosuka Dockyard. Its first modern warship had more horsepower and a higher top speed than the same type of warship built by the Jiangnan Arsenal.[29] Hans van de Ven contends that Japan's modern navy had already demonstrated its superiority during its punitive expedition to Taiwan in 1874, and even more so later after the Sino-French War of 1884–1885, when, allegedly, "the Qing [navy] stood still" because the Manchu state "lacked the financial strength, military power, international significance, and internal cohesion to resist Japan". This view of China's military failures is based on the teleological hindsight that still bedevils most contemporary views of the aftermath of the 1894–1895 Sino-Japanese War. The late Qing regional navies that were built up at great expense at Port Arthur, Weihaiwei, Shanghai, Fuzhou, and Guangzhou were initially far better equipped and larger in scale than the singular Meiji fleet at Yokosuka during the 1870s and 1880s. Yokosuka began to catch up in the 1880s, as van de Ven stresses, but Japan in 1894 — like France in the 1884 — still required the tactic of "surprise attack" without officially declaring war to win the day. The Meiji fleet later benefited substantially from the huge war reparations Japan received from the Qing after 1895, as well as from the addition of captured Qing battleships to defeat Russia's fleet in 1904–1905.

Once shipbuilding was no longer its major task, the Jiangnan Arsenal adapted its machinery to produce the most advanced foreign guns and small arms for military use. As of 1874, the Arsenal had produced a total of 110 cannons and a variety of guns modeled after products from the Armstrong factory in Britain. Three types of large 120 mm, 175 mm, and 200 mm caliber muzzle-loading guns made by the Arsenal were deployed at the Wusong Fort guarding the mouth of the Yangzi River. In the late

---

[29] Meng Yue, "Hybrid Science *versus* Modernity: The Practice of the Jiangnan Arsenal", 17–19. Compare the older, "defeatist" view of the late Qing navy recently revived by Hans van de Ven in his still very enlightening *Breaking with the Past: The Maritime Customs Service and the Global Origins of Modernity in China*, 103–132.

1880s, the Arsenal produced large breech-loading guns that initially used black and then later brown gunpowder. By 1885, Li Hongzhang favored German arms over the British, and the scale of Krupp arms sales to China increased.

Before the Sino-Japanese War, the Jiangnan Arsenal was producing large breech-loading Armstrong guns whose range went from 7,000 to 11,000 yards. They were capable of firing projectiles from 80 to 800 lbs. The Arsenal also became known after 1890 for its success in producing rapid-firing machine guns, which were important in enhancing sea power and coastal defense forts. By 1892 the Jiangnan Arsenal had manufactured ten 40-pound rapid-firing guns. Two years later, the Arsenal was making rapid-firing machine guns capable of launching 40-pound and 100-pound shells. Because annual production in the Arsenal was insufficient to supply the Chinese army, the Qing military still had to purchase such arms from abroad. According to Meng Yue, Japan by comparison did not begin its ambitious artillery program until 1905, during the Russo-Japanese War (1904–1905).[30]

The Fuzhou Navy Yard

Besides the Jiangnan Arsenal in Shanghai, the second major industrial site for shipbuilding and training in the Western sciences and technology was the Fuzhou Naval Yard. When Zuo Zongtang 左宗棠 (1812–1885) submitted his 1866 memorial to establish a navy yard at Fuzhou, the expectation was that after five years the need for foreign experts would be eliminated. The estimated start-up costs of 300,000 taels (417,000 silver dollars) and the 600,000 taels (834,000 silver dollars) for annual operations were to come from maritime customs duties and the inter-provincial trade taxes (*lijin*/likin) collected in Fujian, Zhejiang, and Guangdong provinces. In return, those provinces received naval protection from the "Southern Fleet" based at Fuzhou.

From the start, Zuo and his successor Shen Baozhen 沈葆楨 (1820–1879) relied on French expertise in contrast to the British influence at the

[30] Meng Yue, "Hybrid Science *versus* Modernity: The Practice of the Jiangnan Arsenal", 21–23. See also David Wright, "Careers in Western Science in Nineteenth-Century China: Xu Shou and Xu Jianyin", *Journal of the Royal Asiatic Society*, third series, 5 (1995): 80.

Jiangnan Arsenal. Once the navy yard was established, however, only 400,000 taels (556,000 silver dollars) were raised from the Fujian maritime customs, with another 50,000 (69,500 silver dollars) per month for operations, leaving the venture in a perpetual financial bind. At its peak the shipyard employed 3,000 workers in the navy yard. When later construction was completed the force was dropped to 1,900, with 600 in the dockyard, 800 in workshops, and 500 coolies. Some 500 soldiers guarded the premises. The navy yard had more than 45 buildings on 118 acres set aside for administrative, educational, and production purposes. By comparison, the Jiangnan Arsenal as the largest ordnance enterprise in 1875 had 32 such buildings on 73 acres.[31]

In terms of scale, the Fuzhou Navy Yard was probably the leading industrial venture in late Qing China. Designed as a Westernized enterprise based on machinery and efficiency, the whole plant was served by a tramway with turntables at important workshops and intersections. The Navy Yard's goal was to build a modern Chinese flotilla between 1868 and 1875. Nineteen ships were planned with 80- to 250-horsepower engines. Of these, 13 would be transport ships with 150-horsepower engines. Sixteen ships were finished during this time. Ten transports with 100-horsepower engines, and one corvette as a showpiece, with a 250-horsepower engine, were realized in 1869–1875 while Shen Baozhen was in charge. Nine of the 150-horsepower transports cost over 161,000 taels (224,000 silver dollars) each; five of the 80-horsepower ships cost over 106,000 taels (147,000 silver dollars), with the corvette alone requiring 254,000 taels (353,000 silver dollars).[32]

Like the Jiangnan Arsenal, the Fuzhou Ship Yard also compared favorably with the Yokosuka Naval Yard. The latter began in 1865 with a budget of 1.3 million taels (1.8 million silver dollars) for a four-year period, compared to four million taels (5.6 million silver dollars) allotted to Fuzhou over five years. Actual expenditures at Yokosuka actually doubled the budget, while the Fuzhou Ship Yard expended 5.4 million

---

[31] Knight Biggerstaff, *The Earliest Modern Government Schools*, 200–208, and David Pong, *Shen Pao-chen and China's Modernization in the Nineteenth Century*, 208–209.

[32] Zhang Yufa, *Jindai Zhongguo gongye fazhan shi, 1860–1916*《近代中国工业发展史，1860–1916》.

taels (7.5 million silver dollars) from 1866 to 1874. By 1868, Yokosuka had completed eight ships with 11 more on the way. In comparison, Fuzhou was also at the forefront of naval and technological development. With two major industrial sites in the Yangzi delta and in Fujian province, the Qing was in aggregate ahead of Japanese modernization efforts in the 1860s and 1870s, but such aggregate advantages did not translate into organizational superiority or better training when the Fuzhou naval fleet faced the French flotilla alone, unprepared, and unaided in 1884.[33]

The industrial results in Fuzhou were at first gratifying for the Qing dynasty and praised in the December 10, 1875, *North-China Daily News*. Like the ships built at the Jiangnan Arsenal, however, the Chinese Southern Fleet in Fuzhou harbor were mainly wooden ships and thus vulnerable to European ironclads. Nor were they equipped with the latest compound engines. When faced with war with France in the 1880s and Japan in the 1890s, some Qing officials blamed the French for purposely dumping obsolete equipment and designs on the Chinese navy.[34]

Zuo Zongtang had also suggested opening a school for technical training, which, when established, was called the School for Naval Administration. Foreigners taught English, French, mathematics, and drafting. At the same time, students were expected to master the *Classic of Filial Piety* (*Xiaojing* 《孝经》) and the "Sacred Edict" and "Amplified Instructions" of the Kangxi and Yongzheng emperors, just like candidates for the local civil examinations. The Qing dynasty's long-term goal for the training provided by the French engineers and skilled workmen brought to Fuzhou was to create Chinese naval architects and engineers and to generate modern workmen: carpenters, ironworkers, brass workers, ship construction workers, etc.

---

[33] David Pong, *Shen Pao-chen and China's Modernization in the Nineteenth Century*, 241–243, 261.

[34] Prosper Giquel (1835–1886), a French naval officer, who had joined the Chinese Imperial Maritime Customs as a commissioner of customs at Ningbo in 1861 and later in Hankou until 1866, was the foreign director of the Fuzhou Naval Yard based on the contract he signed in 1866. See Steven Leibo, *Transferring Technology to China: Prosper Giquel and the Self-Strengthening Movement*, David Pong, *Shen Pao-chen and China's Modernization in the Nineteenth Century*, 214–225, and Knight Biggerstaff, *The Earliest Modern Government Schools*, 203–210.

Two divisions of French and English schools were set up. The French division included departments of naval construction, design, and apprentices. In the English division there was a naval academy with departments of theoretical navigation, practical navigation, and engine room training. The naval construction department opened first in February 1867 based on a curriculum of French, arithmetic, algebra, descriptive and analytic geometry, trigonometry, calculus, physics and mechanics. The five-year program suffered a high rate of attrition, however. In the first group of 105 beginning students, only 39 remained at the end of 1873.[35]

To train Chinese officers to operate warships, the English division, headed by John Carroll from England, created a department of theoretical navigation with a curriculum as follows:

> Arithmetic: for knowledge of fractions, proportions, interest, etc.
>
> Algebra: for quadratic equations of second degree, ratios, proportions, progressions, etc.
>
> Geography: used Anderson's *General Features of the Globe*.
>
> Trigonometry: plane and spherical; for solutions of triangles in navigation and nautical astronomy.
>
> Geometry: used Todhunter's Euclid (made up of Euclid's first three books and part of sixth book).
>
> Navigation: used Raper's Correction of Compasses, the Sailings, as usually taught, and the Day's Work.
>
> Nautical Astronomy: finding latitude and longitude methods and errors of the compass.

Besides building the naval yard and training personnel, Shen Baozhen saw to it that 15 ships were launched between June 1869 and February 1874. However, only 19 were completed between 1874 and 1897 when problems in the lower caliber of administration were exacerbated after Prosper Giquel's departure. The yard also faced a curtailment of operating funds due to the decline of interest by Beijing and provincial officials.[36]

---

[35] Knight Biggerstaff, *The Earliest Modern Government Schools*, 203–211. See also Benjamin A. Elman, *A Cultural History of Civil Examinations in Late Imperial China*, 135, 221–222.

[36] Knight Biggerstaff, *The Earliest Modern Government Schools*, 214–219.

A period of Qing self-management from 1874 commenced when operations in the Ship Yard carried on without foreign technicians until 1897, when five new French technicians arrived. Nevertheless, the schools were able to attract native students, mainly from the south, until the late 1880s. After 1874, graduates were sent to Europe, especially England and France, for advanced training to keep up with new technological developments. In 1877 Giquel led a party of 26 students. Twelve students from the English division went to England with five at the Royal Naval College at Greenwich. Nine of the 14 students from the French division studied hull construction and engine principles in France; the other five studied mining and metallurgy.

A second group of eight graduates were sent out in late 1882 for three years of advanced training. Five studied fortifications, defenses, and gunpowder explosives in France; two studied navigation and naval command in England; and one went to Germany for training in naval mines and torpedoes. A third group of 33 graduates were sent in 1886, with ten from the English division, 14 from French division, and nine from the Tianjin yard. Thirty completed their training; 18 studied hydrography, ironclad warship navigation, naval artillery and small arms in England; 12 studied hulls and engines, mathematics and ship construction, river control, bridge and railway construction, and international law in France. A fourth group was scheduled to go to Europe in 1894, but the war with Japan interrupted that.

In 1874, as a 21-year-old graduate, Yan Fu 严复 (1853–1921), for instance, was the acting captain of a small steamer owned by the Fujian–Zhejiang administration but not built by the Fuzhou Navy Yard. As a graduate of the Fuzhou naval division, however, Yan was eligible to receive advanced training in Europe. On his return to China he became a dean and professor of navigation and mathematics for many years at the Fuzhou Navy Yard. In the early 1880s he also became professor of navigation and mathematics in Tianjin Naval Academy where he was a teacher and administrator for nearly 20 years. After the bitter defeat to Japan in the Sino-Japanese War, an 1896 recommendation that foreign teachers should be hired in China rather than sending students to Europe was considered, but the *Zongli yamen* 总理衙门 (General Affairs Office) still wished to send the best naval students to Europe for advanced training. Ten were sent in 1897 for six years of training, but only six went to

France. They were recalled in 1900 after three years due to insufficient funds.[37]

Both David Pong and Knight Biggerstaff have described how industrial decline at the Fuzhou Navy Yard due to financial troubles had set in by 1876–1877. Expenditures totaled 5.35 million taels (7.4 million silver dollars) for the six and a half years to July 1874. This amount significantly exceeded original estimates, partly due to high costs for foreign wages, which used up 12,000 taels (16,700 silver dollars) out of the monthly operation cost of between 50,000 (69,500 silver dollars) and 80,000 (111,200 silver dollars) taels. By contrast, the total wages of 2,000 Chinese workmen amounted to only 10,000 taels (13,900 silver dollars) per month. Corruption and nepotism ate away at rest.

The Chinese staff under Shen Baozhen had to work together with Giquel and his Europeans for construction to remain on schedule. Because the shipyard was financed as a traditional enterprise with numerous sources of income, traditional Qing budgetary practices did not take into account inflation, growth, or retooling. Long-term planning became impossible. After 1880, the Fujian Maritime Customs failed to turn over regularly the full annual allocation of 600,000 taels (834,000 silver dollars). By the 1890s, the allocation fell to between 200,000 (278,000 silver dollars) and 300,000 (417,000 silver dollars) and under 200,000 taels by 1895. As a result, the schools and naval yard were less active in the critical years of the 1890s.[38]

Western Science in Translation

An 1861 proposal by reformist court leaders in Beijing, principally Prince Gong 恭亲王(Yixin 奕䜣, 1833–1898) and Wenxiang 文祥 (1818–1876), called for establishing the General Affairs Office to deal with

[37] Knight Biggerstaff, *The Earliest Modern Government Schools*, 223–241. See also Ting-yee Kuo and Kwang-Ching Liu, "Self-Strengthening: the pursuit of Western technology", 524–525.

[38] Knight Biggerstaff, *The Earliest Modern Government Schools in China*, 53, 220, 239, 271, and David Pong, *Shen Pao-chen and China's Modernization in the Nineteenth Century*, 266–270. See also Ting-yee Kuo and Kwang-Ching Liu, "Self-Strengthening: the pursuit of Western technology", 534.

the unprecedented nature of the Western threat the dynasty faced. The proposal also included a proposal for a School of Foreign Languages in Beijing. Li Hongzhang advocated similar schools in Guangzhou and Shanghai in 1863. His proposal was based on Feng Guifen's 冯桂芬 (1809–1874) 1861 recommendation for establishing an arsenal and shipyard in each Chinese port for better arms and ships in defense. Feng had also stressed establishing schools in Guangzhou and Shanghai for instruction in Western languages and science.[39]

Subsequently in 1866–1867, a proposal that a Department of Mathematics and Astronomy should be added to the School of Foreign Languages in Beijing was approved. Under this traditional focus, which echoed the Jesuit position in the Ming–Qing Astronomy Bureau and the French Jesuit role in the Academy of Mathematics, the goal of teaching students about modern science was realized through instruction in chemistry, physics, and mechanics. When William Martin (1827–1916) returned to Beijing in 1869 to teach physics, after more advanced study in the United States, he assumed the presidency of the School of Foreign Languages.

Science at the Beijing School of Foreign Languages and Elsewhere

Li Shanlan 李善兰 (1810–1882) left Shanghai and the Jiangnan Arsenal when he was appointed professor of mathematics at the Beijing School of Foreign Languages after it was upgraded to a college and a department of mathematics and astronomy was added in 1869. Li taught mathematics at the School for 13 years. A special civil examination in mathematics, however, was opposed in the 1870s, although Li's mathematics examinations at the School of Foreign Languages were influential among civil service candidates.[40]

---

[39] Su Jing, *Qingji Tongwen'guan ji qi shisheng*《清季同文馆及其师生》, *passim*. See also Nancy Evans, "The Canton T'ung-wen Kuan: A Study of the Role of Bannermen in One Area of Self-Strengthening", *Papers on China* (Harvard), 22A (1969): 89–103.

[40] Horng Wann-sheng, "Tongwen'guan suanxue jiaoxi Li Shanlan"〈同文馆算学教习李善兰〉, in Yang Tsui-hua and Huang Yi-long, eds., *Jindai Zhongguo keji shi lunji*《近代中国科技史论集》, 215–259. See also Knight Biggerstaff, *The Earliest Modern Government Schools*, 19–34, and ECCP, 480.

As the capital, Beijing remained a central venue for missionaries such as William Martin (1827–1916) who were linked to the School of Foreign Languages and the foreign scholars and Chinese students there. The missionary Society for Diffusion of Useful Knowledge (*Guangxue hui* 广学会) in Beijing, for example, also produced an illustrated monthly magazine known as *The Peking Magazine* (*Zhongxi wenjian lu* 《中西闻见录》). The *Magazine* was edited by Martin, among others, and printed from 1872 as a monthly for 36 issues before closing in August 1875. Distributed by the Society, the magazine was devoted to Western and international news, but it also included articles on astronomy, geography, and science (*gewu* 格物). William Martin had worked on the journal while heading and teaching at the Beijing School of Foreign Languages. Later, he compiled a selection of articles that were published separately as *Selections from The Peking Magazine* (*Zhongxi wenjian lu xuanbian* 《中西闻见录选编》) in four volumes in 1877.[41]

Hence, *The Peking Magazine* was a voice for the School of Foreign Languages to promote science and missionary concerns. They supported Li Hongzhang and the Self-Strengtheners in their efforts to reform the Qing regime. Altogether 199 articles (55.1%) in the *Magazine* were produced by the School's teachers or students. Moreover, the journal's role in promoting science and technology as a free monthly facilitated its content to be reprinted in the popular *Review of the Times* (*Wanguo gongbao* 《万国公报》), which had originally been called the *Chinese Missionary News* (*Jiaohui xinbao* 《教会新报》).

With Young J. Allen (1836–1907) as editor, the *Review of the Times* was published weekly in Beijing from 1874 and monthly after 1889. Both the *Magazine* and the *Review* addressed issues that concerned those employed in the emerging arsenals and shipyards. Of the 361 essays by the 54 foreign missionaries, traders, and diplomats, 166 (46%) dealt with the sciences and technology. The topics included the technical fields of astronomy, geography, physics, chemistry, medicine, as well as promoting railroads, mining, and the telegraph. Biographies of Western scientists were also added.

---

[41] San-pao Li, "Letters to the Editor in John Fryer's *Chinese Scientific Magazine*, 1876–1892: An Analysis", *Zhongyang yanjiuyuan Jindaishi yanjiusuo jikan* 《中央研究院近代史研究所集刊》 4 (1974): 737–738.

William Martin's use of *The Peking Magazine* to promote science in Beijing was complemented by the numerous contributions of Li Shanlan and his mathematics students in the School of Foreign Languages to the journal. Often the examination papers in the sciences and the mathematics homework of Li's students were included in the *Magazine*, thus confirming that it was the School's journal. The March and June 1875 issues, for instance, carried articles dealing with Martin's reaction to examination papers debating whether the earth or sun was at the center of the world.[42]

Students often replied to the section on "Difficult Questions" and established the precedent for "Answers to readers' queries", which also became a regular feature in later science journals in Shanghai. *The Peking Magazine* was an important model for *The Chinese Scientific and Industrial Magazine* (*Gezhi huibian* 《格致汇编》), which as we saw in Chapter 6 became the scholarly voice simultaneously of the Shanghai Polytechnic and the Jiangnan Arsenal. We have seen in Chapter 6 that a Translation Department was also initiated at the Jiangnan Arsenal.

Besides their use in the increasing number of missionary schools, such translations were also institutionalized as texts within a regional matrix of arsenals, factories, and technical schools that formed the 19th-century roots of the 20th-century industrial revolution in China. Hence, we should also acknowledge the scope and scale of scientific translation and military arsenals elsewhere in China after 1860. A sampling of these empire-wide venues includes the following:[43]

- Anqing Arsenal (1861), set up by Zeng Guofan.
- Beijing Field Force Arsenal (1883).
- Daye Iron Mine (1890), in Hubei.
- Fuzhou Ship Yard (1866), the base for the Southern Fleet, established by Zuo Zongtang.

---

[42] See the mathematics questions and science essays in *Zhongxi wenjian lu* 《中西闻见录》, Vol. 1, 403–404 (1872.2: 10a–b), 411–418 (1872.2: 14a–17b), 481–487 (1873.3: 16a–19b). On the earth versus the sun, see Vol. 4, 189–193 (1875.3: 6a–8a), 351–354 (1875.6: 7a–8b).

[43] See Ting-yee Kuo and Kwang-Ching Liu, "Self-Strengthening: the pursuit of Western technology", 519–537, and K. H. Kim, *Japanese Perspectives on China's Early Modernization*, 3–12.

- Guangzhou Arsenal (1874).
- Hangzhou Arsenal (1885).
- Hanyang Ironworks, in Hubei (1890), established by Zhang Zhidong.
- Hanyang Arsenal (1892).
- Hunan Arsenal (1875).
- Jiangnan Arsenal (1865), set up in Shanghai by Zeng Guofan and Li Hongzhang.
- Jilin Arsenal (1881).
- Jinling Arsenal (1867) in Nanjing used for making breech rifles and steel.
- Lanzhou Arsenal (1871).
- Port Arthur Naval Station (Lüshun, 1881–1882).
- Shandong Arsenal (1875), used for gun purchase, making acid and gun powder.
- Sichuan Arsenal (1877).
- Tianjin Arsenal (1867), under Li Hongzhang, used to manufacture gunpowder and acid.
- Taiwan Arsenal (1885).
- Weihaiwei Shipyard (1882) for the Northern Fleet.
- Yunnan Arsenal (1884).
- Xi'an Arsenal (1869).

Once the destruction of the Fuzhou Ship Yard's fleet during the Sino-French War (1883–1885) demonstrated the vulnerability of the Jiangnan Arsenal and other factories and fleets on the China coast to foreign naval blockade, Zhang Zhidong 张之洞 (1837–1909), then governor-general in Hubei and Hunan provinces in the middle Yangzi region, recognized the need for the Hanyang Ironworks (1890) and Hanyang Arsenal (1892) as protected inland industrial sites. Not funded until 1891–1895, however, and then subject to competing interests of Li Hongzhang's Northern Fleet and the military threat from the Japanese in Korea, the Hanyang Arsenal found that its funds were inadequate for simultaneous development of the ironworks and the arsenal. This problem led to a slowdown in the arsenal, which failed to produce weapons or ordinance in time for the Sino-Japanese War.

Other delays in plant building and a damaging fire in summer 1894 kept the Hanyang project from achieving success in the late 19th century.

Zhang wrestled with the twin goals of strategic industrialization and modern military production in the midst of the emergency diversion of imperial funds and resources to deal with the Russian and Japanese threats. He chose to fund the ironworks for general development rather than the arsenal for military arms. Hence, the Hanyang Ironworks became the hub of China's iron and steel industry during the first half of the 20th century, even though it failed to contribute to the Sino-Japanese War.[44]

If we repopulate this impressive list of modern industrial venues with the human lives and literati careers they instantiated, then we can trace more clearly the post-Taiping successors to the native calendricists and mathematical astronomers that compilation of the *Biographies of Astronomers and Mathematicians* (*Chouren zhuan* 《畴人传》; see Chapter 5) had adumbrated circa 1800.[45] Now, however, a new group of artisans, technicians, and engineers were emerging between 1865 and 1895 who gained an independent position from the fields of classical learning monopolized by the customary scholar-officials. Increasingly, they were no longer subordinate to the dynastic orthodoxy or its official representatives.

Still a necessary part of the cultural, political, and social hierarchies, the new students of the sciences in the arsenals and missionary schools emerged from the older categories of the myriad elite aspirants for official status as a literatus. The "scientist" (*gewu zhe* 格物者) was "one who investigated things", and he now coexisted with the orthodox classical scholar in the bureaucratic apparatus but still at lower levels of political rank, cultural distinction, and social esteem. Meng Yue has perceptively noted that the self-taught students of modern science and technology in the 1850s were successors of the Yangzi delta "astronomers and mathematicians" who had emerged during the rise of mathematics in an age of evidential research.[46]

---

[44] Thomas Kennedy, "Chang Chih-tung and the Struggle for Strategic Industrialization: The Establishment of the Hanyang Arsenal, 1884–1895", *Harvard Journal of Asiatic Studies* 33 (1973): 154–182.

[45] See Benjamin A. Elman, "Jesuit Scientia and Natural Studies in Late Imperial China", *Early Modern History: Contacts, Comparisons, Contrasts* 6, 3 (Fall 2002): 1–24.

[46] Meng Yue, "Hybrid Science *versus* Modernity: The Practice of the Jiangnan Arsenal", 25–28.

In Xu Shou's 徐寿 (1818–1882) case, evidential research could also serve as preparation for mastering Western science. Xu Shou, Li Shanlan et al. were in turn succeeded by the Yan Fus and Lu Xuns 鲁迅 (Zhou Shuren 周树人, 1881–1936) who were drawn to the Fuzhou Navy Yard and the Jiangnan Naval Academy (*Jiangnan shuishi xuetang* 江南水师学堂) for formal training in science, mathematics, and engineering. An examination scandal had affected Lu Xun's family both financially and socially, and Lu was forced to leave his lineage school. Before turning to literature, Lu Xun also trained at the School of Mines and Railways (*Kuanglu xuetang* 矿路学堂), and he later traveled to Japan to study modern medicine at Sendai just before the 1904–1905 Russo-Japanese War.[47]

By going outside the orthodox curriculum of the civil service examination, the newly educated in science, mathematics, and engineering inhabited the unprecedented institutional venues of arsenals, shipyards, and industrial factories that promoted non-degree-oriented engineering, mathematical, and science studies. Once put in place institutionally by the regional leaders of the diffuse Foreign Affairs Movement, technical expertise in engineering and mechanics and specialized knowledge of the modern sciences gathered momentum, albeit slowly, within what Meng Yue describes as "an independent, hybrid, even international field of cultural production".[48]

Eventually, thousands of administrative experts, translators, and advisors — including hundreds of foreigners — served in provincial schools and regional arsenals under the chief provincial ministers of the late Qing, Zeng 曾, Li 李, Zuo 左, and Zhang 张, who were the leaders of the turn toward foreign studies focusing on science and industry. Literati associated with statecraft and evidential studies after the Taiping Rebellion created the intellectual space needed to legitimate natural studies and

[47] David Wright, "The Great Desideratum: Chinese Chemical Nomenclature and the Transmission of Western Chemical Concepts", *Chinese Science* 14 (1997): 35–70. On Lu Xun, see Howard Boorman and Richard Howard, eds., *Biographical Dictionary of Republican China*, 417.

[48] Meng Yue, "Hybrid Science *versus* Modernity: The Practice of the Jiangnan Arsenal", 26–27, and David Wright, "Careers in Western Science in Nineteenth-Century China: Xu Shou and Xu Jianyin", 51–80, 88.

mathematics within the framework of "Chinese studies as fundamental, Western learning as useful" (*Zhongxue wei ti, Xixue wei yong* 中学为体，西学为用).

The promising start made in missionary schools and the empire-wide arsenals accelerated in the 1880s when Shanghai and Beijing took the lead in promoting the new fields associated with the Foreign Studies Movement. China's defeats in the Sino-French and Sino-Japanese wars, unfortunately, produced an intellectual backlash from foreigners in China and Chinese literati that China was doomed unless more radical political initiatives were carried out. In the process, the rhetoric favoring modern science became a key aspect of the political discourses of reformers and revolutionaries (see Chapter 8).[49]

The Impact of Science Translations in Qing China on Japan

Before 1894, Japan had imported many European books on science that had been translated in Qing China after the Japanese expelled the Jesuits for their meddling in the 16th-century civil wars there. Chinese translations of Euclid's geometry and Tychonic astronomy, for example, had made their way to Tokugawa Japan. We have discussed the impact of these works in Ming and Qing China in Chapter 5. Both the late Ming collectanea *Calendrical Studies of the Chongzhen Reign* (*Chongzhen lishu*《崇祯历书》) and the Kangxi era *Compendium of Observational and Computational Astronomy* (*Lixiang kaocheng*《历象考成》) arrived in Japan via the Ningbo–Nagasaki trade after the 1720s. The Japanese also avidly imported physics, chemistry, and botany books from Europe via the Dutch trading enclave in Nagasaki harbor in the early 19th century.[50]

In addition, the importance of 19th-century translations on science prepared under the auspices of Protestant missionaries in the treaty ports and others at the London Missionary Society's Inkstone Press in Shanghai was quickly recognized by the Meiji government. Prominent translations

[49] See the draft study of "Kuang Qizhao and the Significance of the 1870s and 1880s for China's Modernization", by Sam Wong and Valerie Wong, forthcoming.

[50] Jurgis Elisonas (George Elison), "Christianity and the Daimyo", in John Hall, ed., *The Cambridge History of Japan*. Volume 4. *Early Modern Japan*, 301–372.

into Chinese of works dealing with linear algebra, calculus, Newtonian mechanics, and modern astronomy quickly led to Japanese editions and Japanese translations of these works. Dr. Daniel Jerome Macgowan's (1814–1893) 1851 *Philosophical Almanac* (*Bowu tongshu*《博物通书》), for instance, had a Japanese edition, and Dr. Benjamin Hobson's (1816–1873) 1855 *Treatise of Natural Philosophy* (*Bowu xinbian*《博物新编》) from the Guangzhou Hospital came out as a Japanese edition in 1859, for instance. Four other of Hobson's medical works from 1851–1858 quickly came out in Japan between 1858 and 1864.[51]

Issues from the 1850s *Shanghae Serial* (*Liuhe congtan*《六合丛谈》), which included numerous articles introducing European science, were also republished in Japan, along with the translations Fryer et al. completed in the Jiangnan Arsenal and the publications from the Beijing School of Foreign Languages. The translations of algebra (1859/1872), calculus (1859/1872), and Martin's *Natural Philosophy* (1867/1869) were all quickly available to scholars and officials in Meiji Japan. Arguably, these works had greater influence in Japan than China, and they can still be easily found in libraries there, while they are rare in China.[52]

Many Chinese scientific terms were preferred in Meiji Japan over the translations derived from Dutch Learning. The Chinese name for chemistry (*huaxue* 化学), for example, replaced the term *chemie* [*semi* in Japanese] that was derived from Dutch. Similarly, the impact of Jiangnan Arsenal publications can be seen in the choice of Chinese terminology for metallurgy (*jinshi xue* 金石学) used in Japanese publications, which were later changed in Japan and reintroduced to China as a new term for mining (*kuangwu xue* 矿物学).

When the Japanese diplomat Yanagihara Sakimitsu 柳原前光 (1850–1894) visited China several times, he purchased many of the Chinese scientific translations. On his third visit in 1872, for instance, he bought 12 titles on science and technology in 31 volumes from the Jiangnan

---

[51] Wang Yangzong, "1850 niandai zhi 1910 nian Zhongguo yu Riben zhi jian kexue shuji de jiaoliu shulue"〈1850 年代至 1910 年中国与日本之间科学书籍的交流述略〉, *Tōzai gakujutsu kenkyūjo kiyō* 33 (March 2000): 139–152.

[52] Wang Bing, "Jindai zaoqi Zhongguo he Riben zhi jian de wulixue jiaoliu"〈近代早期中国和日本之间的物理学交流〉, *Ziran kexueshi yanjiu*《自然科学史研究》 15, 3 (1996): 227–233.

Arsenal. These included works on chemistry, ship technology, geography, traditional mathematics, mining, and trigonometry. The Japanese government continued to buy Arsenal books until 1877. In 1874, Yanagihara received 21 newly translated books from China. Despite the influence of Dutch Learning and translations from China, the Japanese began teaching modern Western science on a large scale only in 1870s, and thus the Chinese did not borrow many scientific terms from Japan before the Sino-Japanese War.[53]

Naval Warfare and the Refraction of Qing Reforms into Failure

Not until the Sino-Japanese War of 1894–1895, when the Japanese navy, which was tied to Yokosuka military technology, decisively defeated the Qing navy, which was tied to Fuzhou, Port Arthur, and Shanghai technology, did the superiority of Japan in modern science, or so it was interpreted, become common knowledge to Chinese and Japanese patriots. Although the Jiangnan Arsenal and Fuzhou Ship Yard had appeared superior in science and technology to the Yokosuka Dockyard until the 1880s, after 1895 each side read their different fates in the war teleologically back to the early Meiji period, in the case of triumphant Japan, or back to the failures of the Self-Strengthening Movement after 1865, in the case of the defeated Qing.

The Jiangnan Arsenal and the Fuzhou Ship Yard, for example, were generally acknowledged by contemporary Europeans and Japanese to be more advanced than their chief competitor in Meiji Japan, the Yokosuka Dockyard, until the 1880s. David Pong has contended, for instance, that had the Qing navy engaged the Japanese in a naval battle over Taiwan when the Japanese threatened the island in April 1874, Chinese maritime defense preparations would have gained greater support. Due to a policy debate, however, the Chinese sued for peace to avoid hostilities with the result that the budget for the two modern naval fleets in north and south

[53] Wang Yangzong, "1850 niandai zhi 1910 nian Zhongguo yu Riben zhi jian kexue shuji de jiaoliu shulue", 142. Compare David Wright, "The Translation of Modern Western Science in Nineteenth-Century China, 1840–1895", *Isis* 89 (1998): 671

China was cut to four million taels (5.56 million silver dollars), much less than was needed.

We have seen above that the mid-1870s saw a cutback in the production of ships in both the Jiangnan Arsenal and Fuzhou Ship Yard. In the mid- and late 1870s, China's armaments industries were mainly producing ammunition for Zuo Zongtang's army to reconquer Xinjiang in the northwest. Besides financial difficulties, corruption was also rife among leading officials who competed with each other for the remaining funds. Many contended that Zeng, Li, and Zuo built up the armaments industry mainly for their own power bases and to maintain domestic security, not to defend against attacks from foreign aggression.[54]

According to Rawlinson, only three Japanese ships with about 3,600 men were in the 1874 Japanese expedition to Taiwan. The Japanese naval ministry was established in 1872, and by 1874 it had just 17 ordinary ships with an aggregate of about 14,000 tons. Foreign observers thought China's 21 steamers in the 1,000-ton class would be able to handle the Japanese threat, but, as in 1894–1895, the Chinese ships were not organized into a unified fleet. Since it would take time to gather a fleet in Taiwan, and because he wrongly feared that Japan had two ironclad warships, Shen Baozhen as the director-general of the Fuzhou Navy Yard agreed to end the crisis with a financial payment to Japan and de facto recognition of Japanese control over the Liuqiu (Ryūkyū) Islands. By 1879, China had two ironclad steamships, which had been ordered from the Vulcan factory in the Baltic for the Northern Beiyang Fleet (*Beiyang jiandui* 北洋舰队) and were more advanced than anything the Japanese navy had at the time. They were both sunk in the Sino-Japanese War. In gunpowder manufacture, moreover, the machinery used in Germany, interestingly, was not as advanced as that in Shanghai at the Jiangnan Arsenal. China's naval superiority would never again be that manifest.[55]

---

[54] See Thomas Kennedy, *The Arms of Kiangnan: Modernization in the Chinese Ordnance Industry, 1860–1895*, 150–160, and David Pong, *Shen Pao-chen and China's Modernization in the Nineteenth Century*, 292–293, 335.

[55] John Rawlinson, *China's Struggle for Naval Development, 1839–1895*, 60–61, and David Wright, "Careers in Western Science in Nineteenth-Century China: Xu Shou and Xu Jianyin", 81.

The Impact of the Sino-French War

The lack of coordination between the northern and southern navies became the chief disadvantage Chinese fleets faced vis-à-vis their counterpart in Japan, which was a unified fleet stationed in Yokosuka under a central command. This disadvantage became clearer after 1874 when the French claimed Vietnam as protectorate leading to conflict with Qing China in the upper Red River area in northern Vietnam. France then began a naval buildup on the China coast which provoked several naval engagements. France did not win all the battles of the Sino-French War, but it did win the war in 1884–1885 because of the lack of coordination between the vulnerable Chinese fleet based at the Fuzhou Ship Yard and the Beiyang Fleet under Li Hongzhang's control in the north. The irony that a French-sponsored Chinese navy at the Mawei anchorage in Fuzhou would be destroyed — before war had been declared — by a French flotilla using Vietnam as its base suggests the dangers of relying on European aid in an age of imperialism.

The Qing had over 50 modern naval ships in 1884, with more than half built in China. Among the others, 13 were Armstrong gunboats, two were Armstrong cruisers, and two more were German ships with two eight-inch guns each. The latter two pairs were divided equally between the Northern Beiyang Fleet and Southern Nanyang Fleet (*Nanyang jiandui* 南洋舰队). The Qing navy, however, was divided into four fleets: the Northern at Weihaiwei and Port Arthur, one in Shanghai, another in Fuzhou, and the smallest in Guangzhou. Unfortunately, the 1884–1885 war was fought by the Fuzhou flotilla nearly alone in the climactic battle at its home port of Mawei.

At Mawei, the Fuzhou fleet was almost completely destroyed in 15 minutes in part due to the vagueness of international law when war had not yet been declared. This diplomatic nicety had allowed French war vessels to sail past the Min River defenses and approach the Fuzhou dockyard unchallenged. Hence, China, unlike France, had not made preparations for war. The modern fleet at the Mawei anchorage numbered 11 ships on August 23, 1884. All were at least nine years old and made of wood. Eight French vessels were anchored nearby and were on the whole superior, but the Chinese ships had respectable if non-standard armaments. Nor did the Chinese take advantage of the tides to outmaneuver the heavier French vessels. Li Hongzhang only sent two of the ships requested from his Beiyang Fleet, and he withdrew these from the battle by asserting that the Japanese threat in Korea mandated their return north.

The French fleet withdrew to Taiwan, but after a failed landing there it threw a blockade around the west coast of the island. Negotiations then resumed after a surprising Chinese land victory over the French. China's loss, then, was not simply due to French military superiority. Rawlinson has noted that French technological superiority in the 1880s was not as great as England's in the Opium War of 1839–1842 and the Second China War of 1856–1860. The gap between China and Europe had been closed technologically. The remaining problems were: (1) the political and regional disorganization of the empire; and (2) naval personnel were insufficiently trained and had a poor grasp of modern naval strategy.[56]

In the postwar period, progress at the Fuzhou dockyard was limited in scope, while Li Hongzhang sought to purchase naval vessels for his Northern Fleet rather than build them at home. Li also had to supply his Anhui land army. After most of the Fuzhou squadron was destroyed by the French in 1884, a foreign-built ship was purchased and used as a training vessel. The Fuzhou Navy Yard also reduced its engineers and skilled workmen, but it continued to operate in the 1890s despite neglect. One ship each year was launched in 1891, 1892, and 1895. Books and supplies were damaged but restored by 1886.

The rise of the Beiyang Fleet as China's chief fleet after 1885 was the result of the "Disaster in the South". Although demanded by the court, subsequent efforts to create a single command for a unified naval fleet never succeeded. The new Navy Board and Li's Beiyang Fleet competed for financial resources, which were declining due to further naval budget cuts between 1885 and 1894. The Empress Dowager's efforts to garner funds for her ambitious plans to expand the Summer Palace did not use up the imperial treasury and thus leave nothing for the Chinese navy, but

[56]John Rawlinson, *China's Struggle for Naval Development, 1839–1895*, 109–128. Compare Allen Fung, "Testing the Self-Strengthening: The Chinese Army in the Sino-Japanese War of 1894–95", 1010–1015, who also stresses China's lack of preparedness for the Sino-Japanese War when the Japanese navy, like the French, attacked without warning.

inadequate funding did set limits on Li Hongzhang's plans to expand the Northern Fleet.[57]

The apparent strength of the Beiyang Fleet, however, was clear to the Japanese because of stops the Chinese fleet made there in the 1880s after cruises to Vladivostok. Moreover, the inconclusiveness of the Sino-French War, which was reported in Japan, had restored Chinese prestige in Japanese eyes from the low it had reached after the Opium War. In the "Nagasaki Incident" of 1886, for instance, four warships of the Northern Fleet dropped anchor in Nagasaki on their return trip from the Russian port. Reinforced by new ships purchased from Germany, Li Hongzhang sought to make a propaganda statement by showing the Japanese that China's naval equipment was superior. Fights between Chinese sailors, who claimed the right of extraterritoriality while in Japan, and Nagasaki police, who viewed it differently, broke out during the port call. Each side blamed the other.

Japanese hostility was apparently aroused by China's flaunting of its naval superiority. Similarly, the "Kobe Incident" of 1889 was based on Japanese–Chinese fights that became a diplomatic dispute after a Chinese port stop there. Another visit by the Chinese fleet in July 1890 was reported in the newspaper *Citizen's Press* (*Kokumin Shimbun* 国民新闻) as an instance of the Chinese showing off their new ships. Toyama Masakazu (1848–1900), an educator and former president of Imperial Tokyo University, visited the flagship of the Chinese fleet and came away impressed with its large caliber guns and thick steel armor. The Sino-Japanese War put an end to these diplomatic controversies by exploding the notion of Chinese superiority and rejecting Chinese claims of extraterritoriality in Japan.[58]

---

[57] John Rawlinson, *China's Struggle for Naval Development, 1839–1895*, 129–139, and Knight Biggerstaff, *The Earliest Modern Government Schools*, 221–222. See also Kwang-Ching Liu and Richard Smith, "The Military Challenge: The North-West and the Coast", 254–256.

[58] Noriko Kamachi, "The Chinese in Meiji Japan: Their Interaction with the Japanese before the Sino-Japanese War", in Akira Iriye, ed., *The Chinese and the Japanese: Essays in Political and Cultural Interactions*, 69–72. See also Donald Keene, "The Sino-Japanese War of 1894–1895 and Its Cultural Effects in Japan", in Donald Shively, ed., *Tradition and Modernization in Japanese Culture*, 122–123.

The Sino-Japanese War and Its Aftermath

Upon the unexpected outbreak of the Sino-Japanese War on July 24, 1894, the foreign press generally predicted an eventual Chinese victory, even after reports of initial Chinese losses. G. A. Ballard, vice admiral in the British Royal Navy, thought the Beiyang Fleet in the 1890s was in serviceable condition and ready for action. Later comparisons between the naval fleets of China and Japan indicated that China might have won the sea war. Japan's fleet totaled 32 warships and 23 torpedo boats manned by 13, 928 men. Ten were built in Britain, and two in France. The Yoshino from Armstrong's shipyard was regarded as the fastest vessel of its time when it was timed at 23 knots in 1893 trials. China's navy still had a fourfold division into the Beiyang, Nanyang, Fujian, and Guangdong fleets, however. In 1894, these four combined had about 65 large ships and 43 torpedo boats. The strongest was the Beiyang Fleet which more or less equaled Japan's entire fleet.[59]

If general opinion among foreigners favored Li Hongzhang's fleet over Japan's, then Japanese newspapers, magazines and fiction were marked by exhilaration at the prospect of war with Qing China. Many Japanese themselves were not overly confident of victory, however. The publicist Fukuzawa Yukichi 福澤諭吉 (1835–1901) warned against overconfidence, for instance, although he agreed with Japan's just cause in spreading independence and enlightenment to a Korea allegedly subjugated by China. Indeed, Japanese Diet members were surprised at the easy victory, and the Meiji emperor was at first reluctant to begin hostilities. He had refused to send messengers to the imperial shrines at Ise or to his father's grave to announce the war until the news of the initial Japanese victories was communicated to Tokyo.[60]

Another British observer noted, however, that on the Chinese ships engaged in the Sino-Japanese War, Chinese crews were at half-strength but salaries for full crews were paid. The greatest contrast lay in the fact,

---

[59] John Rawlinson, *China's Struggle for Naval Development, 1839–1895,* 163–169. See also Donald Keene, "The Sino-Japanese War of 1894–1895 and Its Cultural Effects in Japan", 132.

[60] Donald Keene, "The Sino-Japanese War of 1894–1895 and Its Cultural Effects in Japan", 127, 132. Compare Shumpei Okamoto, "Background of the Sino-Japanese War, 1894–1895", in *Impressions of the Front: Woodcuts of the Sino-Japanese War, 1894–1895*, 13.

however, that Japan's navy was unified. There was some synchronization between China's four fleets, but in the end the Beiyang navy was left to fight the Japanese principally alone. Li had kept his fleet out of the Fuzhou battle in 1884, and the Nanyang officers now got their revenge on the Northern Fleet by keeping the Southern Fleet out of war with Japan for the most part. No national fleet existed, even on paper.

With the political and economic opening of Korea as the key dispute in Sino-Japanese relations, hostilities commenced when Japan seized the Korean king shortly after Li Hongzhang sent Qing troops into Korea in July 1894 to preserve Korea as a Qing tributary. The Korean king's regent then declared war on China. The first encounter between Chinese and Japanese ships occurred in late July at Fengdao, when China's two warships proved no match against an unprovoked attack by Japan's ships. After that sea battle, the Qing Northern Fleet tried to defend the Chinese coast from Weihaiwei to the mouth of Yalu River and declared war on Japan on August 1st.

Subsequently, the Japanese naval raid at Weihaiwei on August 10, 1894, stunned the Qing court, while Li Hongzhang stalled and made excuses for his inadequacies. The main Beiyang Fleet gathered at the mouth of the Yalu where the great naval battle with Japan for control of the Yellow Sea commenced on September 17. Each side had 12 ships in the clash. China had the advantage in armor and weight in a single salvo, while Japan had a decided advantage in speed of ships and metal thrown in a sustained exchange of salvos. Japan had more quick-firing guns that could fire three times the weight of metal from China's six- to 12-inch guns.[61]

Technology alone, however, was not the key determinant of the outcome, however. Japan proved to be superior in naval leadership, ship maneuverability, fast-firing guns, and the availability/reliability of explosive shells. Some observers described the Fuzhou-trained officers as cowards; they were the dominant Chinese group because of their experience and training when compared to the Tianjin-trained officers, few of whom were captains. In 1892, for example, most engine-room appointments still went to Fuzhou graduates. Nine of the 12 captains of

[61] John Rawlinson, *China's Struggle for Naval Development, 1839–1895,* 169–174, 201.

the Beiyang ships that fought Japan at the mouth of the Yalu were Fuzhou graduates. Cowardice was not the decisive factor, however, because China fired 197 12-inch projectiles at the decisive naval battle of Yalu, with only half of them being solid shot rather than explosive shell. They scored ten hits with six shots and four shells.[62]

From smaller guns, Chinese fired 482 shots and registered 58 hits, 22 on one ship, the Hiyei. They also launched five torpedoes without hits. China scored about 10% of her tries. The Japanese, on the other hand, with their quick-firers scored about 15% of their tries. In addition, the Chinese were hampered by woeful shortages of ammunition especially for her ships' big guns. Some shells were filled through the black market with cement rather than explosives, e.g., the one that struck the Matsushima and the two that passed through the Saikyo. This suggests that there were serious corruption problems in Li Hongzhang's supply command. With hindsight, assuming the same strategic decisions, it was clear that the speed and rapidity of fire were more important at Yalu than the weight of the vessel and its armor.

Shore engagements continued after the battle at the Yalu as the Japanese took advantage of their dramatic victory at sea to launch a land war, which allowed the Japanese First Army to occupy Pyongyang and then cross the Yalu to enter China at the Manchurian border. The Japanese Second Army, formed in September 1894, landed on the Liaodong Peninsula and took Port Arthur. Li Hongzhang now sought to rebuild his navy minus the Weihaiwei naval port. The poor command structure of the Beiyang Fleet and the lack of a court martial system made it impossible to place blame on officers and allocate reward properly, although many were made scapegoats for the defeat. Moreover, the Qing personnel system of naval rewards and punishments was filled with inequity and unpredictability. Many Chinese captains and officers simply committed suicide. No one dared to question the command structure or demand a board of review independent of the navy.[63]

---

[62] Knight Biggerstaff, *The Earliest Modern Government Schools*, 248.

[63] John Rawlinson, *China's Struggle for Naval Development, 1839–1895,* 174–197. See also Louise Virgin, "Japan at the Dawn of the Modern Age", in *Japan at the Dawn of the Modern Age: Woodblock Prints from the Meiji Era, 1868–1912*, 66–72, 86.

The Sino-Japanese War generated intense Japanese self-confidence after 1895. Moreover, Japanese industrialization accelerated after the Qing dynasty was forced to pay a considerable indemnity to the Meiji regime, not to mention the Qing ships added to the Yokosuka fleet. Korea and Taiwan were ceded to Japan as virtual colonies. Wider Western notice of the smaller island kingdom that had defeated the Chinese empire also ensued. The Japanese victory, however, had angered the Russians who feared Japanese expansion on the Asian continent. In concert with Germany and France, the Russians joined in a Triple Intervention after the Treaty of Shimonoseki was signed in April 1895, which forced the Japanese to withdraw from the strategic Liaodong Peninsula in exchange for an additional payment from the Qing government.[64]

For the Japanese public, the war victory developed into the key event that energized the newly emergent Meiji press, and drowned out editorial debate over the war. Public rage, as well as the Meiji emperor's, was directed at the European powers for intervening on the side of China. When Russia later forced the Qing to lease the Liaodong Peninsula to them, the Japanese were primed for war with Moscow over China. Public enthusiasm for military adventures became a common feature when the dissemination of the national news became a central feature of the Japanese press after 1895. There were by then 600,000 newspaper subscribers altogether in Tokyo and Osaka alone. The Japanese victory over China echoed throughout the country and demonstrated to Japanese the preeminence of Meiji Japan in East Asia, and the Japanese naval victory over Russia in 1904–1905 cemented such national exuberance.

The shift to an information press in Meiji Japan, which grew out of news accounts of the Sino-Japanese War, stimulated the demand for news and information in a new, unified Japanese language. The Hakubunkai Publishing House, for example, took advantage of the outbreak of war and quickly published a tri-monthly, illustrated record in September 1894 entitled the *Diary of the Japanese War with Qing China* (*Nisshin sensō jikki* 《日清戦争実記》), which was enormously popular and helped create a cult of Japanese war heroes. Other publishers quickly

[64] See Shumpei Okamoto, "Background of the Sino-Japanese War, 1894–1895", 16, and Louise Virgin, "Japan at the Dawn of the Modern Age", 112.

followed suit, and novels, plays and woodblock printed posters about the war became best-sellers. The *Yomiuri Shimbun*《読売新聞》newspaper initiated a prize competition for the "best" anti-Chinese war songs.[65]

In a completely opposite way for China, the naval disaster at the Yalu River and the decisive Qing defeat in the Sino-Japanese War energized public criticism of the dynasty's inadequate policies and enervated the staunch conservatives at court and in the provinces who had opposed Westernization. The unexpected naval disaster at the hands of Japan shocked many scholars and officials and now led to a new respect for Western studies in literati circles. The renewed success of the Shanghai Polytechnic in 1896, for example, was tied to this event. John Fryer now reported: "The book business is advancing with rapid strides all over China, and the printers cannot keep pace with it. China is awakening at last."[66]

Japanese Science in China after 1895

Unfortunately for Fryer and the Christian missionaries, China increasingly imported science books that were translated or edited in Japan after the Sino-Japanese War. Accordingly, there was a decisive sea change from missionary-based Chinese terminology for science from the 1840s to Japan-based Chinese terminology from 1900 — with the Sino-Japanese War as the key point of change. From 1896 to 1910, Chinese translated science books from Japan that were based on Japan's own translations because, unlike the Chinese, the Japanese no longer worked with foreigners to produce science translations. By 1905, when educational reforms were introduced as the key to the "New Governance" (*Xinzheng* 新政) policies of the Qing dynasty, the new Ministry of Education (*Xuebu* 学部)

---

[65] See Donald Keene, "The Sino-Japanese War of 1894–1895 and Its Cultural Effects in Japan", 121–175. See also James Huffman, "Commercialization and Changing World of the Mid-Meiji Press", 574–579, and Giles Richter, "Entrepreneurship and Culture The Hakubunkai Publishing Empire in Meiji Japan", 591, both in Helen Hardacre and Adam Kern, eds., *New Directions in the Study of Meiji Japan*.

[66] David Wright, "John Fryer and the Shanghai Polytechnic: making space for science in nineteenth-century China", *British Journal of History of Science* 29 (1996): 15, and Ting-yee Kuo and Kwang-Ching Liu, "Self-Strengthening: the pursuit of Western technology", 587.

was staunchly in favor of science education and textbooks based on the Japanese scientific system. Instead of the "West" represented by Protestant missionaries such as Martin or more secular Christians such as Fryer, Japan now became the mediator of West for Chinese literati and officials.[67]

After the Sino-Japanese War, reformers such as Kang Youwei 康有为 (1858–1927) and Zhang Zhidong encouraged Chinese students to study in Japan. Zhang's *A Plea for Learning* (*Quanxue pian* 《劝学篇》), which was presented to the throne and then distributed widely, was particularly influential in this regard. Kang promoted Meiji Japanese scholarship in his *Annotated Bibliography of Japanese Books* (*Riben shumu zhi* 《日本书目志》) and in his reform memorials to the Guangxu emperor. Kang recommended 339 works in medicine and 380 works in the sciences (*lixue* 理学), which now replaced Liang Qichao's 梁启超 (1873–1929) missionary-based list of the best Western books on science. The Guangxu emperor's edict of 1898 encouraged study in Japan. By 1905, 8,000 Chinese were studying in Japan, and this number increased dramatically by 50% in 1906, before declining to 4,000 by the end of the decade. Between 1900 and 1937, about 34,000 Chinese studied overseas in Japan.[68]

The Construction of China's "Backwardness" after the Sino-Japanese War

As we have seen in Chapter 6, the Sino-Japanese War also provoked a dramatic switch in missionary confidence about the future of Qing China. In a May 22, 1895, letter to President Kellogg concerning the chair of Oriental Languages position at Berkeley University, which he would be

[67] Wang Yangzong, "1850 niandai zhi 1910 nian Zhongguo yu Riben zhi jian kexue shuji de jiaoliu shulue", 139–144. See also Tsuen-hsuin Tsien, "Western Impact on China through Translation", *Far Eastern Quarterly* 13 (1954): 323–325.

[68] Wang Yangzong, "1850 niandai zhi 1910 nian Zhongguo yu Riben zhi jian kexue shuji de jiaoliu shulue", 144–145. See also ECCP, 30. Compare Douglas Reynolds, *China, 1898–1912: the Xinzheng Revolution and Japan*, 48, 58–61, and Barry Keenan, "Beyond the Rising Sun: The Shift in the Chinese Movement to Study Abroad", in Laurence Thompson, ed., *Studia Asiatica*, 157.

offered in July, John Fryer explained that his position in China had been strengthened because of China's defeat in the war. A "strong tide of demand for Western learning" was now evident among Chinese literati, who were "becoming aware of their own gross ignorance of modern arts and sciences". He added to Kellogg: "My translations are being bought up as fast as they can be printed, and education conducted on Western principles is becoming the order of the day. It is for this tide that I have waited patiently year after year, and now that it has begun to flow it would seem almost wrong to absent myself from the country that has so long afforded me a home and for those whose enlightenment I have so long been working."

Why then entertain a teaching position at Berkeley University at this promising time? Earlier in 1880 Fryer had rejected the possibility that English would become a universal language or that China would be ruled by foreign powers. In his 1895 letter, however, Fryer explained why he now entertained accepting the Berkeley position: "However necessary it may be for China to have the arts and sciences of the West translated into the native language and disseminated throughout the country in the first instance, it stands to reason that this will only succeed up to a point. Beyond that point no amount of translation can keep pace with the requirements of this age of progress." The "complete education of China" had begun through translation, Fryer quipped, but that was only a first step.

The man who had tirelessly translated several score of works on science and technology into Chinese now assumed a more strident tone. The war had proven to him and the Chinese that their efforts since 1865 had been a failure. Fryer now became a voice of doom for China's future:

> Of course this looks to the gradual decay of the Chinese language and literature, and with them the comparative uselessness of my many years of labor. Their doom seems to be inevitable, for only the fittest can survive. It may take many generations to accomplish, but sooner or later the end must come, and English be the learned language of the Empire.[69]

---

[69] Ferdinand Dagenais, *John Fryer's Calendar: Correspondence, Publications, and Miscellaneous Papers with Excerpts and Commentary*, 1895: 4–6. This overturns Fryer's earlier view in 1880 in his "An Account of the Department for the Translation of Foreign Books at the Kiangnan Arsenal, Shanghai", *North-China Herald*, January 29, 1880: 77–81.

This intriguingly timed Darwinian perspective belied the religious message of a natural theology that Fryer and other missionaries were encoding in their earlier translations of botany and biology for the Chinese.

On the eve of his departure for California, Fryer publicly announced a competition for "new age novels" (*xin xiaoshuo* 新小说) in Chinese that would enhance the morals of China. The Boxer Rebellion (*Yi He Tuan* 义和团) of 1900 confirmed the fears of most missionaries such as the devoted William Martin: "Let this pagan empire be partitioned among Christian powers."[70]

If, however, we look more carefully at the total picture of the Self-Strengthening period from 1865 to 1895, the view that Qing China was irrevocably weak and backward, in contrast to a powerful and industrialized Europe and a rapidly industrializing Japan, is an artifact of the impact of the Sino-Japanese War after 1895 on international and domestic opinion. Impatient perspectives of China's efforts to Westernize after 1865, unfortunately, underestimate the crucial role the missionary translations of science, the industrialization in the arsenals, and the new government schools played in the emergence of modern science and technology in late Qing China. We should deal with the 19th-century arsenals, factories, and translation schools by also considering them as a harbinger of the modern industrial revolution to come in China and not simply as a prelude to the end of the Qing dynasty and imperial China.

---

[70] Ferdinand Dagenais, *John Fryer's Calendar: Correspondence, Publications, and Miscellaneous Papers with Excerpts and Commentary*, 1895: 7–8, 11–12, 1896: 4, and Patrick Hanan, "The Missionary Novels of Nineteenth-Century China", *Harvard Journal of Asiatic Studies* 60, 2 (December 2000): 440–441. See also W. A. Martin, *The Awakening of China*, 177. For the Boxer impact, see Jonathan Spence, *To Change China: Western Advisers in China, 1620–1960*, 158–160.

Chapter 8

Rethinking the 20th-Century Denigration of Traditional Chinese Science and Medicine in the 21st Century

Western scholars and Westernized/Orientalized Chinese scholars have often essentialized the European history of science as the universalist progress of knowledge. When Chinese studies of the natural world, her rich medieval traditions of alchemy, or pre-Jesuit mathematical and astronomical achievements are discussed, as I have tried above, they are usually treated dismissively to contrast with the triumphant objectivity, rationality, or curiosity (see Chapter 3) of science in early modern Europe and the Americas. Most 20th-century scholars remain convinced that imperial China had no industrial revolution and had never produced democratic institutions, or capitalism.

As the evidence of a rich tradition of natural studies and medicine has accrued in volume after unrelenting volume of Joseph Needham's *Science and Civilisation in China* project after 1954, it has become harder to gainsay it all as superstition, irrationality or inductive luck. Ironically, the long-term history of Western science has been decisively refracted when viewed through the lenses chronicling the demise of traditional Chinese natural studies, technology, and medicine. Study of pre-modern Chinese science, technology, and medicine has restored a measure of respect to traditional Chinese natural studies and thereby granted priority to research in areas where the received wisdom was suspect, based as it was on careless speculation about banal generalities. Scholars now build on those contributions to search beneath the surface of self-satisfied discourses about "Western science" and the self-serving appeals to Greek deductive logic upon which they were rhetorically based.[1]

[1] See Geoffrey Lloyd and Nathan Sivin, *The Way and the Word: Science and Medicine in Early China and Greece*.

The beginnings of the "failure narrative" for Chinese science, i.e., why China of her own had not produced science or technology, has paralleled the story of political decline (why no democracy?) and economic deterioration (why no capitalism?) during the late empires of Ming and Qing. During the 20th century, the traditional Chinese sciences were increasingly seen as incompatible with the universal findings of Western science. As part of a complex historical process that began in the late 19th century, the denigration of traditional Chinese natural studies and medicine became a prerequisite among iconoclasts and revolutionaries for the institutional formation of modern science and medicine based on Western models that were decisively mediated by Japan.[2]

The Late Qing Stress on the "Chinese Origins of Western Learning"

Until 1900, however, some cultural conservatives such as Tang Caichang 唐才常 (1867–1900) confidently argued that Zhu Xi 朱熹 (1130–1200) had already enunciated many of the principles of Western science in Song times through his advocacy of the investigation of things (*gewu* 格物). Similarly, Zheng Guanying 郑观应 (1842–1923) contended that the fuller investigation of things section of the *Great Learning* (*Daxue*《大学》) had been lost in antiquity and yielded empty scholarship thereafter. Another popular position among late Qing classical scholars was to appeal to the origins of Western science in the practical and seemingly proto-scientific teachings of the Warring States philosopher Mo Di 墨翟 (470–391 B.C.). Ever since Cantonese scholars such as Zou Boqi 邹伯奇 (1819–1869) and others revived interest in the long forgotten Mo Di in academies such as the Sea of Learning Hall (*Xuehaitang* 学海堂) in Guangzhou, appeals to the *Master Mo's Teachings* (*Mozi*《墨子》) became an important feature of the late Qing claim that Western learning was derived from ancient China.[3]

---

[2] James Reardon-Anderson, *The Study of Change: Chemistry in China, 1840–1949*, 76–78. For background, see Nathan Sivin, "Science and Medicine in Chinese History", in Paul Ropp, ed., *Heritage of China: Contemporary Perspectives on Chinese Civilization*, 164–196.

[3] Tang Caichang, "Zhuzi yulei yiyou Xiren gezhi zhi li tiaozheng"〈朱子语类已有西人格致之理条证〉, in *Tang Caichang ji* 《唐才常集》, 172–176. Compare David Reynolds, "Redrawing China's Intellectual Map: Images of Science in Nineteenth-Century China", *Late Imperial China* 12, 1 (June 1991): 47–48, and Benjamin A. Elman, *From Philosophy to Philology: Social and Intellectual Aspects of Change in Late Imperial China*, 99, 112–113.

The distinguished Cantonese classicist Chen Li 陈澧 (1810–1882), whose teaching career at the Sea of Learning Academy influenced many local students, including Liang Qichao 梁启超 (1873–1929) and Kang Youwei 康有为 (1858–1927), linked the empiricism associated with "searching truth from facts" (*shishi qiushi* 实事求是) to the "investigation of things", and affirmed the Han Learning critique of Cheng–Zhu 程朱Song Learning. Chen also contended that the split between the classic and commentary of the *Great Learning* that Zhu Xi had introduced was unnecessary (see Chapter 2). Nevertheless, Chen Li concluded that Zhu Xi had gotten the meaning of the investigation of things essentially right.[4]

Even liberal critics accepted the late Qing distinction that was drawn between Zhu Xi's positivism and Wang Yangming's 王阳明 (1472–1528) idealism. Yan Fu's 严复 (1853–1921) criticism of Wang Yangming's view of the investigation of things, for example, also allowed him to place Zhu Xi's agenda for investigating things and extending knowledge (*gezhi* 格致) in a more positive light. In this period, Chinese scholars increasingly employed Wang Yangming as a foil to explain the Chinese failure to develop a positivist attitude more amenable to scientific discovery. In the 1890s, classically trained scholars like Yan reassessed the classical tradition for investigating things and where it went wrong.[5]

Some nationalists such as Liu Shipei 刘师培 (1884–1919) and Zhang Binglin 章炳麟 (1868–1936) preferred Later Han glosses for the investigation of things, which had been canonical, they contended, until the Song when Zhu Xi revised the *Great Learning* and placed it within the Four Books. Both Li and Zhang were critical of Zhu Xi for straying from original version of the *Great Learning* in the *Record of Rites* (*Liji*《礼记》). A more positivistic interpretation had emerged, according to Liu, when the Jesuits brought Western learning to China in the late Ming, but because the Yongzheng emperor had limited the impact of such new trends in mathematics and algebra, Zhu Xi's views had remained pervasive. Their views reveal how classical themes pro or contra Zhu Xi

[4] Chen Li, *Dongshu dushu ji* 《东塾读书记》, 9.14.

[5] Wang Hui, "The Fate of 'Mr. Science' in China: The Concept of Science and Its Application in Modern Chinese Thought", *positions: east asia cultures critique* 3, 1 (Spring 1995): 21–23. Mark Elvin's dated account in *The Structure of the Chinese Past*, 203–234, draws on this stock view of Wang Yangming's idealism.

and Wang Yangming could be appropriated by intellectual radicals in the late Qing.[6]

Wang Renjun and the *Ancient Subtleties of Science*

Wang Renjun 王仁俊 (1866–1914), an 1890s partisan of Zhang Zhidong 张之洞 (1837–1909), brought the late Qing "Chinese origins" position to scholarly completion. After arriving in Shanghai in 1897, he edited the reformist journal known as the *Journal of Concrete Learning* (*Shixue bao*《实学报》). Since the Sino-Japanese War of 1894–1895 (*Jiawu zhanzheng* 甲午战争), he had also worked on the *Ancient Subtleties of Science* (*Gezhi guwei*《格致古微》), which followed the traditional "four classifications" (*sibu* 四部; see Chapter 2) for citing native works in natural studies. Another was the *Record of the Essence of Science* (*Gezhi jinghua lu*《格致精华录》), which represented a revised version of the *Ancient Subtleties* and was rearranged according to Western categories for the sciences. Neither was entirely unprecedented, and each drew on earlier works which were included in the *Qing Exegesis of the Classics* (*Huang Qing jingjie*《皇清经解》).[7]

As early as 1870, Liu Yueyun 刘岳云 (1849–1917) had prepared materials for his encyclopedic *Chinese Methods for the Investigation of Things* (*Gewu Zhongfa* 《格物中法》, i.e., "science"), which was completed in its final form in 1899 but not printed in Beijing until after the Boxer uprising (*Yi He Tuan* 义和团) in 1900. Although more complete in comparative scope and Western content than Wang Renjun's *Ancient Subtleties*, the late printing of the *Chinese Methods* limited its influence at a time when the conservative agenda was in complete disarray.

The "Chinese origins of Western learning" (*Xixue Zhongyuan* 西学中源) and Zhang Zhidong's "substance and function" (*tiyong* 体用) dichotomy were also implicit in Wang Renjun's *Ancient Subtleties*. The

---

[6] See Liu Shipei, "Gewu jie"〈格物解〉, in *Zuo'an waiji* 《左庵外集》, in *Liu Shenshu xiansheng yishu*《刘申叔先生遗书》, Vol. 3, 1.23b–24b. See also Zhang Binglin, "Zhizhi gewu zhengyi"〈致知格物正义〉, in *Zhang Taiyan quanji* 《章太炎全集》, Vol. 5, 60–62. For discussion, see Benjamin A. Elman, *From Philosophy to Philology: Social and Intellectual Aspects of Change in Late Imperial China*, 18–23.

[7] Zeng Jianli, "Gezhi guwei yu wan Qing 'Xixue Zhongyuan' lun"〈格致古微与晚清西学中源论〉, *Zhongzhou xuekan* 《中州学刊》 (November 2000): 146–150.

two claims worked as complements in legitimating Western learning as a balance for Chinese learning. Wang stressed the importance of science in the creation of a strong China, and rationalized this mastery by affirming the ancient foundations of Chinese learning. Within the "substance and function" dichotomy, both Chinese and Western learning as a new hybrid were affirmed in the name of preserving China.[8]

Many could see that Wang's use of "Chinese origins" to legitimate Chinese natural studies was based on forced parallels with the specialized fields of Western science. Such obligatory appeals were in turn motivated chiefly by wishful thinking that ancient Chinese masters such as Master Mo or Zhu Xi had already enunciated the principles of modern science. These claims also indicated to those trained in the sciences in China or abroad that Wang had a weak grasp of modern science as a field of technical knowledge. Wang's emotionalist defense of Chinese learning as the basis for adopting Western science appeared to Yan Fu and others as an exercise in self-delusion.[9]

A more sweeping approach to science resulted, which was tied to a more radical political agenda by literati disenchanted by the dynasty's failures in wars with France (1883–1885) and Japan (1894–1895). No longer could the imperial court speak for the literati as a whole as they had in the late 18th century. Domestic public opinion tied to better educated Chinese literati who had worked with the Protestant missionaries could see through the "Chinese origins" schema and declared it intellectually bankrupt. By the time of the May 4th period, circa 1919, the claim that China was the source for Western learning provided proof that late Qing scholars had failed to grasp the essentials of modern science. What was plausible during the Kangxi reign no longer carried as much conviction during the late Qing.[10]

---

[8] See Wang Renjun, "Lueli" 略例, in *Gezhi guwei*《格致古微》, 2b. On the *tiyong* 体用 framework as a rationalization of Western learning, see Joseph Levenson, *Confucian China and Its Modern Fate: A Trilogy*, Vol. 1, 59–78.

[9] See Zeng Jianli, "Gezhi guwei yu wan Qing 'Xixue Zhongyuan' lun", 149–150.

[10] Wang Renjun, *Gezhi guwei*, 5.26b–40a. See also Quan Hansheng, "Qingmo de 'Xixue yuanchu Zhongguo' shuo"〈清末的"西学源出中国"说〉, *Lingnan xuebao*《岭南学报》(June 1935): 89–91.

Chinese Learning vs. Western Learning Mediated through Meiji Japan

Since 1865, literati inside and outside the bureaucracy had distinguished between "Chinese learning" (*Zhongxue* 中学), which represented the whole of native learning, from Western studies (*Xixue* 西学), within which science was prioritized. Neither term was successful as a monolithic designation, but each was politically charged in the 1890s when they were used by conservatives and radicals as a "dominant binary" in the struggle for political modernity. As Western learning gained momentum as a proper model for science and modern institutions, mediating terms for science such as "investigating things and extending knowledge" (*gezhixue* 格致学) were discarded. Educational institutions that used such accommodations were considered old-fashioned.

The increasing numbers of overseas Chinese students in Japan, Europe, and the United States perceived that outside of China the proper language for modern science included a new set of universal concepts and terms that superseded traditionalist literati notions of "Chinese" natural studies. For example, Japanese scholars during the early Meiji period had demarcated the new sciences by referring to *wissenschaft* as science (*kagaku* 科学, "classified learning based on technical training") and natural studies as the "exhaustive study of the principles of things" (*kyūri, qiongli* 穷理). The latter term, long associated with the classical stress on the "investigation of things" popular in early Tokugawa Japan, was reinterpreted in Japan based on the Dutch Learning tradition of the late 18th century, when Japanese scholars interested in Western science still used terms from Chinese learning (*Kangaku* 汉学 = "Sinology") to assimilate European natural studies and medicine.[11]

As accommodational terminology receded from use, so too the multiple identity of science and technology as native and Western disappeared. "Modern science" now meant simply "Western science". When the Qing court launched its "New Governance" (*Xinzheng* 新政) policies in 1901,

---

[11] Albert Craig, "Science and Confucianism in Tokugawa Japan", in Marius Jansen, ed., *Changing Japanese Attitudes Toward Modernization*, 139–142. See also Numata Jirô, *Western Learning: A Short History of the Study of Western Science in Early Modern Japan*, translated by R. C. J. Bachofner, 60–95.

however, this turn could not be described internally as a "Westernizing policy". Just as the early Qing court had refused to call the Jesuit calendar "Western", so now the late Qing court also called its reform policies "new" rather than "Western".

Both Liang Qichao and Zhang Zhidong changed the title of their earlier accounts of "Western learning" to "new learning" (*xinxue* 新学) to follow suit. As in the late Ming, the binary opposition between native and European learning was transformed in favor of the dichotomy between "old" and "new" learning. Lacking the notion of "Chinese origins", however, such parallel moves left the Qing state and its conservative literati bereft of a rationale for compelling study of "ancient learning" (*guxue* 古学) as a complement for "Western learning". Native learning now was simply "ancient learning", with no ties to the new learning from the West.[12]

Science and the Late Qing Reformers

Among the more "Westernizing" literati elites who played important roles in the 1898 Reform Movement, a more assertive approach to modern science was enunciated. Many reform advocates believed that scientific knowledge was a prerequisite for economic productivity in both industry and agriculture. Kang Youwei, like many, believed that the military successes of Meiji Japan served as a model for China. Expanded education in the sciences and industry were required.[13] In his 1905 essay on industrialization, for instance, Kang Youwei emphasized that China like Japan needed to master Western forms of mining, industry, and commerce. Because machines were industrial inventions that had augmented the economic and military power of European states and enhanced the welfare of the people there, Kang contended that the Qing dynasty had to change its goals and educate the people in technology and not just build factories and arsenals based on foreign models. Often, Kang and the

[12] Wang Hui, "The Fate of 'Mr. Science' in China: The Concept of Science and Its Application in Modern Chinese Thought", 1–68.

[13] Kung-chuan Hsiao, *A Modern China and a New World: Kang Yu-wei, Reformer and Utopian, 1858–1927*, 328–346.

reformers demeaned the results that were achieved when the Foreign Affairs Movement had promoted industrialization from 1865 to 1895.[14]

In the post-1895 political environment, reformers claimed they were championing unprecedented policies. In fact their calls for science and industry built on the efforts of their predecessors. Where Kang Youwei et al. did break new ground was in their demand that traditional, subsistence agriculture was outdated and should be replaced by the mechanization of farming. A new industrial-commercial society was the goal. Kang's focus on the required educational transformation that would increase the numbers of those trained to industrialize China through science and technology was on target, however. By 1905, when the civil examinations were abrogated, most literati now faced a new world of career expectations that drew them away from the classical curriculum that had remained prestigious until 1900.[15]

In particular, Kang read late Qing translations of Western political economy. Kang was also influenced by the missionary Timothy Richard (1845–1919), who had more confidence in the Qing reform movement underway. In 1895, Richard had published an influential essay "New Policies" ("Xin zhengce", 〈新政策〉), which had received as much attention as Young J. Allen's (1836–1907) account of the Chinese defeat that year by Japan. Richard also prepared a series of forward-looking essays on policy matters, which the reformers published in Shanghai together under the title *Tracts for the Times* (*Shishi xinlun* 《时事新论》, "new views of contemporary affairs"). Kang Youwei, and Liang Qichao consulted with Richard and drew on his essays for inspiration during the 1898 reforms.[16]

Based on his limited understanding of modern science, Kang Youwei contended that primal *qi* 气 was the creator of heaven and earth. He also

---

[14] See Grace Shen, *Unearthing the Nation: Modern Geology and National Identity in Republican China, 1911–1949*, and Shellen Wu, *Underground Empires: Coal and China's Entry into the Modern World Order*.

[15] Benjamin A. Elman, *Civil Examinations and Meritocracy in Late Imperial China, 1400–1900*.

[16] Kung-chuan Hsiao, *A Modern China and a New World: Kang Yu-wei, Reformer and Utopian*, 306–307, 331–346. See also *Eminent Chinese of the Ch'ing Period*, Hummel, Arthur, ed., hereafter ECCP, 703–704.

believed that *qi* was a sort of pervasive ether (*yitai* 以太) that lent spiritual form to both electricity and lightning. Similarly, Tan Sitong 谭嗣同 (1865–1898) became an amateur mathematician and advocate of science when he established a mathematical academy in 1895, with the help of his teacher. A typical reformer, Tan believed that mathematics was the foundation for both science and technology. Tan also read both traditional and Western mathematical works and became interested in geometry and algebra. In his major published work, *Studies of Benevolence* (*Renxue* 仁学), for example, Tan relied on mathematics as an authority more than as a tool in his writings to reveal the unity of all learning. Moreover, like Kang Youwei, he regarded "ether" as the "element of elements".[17]

Tan was most impressed with a work on psychology by Henry Wood (1834–1909) entitled *Method of Avoiding Illness by Controlling the Mind* (*Zhixin mianbing fa* 《治心免病法》), which later informed Tan's stress on the dynamics of mental power in his *Studies of Benevolence*. There, Tan declared the axiom: "Ether and electricity are simply means whose names are borrowed to explain mental power." Tan's views of the "ether", however bizarre at first sight, were drawn from trends in late 19th-century physics that seem equally bizarre from the standpoint of contemporary physics after 1905 when the nature of electromagnetic fields was better understood.[18]

During this period, a wave of translating Japanese mathematics texts also took hold, which proved to be a convenient shortcut to modern mathematics. The conversion to Western mathematics was also aided by the many Chinese students who returned from studies abroad, particularly from Japan after 1895. Over 10,000 Chinese traveled to Japan to study from 1902 to 1907. Some 90% of the foreign-trained students who joined the Qing civil service after 1905, for instance, graduated from Japanese schools.[19]

---

[17] Tan Sitong, "Xing suanxue yi" 〈兴算学议〉, in *Tan Sitong quanji* 《谭嗣同全集》, 153–194.

[18] See Chan Sin-wai, trans., *An Exposition of Benevolence: The Jen-hsueh of T'an Ssu-t'ung*, 10–20, and Richard Shek, "Some Western Influences on T'an Ssu-t'ung's Thought", in Paul Cohen and John Schrecker, eds., *Reform in Nineteenth-Century China*, 200–201, 237.

[19] Paula Harrell, *Sowing the Seeds of Change: Chinese Students, Japanese Teachers, 1895–1905*, 214, and Barry Keenan, "Beyond the Rising Sun: The Shift in the Chinese Movement to Study Abroad", in Laurence Thompson, ed., *Studia Asiatica*, 157–169.

Because civil examination credentials no longer confirmed gentry status after 1905, sons of gentry turned to other avenues of learning and careers outside officialdom. Du Yaquan 杜亚泉 (1873–1933), Ding Wenjiang 丁文江 (1887–1936), Cai Yuanpei 蔡元培 (1868–1940) and others quickly left behind the typical classical education of a literatus and increasingly traveled to treaty ports such as Shanghai or abroad to seek their fortunes as members of a new gentry-based Chinese intelligentsia that would be the seeds for modern Chinese intellectuals, scientists, doctors, and engineers. Yan Fu, whose poor prospects in the civil examinations led him to enter the School of Navigation of the Fuzhou Ship Yard in 1866, associated the power of the West with modern schools where students were trained in modern subjects requiring practical training in the sciences and technology.

For Yan Fu and the post-1895 reformers, Western schools and Westernized Japanese education were examples that the Qing dynasty should emulate. The extension of mass schooling within a standardized classroom system stressing science courses and homogeneous or equalized groupings of students seemed to promise a way out of the quagmire of the imperial education and civil examination regime, whose educational efficiency was now, in the 1890s, suspect.[20]

Among those affected by the educational changes after 1898, Ren Hongjun 任鸿隽 (1886–1961) would become one of the founders of the Science Society of China (*Zhongguo kexueshe* 中国科学社) in 1914. He had passed the last county civil examinations in 1904, and by 1907 he was a student in Shanghai where he met Hu Shi 胡适 (1891–1962). Ren then traveled to Japan in 1908 where in 1909 he entered the Higher Technical College of Tokyo as a student subsidized by the Qing government. The Qing dynasty had reached a 15-year agreement with the College to send 40 students annually to Tokyo.

While in Tokyo, Ren also joined Sun Yat-sen's 孙中山 (1866–1925) early partisans in the Alliance Society (*Tongmenghui* 同盟会) and rose to

---

[20] Marianne Bastid, *Educational Reform in Early 20th-Century China*, translated by Paul J. Bailey, 12–13, and Y. C. Wang, *Chinese Intellectuals and the West, 1872–1949*, 52–59. See also Benjamin A. Elman, *Civil Examinations and Meritocracy in Late Imperial China*, Chapter 8.

an important position in the Tokyo Sichuan branch. When he returned to China after the 1911 revolution, Ren served in Sun's provisional government, but he received a Qinghua University (*Qinghua daxue* 清华大学) fellowship in 1912 to study chemistry at Cornell. Ren received his B.A. in chemistry from Cornell, where he studied from 1912 to 1916, and his M.A. in chemistry from Columbia in 1917, during which time he assumed a leading role in forming a Chinese science organization that would replace the old-style literary societies (see below).[21]

Influence of Meiji Japan on Modern Science in China

In the late 19th century, an increasing familiarity with Western learning exposed the Chinese to the limits of traditional categories such as "investigating things" for scientific terminology. Increasingly, the claim that Western learning derived from ancient China became unacceptable. Younger literati perceived in the revival of such traditionalistic positions after the Sino-Japanese War, the third stage of the "Chinese origins" argument, a latent conservatism that obstructed the introduction of modern science and technology rather than facilitating it. Hence, those students who studied abroad in Japan and the West after 1895 began to question the use of "investigating things and extending knowledge" as a traditional trope of learning to accommodate modern science. By 1903, state and private schools increasingly used the modern classifications of social science (*shehui kexue* 社会科学), natural science (*ziran kexue* 自然科学), and applied science (*yingyong kexue* 应用科学) as terminology for science, which were borrowed from Japanese translations.[22]

---

[21] Ferdinand Dagenais, "Organizing Science in Republican China (1914–1950)", presented at the "Ideology and Science Symposium", Center for Chinese Studies, University of California, Berkeley, October 19, 2001. See also Douglas Reynolds, *China, 1898–1912: The Xinzheng Revolution and Japan*.

[22] Li Shuangbi, "Cong 'gezhi' dao 'kexue': Zhongguo jindai keji guan de yanbian guiji" 〈从格致到科学：中国近代科技观的演变轨迹〉, *Guizhou shehui kexue* 《贵州社会科学》137 (1995.5): 105–107. See also Victoria Lee, "The Arts of the Microbial World: Biosynthetic Technologies in Twentieth-Century Japan" (Princeton, NJ: Princeton University Ph.D. dissertation in the History of Science, 2014).

Accordingly, there was a decisive sea change from China-based terminology for science from the 1840s to Japan-based terminology from 1900 with the Sino-Japanese War as the key point of change. From 1896 to 1910, Chinese translated science books from Japan that were based on Japan's own translations because unlike the Chinese the Japanese no longer worked with foreigners to produce science translations. By 1905, when educational reforms were introduced as the key to the "New Governance" policies of the Qing dynasty, the new Ministry of Education (*Xuebu* 学部) was staunchly in favor of science education and textbooks based on Japanese scientific system.[23] After the Sino-Japanese War, reformers such as Kang Youwei and Zhang Zhidong encouraged Chinese students to study in Japan. Kang recommended 339 works in medicine and 380 works in the sciences (*lixue* 理学).[24]

The failure of the 1898 reforms, which led to Tan Sitong's martyrdom and the flight of Kang Youwei and Liang Qichao to Japan, did not prevent the mediation of science in China via Japan. Luo Zhenyu 罗振玉 (1866–1940), for example, published the *Agricultural Journal* (*Nongxue bao* 《农学报》) from 1897 to 1906 in 315 issues. He also compiled the *Collectanea of Agricultural Studies* (*Nongxue congshu* 《农学丛书》) in 88 works with 48 based on Japanese books. Du Yaquan edited journals in 1900 and 1901 that translated science materials from Japanese journals. These were the first science journals edited solely by a Chinese. The 1904 translation in Shanghai of a Japanese encyclopedia contained over 100 works with 28 in the sciences and 19 in applied science.

Post-Boxer educational reforms of 1902–1904 were decisive in the transformation of education in favor of Japanese-style science and technology. The last bastion of modern science as "Chinese science" remained the civil examinations, where the "Chinese origins" approach to Western learning remained obligatory. After the examination system was abolished in

---

[23] Wang Yangzong, "1850 niandai zhi 1910 nian Zhongguo yu Riben zhi jian kexue shuji de jiaoliu shulue"〈1850 年代至 1910 年中国与日本之间科学书籍的交流述略〉, *Tōzai gakujutsu kenkyūjo kiyō* 33 (March 2000): 139–144.

[24] Wang Yangzong, "1850 niandai zhi 1910 nian Zhongguo yu Riben zhi jian kexue shuji de jiaoliu shulue", 144–145. See also ECCP, 30. Compare Douglas Reynolds, *China, 1898–1912: The Xinzheng Revolution and Japan*, 48, 58–61, and Barry Keenan, "Beyond the Rising Sun: The Shift in the Chinese Movement to Study Abroad", 157.

1904, Japanese science texts finally became models for Chinese education at all levels of schooling. In 1886–1901, for instance, Japan officially approved 11 different texts on physics. Eight of those produced after 1897 were translated for Chinese editions. In 1902–1911, 22 different physics texts were approved in Japan, and seven were translated into Chinese.

Similarly, in chemistry from 1902 to 1911, 71 Japanese texts were translated into Chinese. Most were produced for middle schools and teachers' colleges. Twelve middle school chemistry texts were produced in Japan between 1886 and 1901. Of these, six were translated into Chinese. Eighteen Japanese middle school chemistry textbooks were produced between 1902 and 1911. Five were translated into Chinese. Japanese scientists were also invited to lecture in China. More technical physics and chemistry works were also translated from the Japanese. Iimori Teizō's 飯盛挺造 (1851–1916) edited volume on *Physics* (*Wulixue*; *Butsurigaku* 《物理学》) was completed in Chinese at the Jiangnan Arsenal from 1900 to 1903. Iimori's influence on Chinese physics grew out of this project.[25]

Updated Sino-Japanese dictionaries such as the 1903 *New Progress toward Elegance* (*Xin Erya* 《新尔雅》), which modernized ancient lexicographies (*xunguxue* 训诂学), were also compiled. By 1907, the use of Japanese scientific terms was formally approved by Yan Fu when he was in charge of the Ministry of Education committee for science textbooks. During the last decade of the Qing dynasty, the terminology in science journals in China shifted dramatically away from the Protestant terminology associated with investigating things and extending knowledge to Japanese terms in science texts such as: *World of Science* (*Kexue shijie* 《科学世界》, 1903–1904); *First Level in Science* (*Kexue yiban* 《科学一班》, 1907); *Science Journal* (*Lixue zazhi* 《理学杂志》, 1906–1907); and *Science* (*Kexue* 《科学》, 1908–1910).

The historical importance of Japanese translations for the development of modern science in China should not be underrated. Japanese translations were much more widely available in China than those produced earlier by the Jiangnan Arsenal or Fuzhou Ship Yard had been.

---

[25] Wang Yangzong, "1850 niandai zhi 1910 nian Zhongguo yu Riben zhi jian kexue shuji de jiaoliu shulue", 146–147. See also Xiaoping Cong, *Teachers' Schools and the Making of the Modern Chinese Nation-state, 1897–1937*.

In addition, the new Japanese science textbooks contained newer content than the 1880s Arsenal and missionary translations, which were already outdated by European standards in the 1890s. The introduction of post-1900 science via Japan, which included new developments in chemistry and physics, went well beyond what Fryer et al. had provided to the emerging Chinese scientific community (see Chapter 6).

Changes in the classical language used to make the translations more easily understood also helped produce a new literary form for presentation of the sciences that contributed to the rise of the vernacular for modern Chinese scholarly and public discourse. The new texts translated from Japanese made modern science available to a wider audience and raised the level of knowledge among students and teachers. Among urbanites, especially in Beijing and Shanghai, the first decade of the 20th century provided the basic education in modern science via Japanese textbooks for the generation that matured from the New Culture Movement (*Xinwenhua yundong* 新文化运动) of 1915 to the May 4th Movement of 1919 (*Wusi yundong* 五四运动).[26]

The Institutional Formation of Modern Science in Republican China

Many university and overseas students were by 1915 as radical in their political and cultural views, which carried over to their convictions about science. Traditional natural studies became part of the "failed" history of traditional China to become "modern", and this view now included the assertion that the Chinese had never produced any science. The earlier claim for the "Chinese origins" of Western science, so prominent until 1900, was now deemed superstition (*mixin* 迷信). Both "modernists" and "socialists" and even the "New Confucians" (*Xinru* 新儒) in China accepted the West as the universal starting place of all science, which was diametrically opposed to Chinese superstitions.[27]

---

[26] Wang Yangzong, "1850 niandai zhi 1910 nian Zhongguo yu Riben zhi jian kexue shuji de jiaoliu shulue", 147–150.

[27] Compare Hildred Geertz, "An Anthropology of Religion and Magic, I", *Journal of Interdisciplinary History* 6, 1 (Summer 1975): 71–89. She notes: "It is not the 'decline' of the practice of magic that cries for explanation, but the emergence and rise of the label 'magic.'"

After 1911, some scientists such as Ren Hongjun linked the necessity for Chinese political revolution to the claim that a scientific revolution was also mandatory. Those Chinese who thought a revolution in knowledge based on universal Western learning was required not only challenged classical learning, or what they now called "Confucianism" (*Kongjiao* 孔教), but they also unstitched the patterns of traditional Chinese natural studies and medicine long accepted as components of imperial orthodoxy.[28]

Chinese students educated abroad at Western universities such as Cornell University or sponsored by the Rockefeller Foundation after 1914 for medical study in the United States, as well as those trained in the sciences locally at higher-level missionary schools and the new universities, regarded modern science in light of the Japanese appropriation of Western science, and not investigating things and extending knowledge.

The belief that Western science represented a universal application of scientific methods and objective learning to all modern problems was increasingly articulated in the journals associated with the New Culture Movement. The journal *Science* (*Kexue* 《科学》), which was created by the newly founded Science Society of China in 1914, assumed that an educational system based on modern science was the panacea for all of China's ills because its universal knowledge system was superior.[29] Patterned after the journal *Science* published by the American Association for the Advancement of Science, the first issue of the Chinese version included an article on why China lacked science. Ren Hongjun, who pioneered the promotion of modern science in China, helped organize the Society at Cornell in 1914 and became its first president from 1914 to 1923. Later he served two more terms as president in 1934–1936, and 1947–1950. He contended that China had failed to develop the research methodology of modern science.[30]

---

[28] See Benjamin A. Elman, "The Formation of 'Dao Learning' as Imperial Ideology During the Early Ming Dynasty", in T. Huters, R. Bin Wong, and Pauline Yu, eds., *Culture and the State in Chinese History*, 58–82.

[29] Ferdinand Dagenais, "Organizing Science in Republican China (1914–1950)", 1–31.

[30] Yang Tsui-hua, "Ren Hongjun yu Zhongguo jindai de kexue sixiang yu shiye"〈任鸿隽与中国近代的科学思想与事业〉, *Jindaishi yanjiusuo jikan*《近代史研究所集刊》24 (June 1995), 297–326. Compare Howard Boorman and Richard Howard, eds., *Biographical Dictionary of Republican China*, Vol. 1, 219–222.

The Science Society of China

When the Science Society was founded, the choice of the Japanese term for modern science — *kexue* — for the society and its journal represented a break with tradition and was seen as transcending the limits of the past focus on the investigation of things and the "Chinese origins". Too often, however, the notion that Chinese scientists shared the iconoclasm of their intellectual peers after the 1911 revolution has been exaggerated. Many of the members of the Science Society, once it moved to China, were not in favor of the cultural iconoclasm or political revolution associated with the New Culture or the May 4th movements. As cultural reformers, many scientists actually favored a blending of modern science with traditional values. Many scientists, unlike revolutionaries who championed science such as Chen Duxiu 陈独秀 (1880–1939), were traditionalistic in their impulses.[31]

Although Ren et al. may not have intended to equate science with traditional learning, they were referring to late Ming and early Qing trends toward empiricism and practical learning that they found amenable with modern science. Like Hu Shi, they at times equated the spirit of critical inquiry found in Qing evidential studies (*kaozhengxue* 考证学) with the methodology of modern science. In the same way, the scientist Ding Wenjiang often appealed to the travel diary of the late Ming adventurer Xu Xiake 徐霞客 (also known as Xu Hongzu 徐宏祖, 1586–1641) as a model and precedent for his own intellectual pursuits in geology.[32]

Hu Shi published an article in 1921 that presented the scholarly methodology of Qing dynasty textual scholars as a Chinese precedent for contemporary scientific research. The English version of this essay was reworked in 1962 when it was published under the title "The Scientific Spirit and Method in Chinese Philosophy". The essay in Chinese was

---

[31] David Reynolds, "The Advancement of Knowledge and the Enrichment of Life: The Science Society of China and the Understanding of Science in the Early Republic" (Madison, WI: University of Wisconsin Ph.D. dissertation in History, 1986), 15, 107, 116–117, 294, 300–301. See also Ssu-yü Teng and John Fairbank, *China's Response to the West: A Documentary Survey, 1839–1923*, 244–246.

[32] Compare David Reynolds, "The Advancement of Knowledge and the Enrichment of Life: The Science Society of China and the Understanding of Science in the Early Republic", 84, 124, 147. See also ECCP, 314–315.

followed in 1922 by the publication in Shanghai of Hu's 1917 Columbia University dissertation entitled *The Development of the Logical Method in Ancient China*.[33]

An uncritical faith in science, i.e., "scientism" (*kexue zhuyi* 科学主义), on the part of some Chinese scientists trained abroad — many from Cornell — did influence other intellectuals such as Chen Duxiu, who argued in the issues of the journal *New Youth* (*Xin qingnian* 《新青年》), which he helped found in 1915, that science and democracy were the twin universal pillars of a modern China that must dethrone the Chinese imperial past. In the process, post-imperial scholars and novelists such as Ba Jin 巴金 (also known as Li Feigan 李芾甘, 1904–2005) in his 1931 novel *Family* (*Jia* 《家》), for example, initiated an assault on pre-modern Daoist religion and traditional medicine as a haven of superstition and backwardness.

Although many iconoclasts such as Hu Shi moderated their views after the First World War, they remained public advocates of the West and saw modern science as the universal model. When China conformed to that model, as with aspects of Qing evidential studies, Hu Shi and others praised China; when it did not, then China had to change. Despite their rhetorical appeal to the empirical tradition for the investigation of things, the scientists who joined the Science Society were reading the presuppositions of modern science from Bacon (1561–1626) to Newton (1643–1727) into a popular slogan from traditional learning. Ren Hongjun maintained that the goal of science was to open up the secrets of nature by investigating things and extending knowledge, which depended on the search for truth.[34]

Peter Buck has described the role of Cornell University and American science in influencing the members of the Chinese Science Society, particularly their confidence about the proper relations between science and society and the value of the scientific method for educational, intellectual, and cultural enlightenment. The Rockefeller Foundation simultaneously sponsored the professionalization of American medicine in China by

---

[33] Hu Shi, "Qingdai xuezhe de zhixue fangfa" 〈清代学者的治学方法〉, in *Hushi wencun* 《胡适文存》, Vol. 1, 383–412. See also Hu Shi, "The Scientific Spirit and Method in Chinese Philosophy", in Charles A. Moore, ed., *The Chinese Mind*, 199–222.

[34] Compare Jia Sheng, "The Origins of the Science Society of China, 1914–1937" (Ithaca: Cornell University Ph.D. dissertation in History, 1995).

building a leading medical center for research and teaching in Beijing. The medical center attempted to raise the prestige of modern medicine in China by linking it with laboratory research in the 1920s.[35]

Nathan Sivin has noted that the scientism the Chinese imbibed in America fared very poorly when the returned students faced a Republican China in an era of rampant warlordism from 1915 to 1927. The refraction of American scientific values in the Chinese context could not replicate the social and scientific changes in late 19th- and early 20th-century America. Moreover, the American myth that the scientific method could be used to justify social change was tempered in China by the many more Chinese scientists, physicians, and engineers who were trained in Japan than in the United States.[36] From 1900 until 1937, when the Second Sino-Japanese War (1937–1945) commenced, some 34,000 Chinese studied in Japan, while up to 1953 only about 20,000 trained in the United States. Science and state power were indelibly linked in Japan and China. Few of the returned scientists from Japan entertained the sort of scientism that the iconoclasts trained in the United States envisioned.[37]

At its inception in 1915, the Science Society of China had only 115 members on three continents. By 1920, the Society had some 500 members in China alone. It grew to 1,000 members in 1930 and reached 1,500 by 1935. There was a comparable increase in the number of publications associated with the new fields of science from 1920 to 1930. Initially, the Society's journal promoted the research of its founders: Zhao Yuanren 赵元任 (1892–1982, mathematics, physics, and psychology); Hu Da 胡达 (1891–1927, physics and mathematics); Bing Zhi 秉志 (1886–1965, biology and agricultural science); as well as Jin Bangzheng 金邦正 (1887–1946?, forestry), president of Qinghua University from 1920 to 1922. Altogether the 12 issues in the first volume ran to almost 1,400 pages. Approximately 1,000 copies were printed of each issue, which appeared

---

[35] See Peter Buck, *American Science and Modern China: 1876–1936*, 171–185, and D. W. Y. Kwok, *Scientism in Chinese Thought, 1900–1950*.

[36] See the Nathan Sivin review of Peter Buck, *American Science and Modern China: 1876–1936*, in *Journal of Asian Studies*, 1981, 231.

[37] Barry Keenan, "Beyond the Rising Sun: The Shift in the Chinese Movement to Study Abroad", 157–169.

monthly from 1915 to 1940, occasionally from 1940 to 1946, and again monthly until 1950, for a total of 32 volumes.[38]

In March 1917, Beijing University under the leadership of Cai Yuanpei began contributing a monthly subsidy to the Society. When it received additional funds from the China Foundation after 1926 to supplement its income from subscriptions, its activities included: (1) publishing the Society's journal, transactions, monographs, and memoirs; (2) maintaining a library that had 37,000 volumes by 1929; (3) operating a biological laboratory in Nanjing from 1922; and (4) promoting science education through public lectures, standardizing terminology, and participation in international scientific congresses. Boxer indemnity funds that were returned to China were also used to support the Science Society. In addition, from 1930 the Society served as the Bureau of Scientific Information for the Nanjing government.

Thereafter, many other scientific societies were formed in China between 1915 and 1927:

- National Medical Association: 1915 in Beijing.
- The Geological Society of China: 1922 in Beijing.
- The Wissen and Wissenschaft Society: founded in 1916 by Chinese students in Tokyo.
- The Société Astronomique de Chine: 1922 in Beijing.
- The Chinese Meteorological Society: 1924 in Qingdao.
- The Chinese Chemical Industry Society: 1922 in Beijing.
- Academia Sinica: 1927 in Nanjing.
- Chinese Engineering Society and Engineer's Society: 1912 in Beijing.

Several non-missionary foreign societies were also founded in China:

- The Anatomical and Anthropological Association of China: 1920 in Beijing.
- The China Society of Science and the Arts: 1923 in Shanghai.

---

[38] Ferdinand Dagenais, "Organizing Science in Republican China (1914–1950)", 25a. See also David Reynolds, "The Advancement of Knowledge and the Enrichment of Life: The Science Society of China and the Understanding of Science in the Early Republic", 152 ff.

- Shanghai Chemical Society: 1922.
- Peking Natural History Society: 1925.[39]

Modern Medicine in China

Traditional Chinese medicine (*Zhongyi* 中医), which was the strongest field of the Chinese sciences during the transition from the late Qing to the Republican era — 1895–1911 — was also derided by those trained in modern, Western medicine, although *Zhongyi* was more successful in retaining its prestige than Chinese astrology (*tianwen* 天文), geomancy (*fengshui* 风水), and alchemy, which were dismissed by modern scholars as purely superstitious forms of knowledge. Chinese scholars increasingly called for a cosmopolitan synthesis of Western experimental procedures with traditional Chinese medicine.

Chinese physicians, however, remained very large in numbers and very influential despite the inroads of missionary physicians, Western hospitals, and the success of anatomy in mapping the internal venues for bodily illnesses. Coexisting for several decades since 1850, Western-style doctors and Chinese physicians had remarked upon the limitations in each other's theories of illness and therapeutic practices, but for the most part each was practiced in its own institutional matrix of care-giving traditions. Moreover, until the "miracle drugs" were discovered and anesthesiology was introduced early in the 20th century, the curative power of Western medicine, especially surgery, remained problematic when compared to the non-invasive pharmacopoeia traditions of Chinese physicians.[40]

One of the aspects of the "New Governance" policies after 1901 was the increasing involvement of the Qing state in policing public health. Unlike the Song dynasties, when the government created local medical bureaus to deal with epidemics, state involvement in local health issues since the late Ming had markedly diminished. Local elites filled the

---

[39] See Lin Mousheng, *A Guide to Chinese Learned Societies and Research Institutes*, and David Reynolds, "The Advancement of Knowledge and the Enrichment of Life: The Science Society of China and the Understanding of Science in the Early Republic", 312–314.

[40] Yili Wu, *Reproducing Women: Medicine, Metaphor, and Childbirth in Late Imperial China*.

vacuum by making charitable contributions to deal with health emergencies at a time when literati also took an increased interest in medical knowledge. Late Qing public health policies signified a decisive break with this long-term secular decline of dynastic interference in local affairs since the middle of the Ming dynasty, a devolution that had allowed medical issues to become the prerogative of local gentry and Chinese literati-physicians.[41]

The late Qing government increasingly saw its role in more statist terms. Using Germany as a model, which Meiji Japan had also emulated, the New Governance courses of action encouraged the Qing government to carry out new disease control methods such as quarantine and isolation hospitals to deal with epidemics caused by infectious diseases. During the Ming and Qing, when local physicians had faced southern epidemics they associated with "heat factor" (*rebing* 热病) causes, the state was minimally involved in dealing with illnesses when they broke out. The emergence of the modern Chinese state during the Qing–Republican transition was increasingly tied to the extension of Western medicine and the appropriation of Western models for government-run public health systems.[42]

One by-product of the expansion of state involvement in public health, however, was that the state was now drawn into the contest between Western-style physicians and traditional Chinese doctors who organized into separate medical associations. Hence, the modernizing state was progressively tied to Western medical theories and institutions, while Western-style doctors controlled the new Ministry of Public Health (*Weisheng bu* 卫生部). When the Guomindang-sponsored Health Commission proposed to abolish traditional Chinese medicine in February 1929, however, traditional Chinese doctors immediately responded by calling for a national convention in Shanghai on March 17, 1929, which was supported by a strike of pharmacies and surgeries nationwide on that day. The protest succeeded in having the proposed abolition

[41] Angela Leung, "Medical Learning from the Song to the Ming", in Paul Smith and Richard von Glahn, eds., *The Song–Yuan–Ming Transition in Chinese History*, 374–398.

[42] Carol Benedict, "Policing the Sick: Plague and the Origins of State Medicine in Late Imperial China", *Late Imperial China* 14, 2 (December 1993): 60–77, and Ruth Rogaski, *Hygienic Modernity: Meanings of Health and Disease in Treaty-Port China*.

withdrawn, and the Institute for National Medicine (*Guoyi guan* 国医馆) was subsequently established. One of its objectives, however, was to reform Chinese medicine along Western lines.[43]

The consequences of increased state involvement in Chinese medical policy after 1901 were significant for both Western and Chinese medicine. After 1929, two parallel institutions, one Western and one Chinese, emerged in the political and educational frameworks that the government established to monitor both public health and medical instruction. This dichotomy between Western and traditional Chinese medicine became de rigueur throughout the 20th century under both the Guomindang Republic and the Communist People's Republic. The bifurcation also entailed the modernization of traditional Chinese medicine, or what Bridie Andrews has called the "re-invention" of Chinese medicine in the early decades of the century.[44]

The influence of Western medicine in early Republican China presented a substantial challenge to traditional Chinese doctors. The practice of Western medicine in China was assimilated by individual Chinese doctors in a number of different ways. Some defended traditional Chinese medicine, but they sought to update it with Western findings. Others tried to equate Chinese practices with Western knowledge and equalized their statuses as medical learning. The sinicization of Western pharmacy by Zhang Xichun 张锡纯 (1860–1933), for example, was based on the rich tradition of pharmacopoeia in the Chinese medical tradition. Another influential group associated with the Chinese Medical Association, which stressed Western medicine, criticized traditional Chinese medical theories as erroneous because they were not scientifically based.[45]

---

[43] See Bridie Andrews, *The Making of Modern Chinese Medicine, 1850–1960*, and Hsiang-lin Lei, *Neither Donkey nor Horse: Medicine in the Struggle Over China's Modernity*. See also Bridie Andrews, "Tuberculosis and the Assimilation of Germ Theory in China, 1895–1937", *Journal of the History of Medicine and Allied Sciences* 52, 1 (1997): 142–143.

[44] See Ralph Croizier, *Traditional Medicine in Modern China: Science, Nationalism, and the Tensions of Cultural Change*, 151–209. Compare Bridie Andrews, "Traditional Chinese Medicine as Invented Tradition", *Bulletin of the British Association for Chinese Studies* 6 (1995): 6–15.

[45] Bridie Andrews, "Tuberculosis and the Assimilation of Germ Theory in China, 1895–1937", 114–157.

In this cultural encounter, techniques such as acupuncture were modernized by Chinese practitioners such as Cheng Dan'an 承澹盦 (also known as 承淡安, 1899–1957), whose research on acupuncture enabled him to accept Japanese reforms by using Western anatomy to redefine the location of the acupuncture points. Cheng's redefinitions of acupuncture thus revived what had become from his perspective a moribund field that was rarely practiced in China and, when used, mainly served as a procedure for "primitive" blood-letting. In this manner, acupuncture survived mainly as an artisanal procedure among the populace, but the Imperial Medical Academy (*Taiyi yuan* 太医院) had banned it in court, along with moxibustion, since 1822.

This Japanese and then Western reform of acupuncture, which included replacing traditional coarse needles with the filiform metal needles in use today, ensured that the body points for inserting needles were no longer placed near major blood vessels. Instead, Cheng Dan'an associated the points with the Western mapping of the nervous system. A new, "scientific acupuncture" sponsored by Chinese and Japanese research societies emerged, which presented a better map of the human body that would enhance diagnosis of its vital and dynamic aspects.[46]

The Legacy of the 1923 "Debate on Science and Philosophy of Life"

After 1915, the teleology of a universal and progressive "science" first invented in Europe replaced the Chinese notion that Western natural studies had their origins in ancient China. The dismantling of the traditions of natural studies and natural history (*bowuxue* 博物学), among many other categories, which had linked the pre-modern sciences

---

[46] Bridie Andrews, "Tailoring Tradition: The Impact of Modern Medicine on Traditional Chinese Medicine", in Viviane Alleton and Alexeï Volkov, eds., *Notions et Perceptions du Changement en Chine*, 154–155, 158–159. For other fields, see Laurence Schneider, "Genetics in Republican China", in John Z. Bowers, J. William Hess, and Nathan Sivin, eds., *Science and Medicine in Twentieth-Century China: Research and Education*, 3–29, and Yang Tsui-hua, "The Development of Geology in Republican China, 1912–1937", in Lin Cheng-hung and Fu Daiwie, eds., *Philosophy and Conceptual History of Science in Taiwan*, 221–244.

and medicine to classical learning from 1370 to 1905 climaxed during the New Culture Movement. Through their opposition to classical learning and its traditions of natural studies, New Culture advocates helped replace the imperial tradition of natural studies with modern science and medicine.[47]

As elites turned to Western studies and modern science, fewer remained to continue the traditions of classical learning (Han Learning, *hanxue* 汉学) or Cheng–Zhu 程朱 moral philosophy (*lixue* 理学), increasingly called "Neo-Confucianism" (*Xin Ruxue* 新儒学) in the 20th century, which had been the basis for imperial orthodoxy and literati status before 1900. Thereafter, the "traditional Chinese sciences", classical studies, "Confucianism" (*Ruxue* 儒学), and "Neo-Confucianism" survived as vestigial native learning in the new public schools (*xuetang* 学堂) established by the Ministry of Education after 1905. They have endured as contested scholarly fields taught in the vernacular in universities since 1911.[48]

Even the protagonists of Chinese moral philosophy involved in the 1923 "Debate on Science and Philosophy of Life" (*Kexue yu rensheng guan* 科学与人生观), which will be our final topic, accepted the West as the repository of universal scientific knowledge. They sought to complement such knowledge of nature with Chinese moral and philosophical purpose.[49]

The Impact of the First World War

The "Great War" from 1914 to 1919 acted as a profound intellectual boundary between those in Republican China who still saw in modern

---

[47] See Min-chih Maynard Chou, "Science and Value in May Fourth China: The Case of Hu Shih" (University of Michigan Ph.D. dissertation in History, 1974), 23–35, and James R. Pusey, *China and Charles Darwin, passim.*

[48] Benjamin A. Elman, *Civil Examinations and Meritocracy in Late Imperial China*, Chapter 8.

[49] Wang Hui, "The Fate of 'Mr. Science' in China: The Concept of Science and Its Application in Modern Chinese Thought", 14–29. See also Charlotte Furth, *Ting Wen-chiang: Science and China's New Culture*, and Roger Hart, "Beyond Science and Civilization: A Post-Needham Critique", *East Asian Science, Technology, and Medicine* 16 (1999): 88–114.

science a universal intellectual model for the future and the "New Confucians", such as Zhang Junmai 张君劢 (1886–1969), who showed renewed sympathy for distinctly Chinese moral teachings after the devastation visited on Europe. The former reformer and now scholar-publicist Liang Qichao, who was then in Europe leading an unofficial group of Chinese observers at the 1919 Paris Peace Conference, visited a number of European capitals. Both Zhang Junmai and Ding Wenjiang, the future antagonists in the 1923 "Debate on Science and Philosophy of Life" were part of Liang's traveling group. Each witnessed the war's deadly technological impact on Europe. They also met with leading European intellectuals, such as the German philosopher Rudolf Christoph Eucken (1846–1926), Zhang Junmai's teacher, and the French philosopher Henri Bergson (1859–1941), to discuss the moral lessons of the war.[50]

In his influential *Condensed Record of Travel Impressions While in Europe* (*Ouyou xinyinglu jielu*《欧游心影录节录》), Liang Qichao related how the Europeans they met regarded the First World War as a sign of the bankruptcy of the West and the end of the "dream of the omnipotence of modern science". Liang found that these Europeans now sympathized with what they considered the more spiritual and peaceful "Eastern civilization" and bemoaned the legacy in Europe of an untrammeled material and scientific civilization that had fueled the world war. Liang's account of the spiritual decadence in post-war Europe included an indictment of the materialism and the mechanistic assumptions underlying modern science and technology. A turning point had been reached, and the dark side of "Mr. Science" (*Sai xiansheng* 赛先生) had been exposed. Behind it lay the colossal ruins produced by Western materialism.[51]

Liang Qichao was still careful to criticize the mythology surrounding science and not science itself. He added a note: "Readers, please do not

---

[50] See Ding Wenjiang, *Liang Rengong xiansheng nianpu changbian chugao*《梁任公先生年谱长编初稿》, 551–574.

[51] See Liang Qichao, *Ouyou xinyinglu jielu*《欧游心影录节录》, in Liang Qichao, *Yinbingshi zhuanji*《饮冰室专集》, Vol. 7, 10–12. For discussion, see Chow Tse-tsung, *The May 4th Movement: Intellectual Revolution in Modern China*, 327–329, and Jerome Grieder, *Hu Shih and the Chinese Renaissance: Liberalism in the Chinese Revolution, 1917–1937*, 129–135. Compare Theodore Huters, *Bringing the World Home: Appropriating the West in Late Qing and Early Republican China.*

misunderstand this as an attack on science. I definitely do not acknowledge the bankruptcy of science. However, I do not acknowledge the omnipotence of science either." It was clear from Liang's account that the caliber of the West that had produced science now troubled him. To remedy its excesses, he appealed to the spiritual resources that traditional Chinese civilization could provide. Liang made no mention of the pre-modern scientific achievements of Chinese civilization in 1919 (he had in some of his earlier writings), because his measure of science and technology was modern science and not traditional natural studies.[52]

In the comparison between China and Liang's "imagined" West, China's pre-modern science was not worth taking seriously. Earlier in 1902, while Liang was living in exile in Japan, he had composed a three-part article surveying the history of science in the West, based on Japanese translations, which he entitled "Synopsis of the Vicissitudes in the History of Science" ("Gezhixue yange kaolue" 〈格致学沿革考略〉). There Liang noted that 200 to 300 years ago, except for "modern science", i.e., *gezhixue*, China had been comparable to the West in all other fields of learning. Liang's article presented the scope of science in the West and its relation to other specialized fields since the ancient Babylonians and Greeks. He then discussed the Arab transmission of Greek science to Europeans.

Liang added, ironically, that printing, gunpowder, and the compass, which all came from China to Islam and were then transmitted to Europe, had enabled the scientific revolution in 16th-century Europe but not in China. Although Liang occasionally used the term "investigating and extending knowledge" for modern science in the essay, in addition to the Japanese term *kexue*, the former no longer evoked memories of the patient "investigation of things" that late Qing literati had inscribed in its semantic life as a translation for "science". Liang's intellectual transition from parochial science — *gezhixue* — in 1902 to universal science — *kexue* — in 1919 was not mentioned in the *Condensed Record of Travel Impressions While in Europe*.[53]

Liang Qichao's postwar disillusionment with Western civilization and its belief in the omnipotence of science and technology impacted

[52] Liang Qichao, *Ouyou xinyinglu jielu*, 12.

[53] Liang Qichao, "Gezhixue yange kaolue" 〈格致学沿革考略〉, in *Yinbingshi wenjii*《饮冰室文集》, Vol. 2, 3–14.

Chinese students and scholars when his travel impressions of Europe were syndicated in China in 1919 and again when published as a separate work. Subsequently, Liang Shuming 梁漱溟 (1893–1988) presented a series of lectures at Beijing University and elsewhere in 1920 and 1921 that addressed the subject of Eastern and Western civilizations, specifically comparing the West, China, and India. Liang Shuming's lectures reopened the "cultures controversy" that Liang Qichao's travel impressions had initiated. After it was published in late 1921, Liang Shuming's book on the subject went through eight printings in four years, signaling that the nativist backlash against the excesses of the 1915 New Culture Movement and its faith in "Mr. Science" was gaining a wider audience.[54]

Evoking the legacy of the German romantics and their anti-materialist appeals to human voluntarism and vitalism, Liang Shuming in many ways reiterated for Chinese what Eucken and Bergson were arguing in Europe in the aftermath of the First World War. Oswald Spengler's (1880–1936) *Der Untergang des Abendlandes* (*The Decline of the West*), for example, was first published in 1918, at the moment of Germany's bitter defeat, and became a bestseller in Europe. Its central premise was that all cultures had their peculiar configurations and therefore must be studied to understand their unique strengths and weaknesses. The cultural life of all nations and peoples followed cultural cycles that by necessity must run their course, as Western civilization now realized. Liang Shuming revived China as an antidote.[55]

This approach struck a responsive cord among many Chinese intellectuals who thought that cultural iconoclasm in China under the Republic had gone too far. Traditional Chinese culture and values could now be salvaged intact because their anti-materialist spiritual foundations were the remedy for modern European excesses. The Beijing University philosophy professor Feng Youlan (Fung Yu-lan, 冯友兰,

[54] See Guy Alitto, *The Last Confucian: Liang Shu-ming and the Chinese Dilemma of Modernity*, 77–81.

[55] See Benjamin A. Elman, "Wang Kuo-wei and Lu Hsun: The Early Years", *Monumenta Serica*, 34 (1979–1980): 389–401, Guy Alitto, *The Last Confucian: Liang Shu-ming and the Chinese Dilemma of Modernity*, 82–125, and Jerome Grieder, *Hu Shih and the Chinese Renaissance: Liberalism in the Chinese Revolution, 1917–1937*, 137–144.

1895–1990), for instance, while at Columbia University in 1922 published an article in English explaining "Why China Has No Science". Contributors to the journal *Science* published by the Science Society of China had thought that China was three centuries behind the West in terms of science.

Feng argued instead that according to her own traditional standard of values China did not need science. Daoism had emphasized the return to nature against human artificiality. The pre-Han philosopher Master Mo had stressed utilizing the past to control the future, which included a system of logic or definitions. Later literati had debated whether knowledge of external things took priority (following Zhu Xi), or the internal stress on the mind (following Wang Yangming) was more important than mastering the world. Since Chinese did not regard life as a search for power, Feng continued, they stressed human and practical affairs and thus had no need of scientific certainty. The lesson Feng Youlan drew, however, was that Wang Yangming had been wrong. Chinese must stop searching for the truth in the "barren land of the human mind". The sciences of the outside world must be studied.[56]

Zhang Junmai vs. Ding Wenjiang

The cultures controversy boiled over, however, when Zhang Junmai presented a lecture at Qinghua University in Beijing on February 14, 1923, before a group of science students. There Zhang laid down the gauntlet to those who championed science in China, notably at Qinghua and Beijing universities. Borrowing ideas from Rudolf Eucken, Zhang contended that science must be secondary to and complement a viable "philosophy of life" (*rensheng guan* 人生观). Science of itself, Zhang contended, could not provide a vision of life that people could follow because its materialist assumptions ruled out a spiritual vision of human values. It was fascinating that Zhang based his defense of spirituality and moral conscience on the subjectivity of human values, which in effect

[56] Feng Youlan (Fung Yu-lan), "Why China Has No Science — An Interpretation of the History and Consequences of Chinese Philosophy", *International Journal of Ethics* 32, 3 (April 1922): 237–263.

jettisoned the universalistic pretensions prominent in traditional Chinese classical learning.[57]

As one of Liang Qichao's travel companions in Europe in 1919, Zhang shared Liang's views of Europe. In addition, he had studied abroad in Japan and later in Britain and Germany, which put him in touch with European thinkers such as the philosopher Eucken at the University of Jena. For Zhang, science must be complemented by a spiritual vision giving it moral direction and purpose. China's spiritual legacy was rich and must be restored, if China was to avoid the materialist excesses of Europe.[58]

Another of Liang Qichao's 1919 travel companions in Europe, Ding Wenjiang, found Zhang's views outrageous. Two months after Zhang Junmai's lecture at Qinghua, Ding picked up the gauntlet. A noted scientist who had trained for seven years in London and Glasgow and received degrees in biology and geology, Ding Wenjiang published a rejoinder in April 1923 to Zhang's lecture. Lasting a year, the controversy led to the publication of some 250,000 to 300,000 words on both sides. All the rejoinders and surrejoinders were then published in a memorable collection that went through three printings by 1928 and successfully aired in public the misgivings scholars such as Zhang Junmai had concerning the modernist agenda that scientists and radicals had promoted as in China's best interests.[59]

---

[57] See Zhang's "Rensheng guan" 〈人生观〉, in *Kexue yu rensheng guan* 《科学与人生观》, I, 4–10. See also Saitō Tetsurō 齊藤哲郎, "Chi no ryōan — Chūgoku no 'kagaku to jinseikan' ronsō o chūshin ni" 〈知の兩岸 — 中國の '科學と人生観' 論爭を中心に〉, *Chūgoku — shakai to bunka* 《中國 — 社會と文化》 8 (1993): 133–155, and Lin Yü-sheng, "The Origins and Implications of Modern Chinese Scientism in Early Republican China: A Case Study — The Debate on 'Science vs. Metaphysics' in 1923", in *Zhonghua minguo chuqi lishi yantaohui lunwenji, 1912–1927* 《中华民国初期历史研讨会论文集, 1912–1927》, 1183.

[58] See D. W. Y. Kwok, *Scientism in Chinese Thought, 1900–1950*, 140–160, Chow Tse-tsung, *The May 4th Movement: Intellectual Revolution in Modern China*, 333–334, and Jerome Grieder, *Hu Shih and the Chinese Renaissance: Liberalism in the Chinese Revolution, 1917–1937*, 145–150.

[59] See *Kexue yu rensheng guan*. Another series of essays prefaced by Zhang Junmai appeared in Shanghai in 1924.

Ding's initial reply to Zhang Junmai was entitled "Metaphysics and Science" ("Xuanxue yu kexue" 〈玄学与科学〉, "Dark studies and science") and appeared in April 1923 in a journal that Ding and Hu Shi had established in Beijing as a forum for political discussion. A distinguished scientist at Beijing University, Ding had initiated the Chinese Geological Survey while chief of the geology section of the Ministry of Industry and Commerce after his return from his studies in Britain. Hence, his credentials as a scientist were impeccable. At the outset, he accused Zhang of resurrecting the "ghost of metaphysics" in Zhang's fanciful and relativistic account of traditional spirituality, human intuition, and humanistic values:

> Metaphysics is really a worthless devil — having scraped along in Europe for something over two thousand years, until he is now coming to find himself with no place to turn and nothing to eat, suddenly he puts up a false trade mark, hangs out a new signboard, and comes swaggering along to China to start working his swindle. If you don't believe it, please just take a look at Zhang Junmai's "The Philosophy of Life" in the *Qinghua Weekly*.[60]

Ding accused Zhang Junmai and others of trying to turn the tables and claim that science and technology via materialism had bankrupted Europe and would do the same in China. According to Ding, the European debacle had been the result of international politics and not science per se. To blame science and its constant search for the truth for the First War was misguided. Its technologies had been misused by European politicians. The problem was politics, not science.

Zhang Junmai countered with a long article on the philosophy of life in which he invoked European thinkers such as Eucken and Kant to show that scientific knowledge was limited to phenomenal experience and thus could not reach the higher levels of human feelings, art, and religion. These were separate domains of human experience that the scientism of Ding Wenjiang refused to acknowledge. Ding replied immediately in May 1923 that Zhang was confusing the difference between spiritual and material matters, which was neither absolute nor inaccessible to human

---

[60] See *Kexue yu rensheng guan*, I, 1, translated in Jerome Grieder, *Hu Shih and the Chinese Renaissance: Liberalism in the Chinese Revolution, 1917–1937*, 150. Compare Saitō Tetsurō, "Chi no ryōan — Chūgoku no 'kagaku to jinseikan' ronsō chūshin ni", 140–142.

reason. Only the scientific method, Ding argued, was the universal means to solve the quandaries of human life.[61]

Liang Qichao tried, unsuccessfully, to mediate. He repeated that he had never stated that science per se was bankrupt, but he added that human feelings did go beyond science and reason and must be addressed through art, religion, and philosophy. The debate, however, had polarized Chinese intellectuals in Beijing and elsewhere. Hu Shi, who had been ill for much of 1923, had earlier written a critical review of Liang Shuming's book on Eastern and Western cultures. He noted that the only way to solve human problems was to apply the scientific method to them. According to Hu Shi, Liang Shuming and Zhang Junmai wrongly challenged the universal basis of science by appealing to the subjective world of human feelings, art, and religion.[62]

In his preface to the volume devoted to the controversy, Hu Shi added that Liang Qichao's 1919 travel impressions had initiated the challenge to the materialistic foundations of science. The discussions and debates had been useful, Hu added, but in the end he sided with the Guomindang spokesman Wu Zhihui 吴稚晖 (1864–1953), who in the second stage of the debate went beyond even Ding Wenjiang by acknowledging that science could only provide a philosophy of life that was purely materialist and mechanistic. Both Hu and Wu held that a "naturalistic conception of life" was the only position that science could uphold. China's pretensions to spiritual superiority would hide the country's material and spiritual backwardness. Universal science was the only solution.[63]

---

[61] See D. W. Y. Kwok, *Scientism in Chinese Thought, 1900–1950*, 142–148, Chow Tse-tsung, *The May 4th Movement: Intellectual Revolution in Modern China*, 334–335, and Jerome Grieder, *Hu Shih and the Chinese Renaissance: Liberalism in the Chinese Revolution, 1917–1937*, 150–151.

[62] For Hu's review of Liang's book, see *Hushi wencun*, Vol. 2, 158–177.

[63] *Kexue yu rensheng guan*, Hu Shi "Xu" 胡适〈序〉, 10–13. The preface can also be found in *Hu Shi wencun*, Vol. 2, 120–147. See also D. W. Y. Kwok, *Scientism in Chinese Thought, 1900–1950*, 154–155, Chow Tse-tsung, *The May 4th Movement: Intellectual Revolution in Modern China*, 336–337, Guy Alitto, *The Last Confucian: Liang Shu-ming and the Chinese Dilemma of Modernity* 126–129, Jerome Grieder, *Hu Shih and the Chinese Renaissance: Liberalism in the Chinese Revolution, 1917–1937*, 151–152, and Saitō Tetsurō, "Chii no ryōan — Chōgoku no 'kagaku to jinseikan' ronsō o chūshin ni", 142–144.

The "Debate on Science and Philosophy of Life" continued for several years, and it is usually argued that the advocates of science gained the upper hand in this brouhaha. Many others joined in the fray, including leading members of the newly established Communist Party, who saw the debate as a chance to promote the scientific pretensions of Marxist socialism. During the third stage of the debate, which lasted from December 1923 to August 1924, Qu Qiubai 瞿秋白 (1899–1935) prepared an essay for *New Youth* entitled "Freedom and Necessity" ("Ziyou shijie yu biran shijie"〈自由世界与必然世界〉) in which he stressed the social realities conditioning human agency. Similarly, Chen Duxiu, one of the founders of the Chinese Communist Party, denied that human agency depended on individual subjectivity in an August 1923 essay for *New Youth*, which gainsaid the claims made by Zhang Junmai and Liang Qichao. Human choice was based on social and economic factors that could be scientifically delineated.[64]

What is significant here is that the followers of scientism and Marxism both championed modern science and materialism. Moreover, the humanist appeal to China's spiritual resources never questioned that "science" meant "Western science and technology". When Wu Zhihui presented a materialist philosophy of life to complement his view of science, he revealed that the significance of modern science carried over to human agency, which the Marxists also readily accepted in their appropriation of science to serve socialism.

Accordingly, the 1923 debate was premised on the value of universal science in its modern Western form. Neither side wished to appeal to traditional Chinese achievements in astronomy, mathematics, or medicine because for each side China had "failed" to develop science. When Liang Qichao appealed for a new unity of Chinese cultural values and modern science, his position required the amalgamation of European science and technology, that is, what China lacked, with Chinese culture, i.e., what China already had.[65]

---

[64] Saitō Tetsurō, "Chii no ryōan — Chūgoku no 'kagaku to jinseikan' ronsō o chūshin ni", 144–146.

[65] D. W. Y. Kwok, *Scientism in Chinese Thought, 1900–1950,* 160–162, and Saitō, Tetsurō, "Chii no ryōan — Chūgoku no 'kagaku to jinseikan' ronsō o chūshin ni", 138.

Those who saw in modern science the intellectual revolution of the future, such as Chen Duxiu, would march on in the name of Communism and modern science. Mao Zedong 毛泽东 (1893–1976) and other Communists would leave Chen Duxiu behind after 1927, but they always carried modern science along with them in their quest for socialism. Many of those who saw modern science as a rival, such as Zhang Junmai, would continue to appeal to the preservation of traditional philosophical values from the borderlands in Hong Kong and Taiwan. Like German romantic philosophers, such as Heidegger (1889–1976) and Gadamer (1900–2002), who also questioned the moral meaningfulness of modern science, however, "New Confucians" after 1949 would have to find a middle ground that would allow "Confucian hermeneutics" to continue to affirm scientific and medical research.[66]

Final Comments

In the early 21st century, we tend to forget the degree of skepticism that Joseph Needham's remarkable *Science and Civilisation in China* series initially provoked six decades ago. Many dismissed the great embryologist's foray into the history of Chinese science as a deadend, a project they felt revealed Needham's wishful thinking about pre-modern China.[67] While Needham granted that China lacked capitalism, he did not stop there. Few besides Needham, his collaborators, and Nathan Sivin stopped to consider what the rich archives in Taiwan, China, and Japan might yield if someone bothered to go through them. As the evidence of a rich tradition of natural studies and medicine accrued in volume after unrelenting volume of the *Science and Civilisation in China* series since 1954, it became clear that the largest archive of pre-modern records for the study of nature most likely remained in China.

By better understanding the history of imperial Chinese natural studies, technology, and medicine, we are more perceptive about

[66] See Benjamin A. Elman, "Rethinking 'Confucianism' and 'Neo-Confucianism' In Modern Chinese History", in Benjamin Elman, John Duncan, and Herman Ooms, eds., *Rethinking Confucianism: Past and Present in China, Japan, Korea, and Vietnam*, 518–554.

[67] For more informed accounts, see the "Review Symposia" on Needham's work in *Isis* 75, 1 (1984): 171–189.

ourselves and the mystifications that undergird our contemporary versions of modern science. Now that "Chinese science" has grown in respectability among the members of the "international society for the study of East Asian science, technology, and medicine" worldwide, the romanticized story of European science, whether capitalist or socialist in genre, has slowly unraveled under the onslaught of Thomas Kuhn and his historicist successors.

No longer reified and automatically placed on a pedestal as the successor to our pre-1900 religions, the story of Western science has become more complicated, more tragic, and more tense. The boyhood dreams of invention among Europeans, and their shrieks of "eureka" after new discoveries, have been revisited. The historical accounts of Newton's mystical fascination with alchemy, Copernicus's (1473–1543) devout religiosity that informed both his curiosity and his astrology, and the deadly serious but ungentlemanly competitive nature of "priority" in science and invention have challenged earlier fictions about the allegedly apolitical scientific revolution and its allegedly "neutral spaces for public discourse" in Europe.

We forget the brouhaha that greeted Thomas Kuhn's 1962 claims that science was historically contingent and grounded in its time and place. As a trained scientist, Kuhn had seen through the scientism of his peers; as a trained historian he unraveled the allure of the philosophy of science, which allegedly explained the universal "logic" of scientific discoveries. Instead of Karl Popper's conjectures versus refutations, Kuhn argued that the claims of science were always tied to what we might call a zeitgeist or *weltanschauung*, but that he defined in terms of a "paradigm", a bloated concept that he later regretted coining and retracted.

Unraveling a consensus of scholarly opinion about the "failed" history of science in China and the "victorious" history of European science reveals that both histories were pieces in the larger global narrative of Western success, Japanese prowess, and Chinese failure. With the exception of a reformed version of traditional Chinese medicine that has survived and is now thriving as one version of "holistic" medicine, the traditional fields of natural studies in imperial China were derailed by the impact of modern science. The historical construction of Western success in science, technology, and medicine was confirmed simultaneously by the delineation, unfairly and prematurely, of imperial China's scientific failure during the Ming and Qing dynasties.

When the history of Chinese science is rewritten in an age of Chinese landings on the moon and Mars later this century or next, it will be presented more positively and triumphantly in Beijing, Taipei, Singapore, and Hong Kong from a new linear framework whose teleology will announce Chinese success and not failure. The 21st century will mark the end of the triumphal narrative of Euro-American success at the expense of much of the world. By mid-century there will likely be more scientists and engineers in China and India than the rest of the world combined.

Appendices

Age of the Return of Rationality — Elman on Breaking Through the Crisis in Historiography

Interview by Reporter Zhang Quan of *Life* Magazine《生活》, Originally in Chinese (2008)

To break through the crisis of historiography caused by the impact of Post-Modernism, Benjamin A. Elman ventured into case studies of particular eras with the concept of a "New Cultural Historicism", to understand China more from the Chinese perspective. He narrates the age of rationality in China by placing intellectual history within the context of social history — the academic world, technology, and the civil examinations in the late imperial China — to discover the missing "silent majority".

Blooming Late

Elman removes his huge silver watch from his wrist and places it down properly, lowering his head to monitor the time. As the dial rotates, sunlight comes into the spacious room. The loud ticking of the second hand of the watch leads us to embark on a journey through time and space, from Elman's legendary life story, to Jiangnan academia in late Imperial China that he has been pursuing in research relentlessly. We pass through the borderless forest of Thailand in the 1960s, pay respect to the Chinese pioneer of technological education Ruan Yuan 阮元 (1764–1849), and depart from Lord Macartney's Embassy that Qianlong emperor loathed and the self-cocooning "Needham Puzzle", finally heading towards the sunset of the empire, finding our way back to Changzhou, the strategic center of the Qing dynasty academia. What excitement it induces — Elman's multi-faceted studies of the late empire!

Sinologists' interest in late imperial China has not waned, an obvious reason being that it was of recent times to the Sinologists and materials are in abundance, but more importantly, so much doubt accumulated internally, such as why China was left behind by the West at the modern turning point in history, and why the many revolutions in the West did not occur in China concurrently. From the travelers in China in the late Qing era, to the academic critics, the inquiry has never stopped, and Elman is one of the most accomplished scholars questioning these issues.

Before becoming a Sinologist, Elman was an officer of public health. In 1968, he went to Thailand with the World Health Organization on a malaria disease eradication mission. He returned home in 1972, continuing his service as an officer of public health. Elman has a dramatic account on this extraordinary experience. He desperately wanted to go to China back then, but America had yet to establish diplomatic ties with China, so he could only choose China's neighbor Thailand, which he heard had many Overseas Chinese that would allow him to continue honing his spoken Chinese. However, as Elman recollected, "Unexpectedly, they all spoke the Southern Min dialect (*minnan hua* 闽南话) …" while staring at my dumbfounded face with a stern look. In the 1980s, Elman finally arrived in China, cycling to the Shanghai Library everyday to read. This triggered my curiosity and I asked, "You must have drawn a lot of attention to be cycling like any other Chinese at that time right?" Elman replied plainly, "No. They thought I was from Xinjiang, and they said I really looked like one." This is Elman, you can never tell how emotional he is through his decorum, but humor and wisdom are tugged under his steady tone, catching you unaware and leaving you dumbfounded. There are similarities between his character and academic studies, and this makes me wonder if it is this character that influences the choices he makes. While most Sinologists were intrigued by the ancient poetry of China, or the bizarre state of the Chinese politics and business, Elman kept a distance, and turned to the heavier topics such as academics, technology and civil examinations in late imperial China. Perhaps, only he knows that under the surface of these monotonous topics lies a certain dynamic wisdom that can break new grounds anytime.

Although having maintained an interest in China for a long time, Elman actually started his studies in Sinology very late, only having embarked onto the journey in his 30s, pursuing his Ph.D. in Oriental Studies at the University of Pennsylvania. However, in just five years upon graduation, he shocked his peers with his debut *From Philosophy to Philology: Intellectual and Social Aspects of Change in Late Imperial China*, which was praised by the *Journal of Asian Studies* (*Yazhou yanjiu xuekan* 《亚洲研究学刊》) as "never has the interaction of Chinese thought been studied in such detail and clarity". Not long after its publication, this book was nominated for but did not win the John K. Fairbank Prize, the highest accolade in the field of Sinology, despite the fact that the direction Elman took in his research was opposite of Fairbank.

Prospective

John K. Fairbank, pioneer of the Fairbank Center for East Asian Research at Harvard University, and his student Joseph Levenson, dominated the ideological trend in Sinology studies after World War II. They expected Sinologists to extract themselves out of the ancient archives, and pay attention to contemporary China, studying its society, politics and economy. They advocated the "Impact-Response Model" and "Tradition-Modernity Model", differentiating them clearly from the Sinology in European tradition that was influencing the overall development of Sinology at that time. However, in the early 1970s, a new trend of thinking emerged, becoming the mainstream in the 1980s, where another of Fairbank's brilliant disciples, Philip A. Kuhn, was at the forefront. In Paul A. Cohen's renowned work *Discovering History in China: American Historical Writing*, this new trend was named the "China-centered" approach, in which historians understood Chinese history on its own terms, rather than through the relatively skewed set of expectations derived from the "Western-centered" one. Cohen paid special attention in analyzing Elman's *From Philosophy to Philology,* perceiving it as another important result of the "China-centered" approach.

Elman's research methodology can be summarized as the "New Cultural Historicism", advocating "contextualization" in intellectual history, placing intellectual history in the context of social history, in search of the realistic values in intellectual history. He forecasted confidently that "once the studies of intellectual history, political history and social history combine, the studies of Chinese academic history will be so rich".

This measure taken by Elman was to counter the impact from Post-Modernism that caused the crisis in historiography. He investigated this crisis from two different angles, internally and externally: the modernization narrative was simplifying history extensively, misplacing eras, covering up the historical phenomenon irrelevant to modern times; while within historiography, "it has actually sunk into its own pretension and exaggeration, unable to reinvent itself in a new environment." Hence, Elman took the decline of Confucius Classical Studies in the 20th century as an example to warn the academia to be aware of "what will happen to a field of studies once it has lost its life and vibrancy."

By deploying an improved methodology in his physical research on Chinese intellectual history, Elman accurately pointed out that previous studies on "Chinese intellectual history" were usually "Chinese philosophical history", which he tried to break through. This enabled him to venture into case studies under certain backgrounds in particular times.

Thread of Thought

The scope of Elman's research spreads across the academic world, civil examinations and technology in late imperial China, and within his thread of thoughts, these three independent fields are somehow linked as a complete network. Six years after *From Philosophy to Philology* was published, Elman wrote yet another quality work that won him even more prestige — *Classicism, Politics, and Kinship: The Ch'ang-chou School of New Text Confucianism in Late Imperial China*. This time, he approached this case study from a micro perspective, focusing on the center of *kaojuxue* 考据学 (textual criticism) in the Qian–Jia era 乾嘉 — Changzhou (Ch'ang-chou 常州). Results of early studies in this area had been authoritative for years, forming an overwhelming cultural awe and habitual cogitation, but Elman discovered the misconception in this cogitation paradigm, "that academia overlooked the difference in era, using the *kaojuxue* in the 20th century to interpret the themes of the 17th- and 18th-century Confucian works". Therefore, Elman emphasized that "to replace the end with the beginning," historical narrative should switch from "Kang–Liang 康梁"

(Kang Youwei 康有为, 1858–1927, and Liang Qichao 梁启超, 1873–1929) to "Zhuang–Liu 庄刘" (Zhuang Cunyu 庄存与, 1719–1788, and Liu Fenglu 刘逢禄, 1776–1829). Zhuang and his maternal grandson Liu Fenglu drove the formation of the *Jinwen xuepai* 今文学派 (New Text Confucianism). However, Elman realized that the Zhuang and Liu families were eventually buried in history due to the Utilitarian approach in Historicism — "who then is Zhuang Cunyu? In the writings of a historian who has accepted the saying of the revolutionary spirit represented by Wei Yuan 魏源 and Gong Zizhen 龚自珍 in the 19th century, he is usually just a mention in the footnotes. Who is Liu Fenglu? In historical narrative, he is usually just regarded as the teacher of Wei and Gong." However, upon having traced the historical details, Elman discovered that "Zhuang Cunyu and Liu Fenglu stood in the center of the late imperial political arena; Wei Yuan and Gong Zizhen were relatively marginalized instead, as the majority of their historical importance actually came from the common understanding of the 20th century scholars." Elman thus advocated to relook at and recognize the historical status of Zhuang and Liu, and from there triggered a repercussion, addressing the "three-way interaction of Classicism, kinship and politics", and reconstructed the late imperial Jiangnan, which was the center of official studies, colleges, publishing and personal libraries in those thriving times.

Since then, Elman shifted his focus to those factors that had influenced the development of academics — the civil examinations — and published *A Cultural History of Civil Examinations in Late Imperial China*, which thereafter zoomed in on the failures of the examinations, which constituted about 95% of the entire cohort, with some of them seeking alternative careers to become physicians, astronomers, or teachers. This change in perspective resulted in the publication of *On Their Own Terms: Science in China, 1550–1900* in 2005. Elman uncovered loads of interesting historical details, such as the limitations of the Jesuits that bore a negative influence on the Chinese. He noted that when Macartney's Embassy met the Qianlong emperor, they stubbornly did not offer the latest technological advancements from Europe, and only showed him the outdated astronomical apparatus, resulting in the emperor's lack of interest in Europe and not valuing it rightfully. The more important significance of this book is its firmness in following up the "China-centered" approach. To the question "Why did modern science not happen in China?", Elman responded that the "Needham Puzzle" was no longer worthy of debate, as China has its own system of natural science that is parallel to that of the Europeans. Elman felt that "we should analyze why the Chinese were the way they were, why the Europeans had their ways, and not why the Chinese could not be like the Europeans." This questioning was the result of following the "China-centered" approach, and what intrigued him more was that "the modern science, medicine and technical skills reconstructed by the Chinese in their own ways are achievements not to be overlooked." He even had the foresight of seeing that "the problem does not depend on who invented science, but instead, on what is the interest of the Chinese after the popularization of science, and after the rapid development of science in the 21st century, are China, Japan and India able to surpass Europe and the United States?"

"Gold-digging"

While the photographer is shooting Elman, I pose a question that was not pre-planned for the interview: *Classicism, Politics, and Kinship* was published for almost 20 years, has he been to Changzhou? Elman gives a sudden grin and catches me unaware again, saying, "The Zhuang lineage there had just published their Genealogy, and I was invited to write a Foreword." He is so sincere, looking like he is talking about his own family.

In a way, it can be considered his family, as without the discovery and promotion from Elman, the Zhuang and Liu families would have long been buried in history. Whenever one mentions the Changzhou School, it is instinctive to just quote Qian Mu's 钱穆 critique in his *Zhongguo jin sanbai nian xueshu shi* 《中国近三百年学术史》(*Academic History of China in the Recent Three Hundred Years*), which rated the academic achievement as overshadowing all others in the late Qing, but not able to give much thought to what this statement truly means. The Zhuang and Liu families once had the narrative authority, pushing the culture of Jiangnan to new heights, but had to settle for the fate of falling into oblivion after the era had passed. Life has its limits, and is unpredictable, forming an indescribable tone of melancholy.

Deep down in my heart lies another kind of sadness. I had once searched the name of Zhuang Cunyu on the Changzhou website, but under the section of prominent figures of Changzhou, I could only see Qu Qiubai 瞿秋白 (1899–1935), Yun Daiying 恽代英 (1895–1931) and Zhang Tailei 张太雷 (1898–1927); Zhuang Cunyu and Liu Fenglu were obviously chucked boorishly within the entry of "Changzhou School". I was left with a feeling of pessimism when I saw this hyperlink is broken.

Elman had once expressed confidently, "Dreams will survive, even after the system is demolished." However, what I can only see is the new positioning of Changzhou blinking on the website — "hotspot for investment, precious land for gold-digging." In this passive era, to unravel the historical details from our buried and sealed time, this "gold-digging" work is actually substituted by a foreigner who loves the Chinese civilization. How can we explain this?

~ · ~

***Life*:** In your opinion, whilst the West was developing its modern technology, China actually had a system of its own.

Elman: That was another kind of system, which should be regarded as the system of natural studies (*ziranxue* 自然学) (referred to in Chapter 2 as *gezhi* — to investigate things, which referred to natural studies as well, as many other terms). The early modern natural studies, *gezhi* 格致 (investigation and inquiry) and *bowu* 博物 (broad learning concerning the nature of things), are not modern science. China has its own tradition of natural studies; Europe has that of its own. The modern science that flourished in the 18th-century Europe was brought to China by the Jesuit missionaries.

***Life*:** You do not seem to hold a positive view of the Jesuits, thinking that certain influence they had on China was negative.

Elman: That is correct. They have their own limitations, such as their views of Copernicus could not be shared with the Chinese. Jesuits who went to China in the 17th and 18th centuries brought a lot of advanced knowledge and technology, but due to religious considerations they were against Copernicus (1473–1543). They amended his views on purpose in their books and added their own opinions; of course, there were parts that did not interest or were opposed by the Chinese, which made the Jesuits edit accordingly to suit them. Therefore, these Jesuits were not able to deliver the latest advancements to China. Another scenario could be that the Chinese in the 18th century had mistaken these Jesuits as the most notable and knowledgeable in Europe, and thought they could represent the latest advancements of the whole of Europe, but this was not the case. In fact, in 18th century Europe, the religious traditions were collapsing; the Roman Jesuits could not carry on their missions anymore, and had no choice but to travel all the way to China. Their arrival should have been beneficial to China, but they could not provide positive influence on the Chinese. It was impossible for China to understand Europe accurately looking at this disintegrating Aristotelian organization of knowledge to be representative, one that opposed the latest scientific and philosophic knowledge. And sure, this misconception was of no fault of the Chinese.

***Life*:** Prior to your studies on technology in China, there was already quite an authoritative work in the West — *Science and Civilisation in China* by Joseph Needham.

Elman: This book has made great contributions. Needham analyzed the development of natural studies in China from ancient to modern times, which included mathematics and geography. However, there was a problem about his most fundamental attitude — he questioned why China did not invent modern science despite the presence of so many scholars and inventions in China? Why did the Europeans succeed? Why did the Chinese fail? I think this is not right. We should be analyzing why the Chinese were like this, and why the Europeans like that, and not why the Chinese could not be like that. This attitude is the same as that of the Chinese after the *Jiawu zhanzheng* (甲午战争 Sino-Japanese War) and the May 4th Movement: Why did we not invent science? Eight thousand plus years ago, why did those in the Middle East invent agriculture, but those inhabiting other areas remained backward? We have now lost interest in questions of this sort. The point here is not who invented science, but how interested were the Chinese in science after it was popularized. Also, after the hyper-speed development of science in the 21st century, will China, Japan and India surpass Europe and the United States? Chinese students have a better understanding of science nowadays, and in contrast, the American students know less. I therefore think Needham's contribution was huge, but he had his limitations, he channeled many people's thinking towards why the Chinese had failed. I think it is time to move away from this attitude, as failure and success are no longer the most important point of view.

***Life*:** You had mentioned in one of your lectures that the tradition of natural studies in China reflected more the relationship among mentors and factions, whereas it is traditionally more open and public in the West.

Elman: The syllabus for the civil examinations in China was the Four Books and Five Classics. Those qualified would normally choose to become a civil servant, while the unsuccessful candidates would seek alternatives, such as becoming physicians or mathematicians, etc. This difference in choices determined that Chinese scientists and natural studies were not considered mainstream and occupied second and third positions, respectively. The attrition rate of the examinations was 95%, only the remaining 5% had the chance to secure a place in history. In my opinion, the civil examinations made us overlook the 95% of literates who failed — we do not know the careers they chose after the failure. They probably became physicians, researchers of natural studies, or novelists, like Pu Songling 蒲松齡 (1640–1715). These personages should not be overlooked. And also, most of them had genealogical ties, for example, a mathematician would most likely have a tradition of mathematics running in the family. The cause of natural studies being regarded in a lower position is inseparable from the civil examinations.

***Life*:** We also noticed a fairly weird phenomenon — the Chinese scholars emphasized a lot on establishing values, words and deeds, but majority of the scientists did not leave any legacy behind. China does have scientific works but very minimal. What do you think is the reason behind this?

Elman: You are right. The problem is that the available books now were written by famous figures. And in the Ming and Qing eras, only the rich and influential, and the temples, could print books; the general public did not have the right, possibly only a handful of handwritten manuscripts managed to survive the times. There were after all many who did not publish any works, and those that we get to see are mostly written by men, hardly any by women. It was of course very popular for women to write poetry in the Ming and Qing eras. I discovered many poetry collections by Zhuang family's poetesses, but the problem is in finding out who owned these collections, and how did they get printed? If only the lineage was networked and influential enough, they would have been capable of printing the works and genealogies, but such a possibility was rare. Historical information hence became elitist information.

Life: Ruan Yuan in his compilation of the *Chouren zhuan*《畴人传》(*Biographies of Astronomers and Mathematicians*) advocated the nurturing of *chouren*, i.e., the astronomers and mathematicians, and established the Sea of Learning Academy (*Xuehaitang shuyuan* 学海堂书院), which was described by Susan Mann Jones and Philip Kuhn as "[t]his tiny enclave of military strategists who were also poets and scholars teaching in a frontier commercial city".[1] How do you perceive this method of Ruan's?

Elman: Ruan Yuan already started teaching mathematics and astrology when he was in the Imperial College, and collected abundant materials. He found the long lost *Suanxue qimeng*《算学启蒙》(*Primer of Mathematical Calculations*) written by a Ming mathematician. Though the main content was astrology-based, there was much relevance for mathematics. Although the Chinese did not invent modern science, they picked up the tradition of mathematics in China once again, having found lots of materials on mathematics and astrology, eventually on differential (*weifen* 微分) and integral (*jifen* 积分) calculus as well. On the one hand, this improved the knowledge among mathematicians, and on the other, it helped to absorb new directions and new knowledge from Europe in the future. This is the main contribution by Ruan and his associates. Also, his compilation included many biographies of Westerners, such as Newton, that showed how widely read and how broad his horizon was.

However, he had an opinion which we disagree with — he thought that the roots of Western knowledge could be traced back to China. Whether

---

[1] Susan Mann Jones and Philip A. Kuhn, "Dynastic decline and the roots of rebellion", in Denis Twitchett and John Fairbank, eds., *The Cambridge History of China*, Volume 10, *Late Ch'ing, 1800–1911*, Part 1, 160.

this opinion is correct or not is another issue, but Ruan made use of the biographies of Western scientists to show that China already possessed this "new" knowledge, just that their works were lost or their interest eventually diminished, and so reviving such knowledge would be very easy. This would also imply that Europe's invention of modern science was not better than China. It only reaffirms that Europe was continuing the tradition which China had begun and so China's later success would be inevitable. Opinions with political inclinations as such were mainstream back in the Qing dynasty. Even though this view eventually lost its influence towards the late Qing era, it was very influential during Ruan Yuan's time.

***Life*:** You opined that the popularization of *kaojuxue* was due to the doubt in the values Chinese intellectuals had been upholding. However, throughout history, they had limitations of their own, being more of vassals to politics. What roles do you think the Chinese intellectuals should play?

Elman: This is not the problem I want to solve, but it is yours to tackle. Intellectuals secured civil positions by studying, obtaining power from there. Nowadays, intellectuals can only teach in universities, they do not join the government, and so are powerless. This applies to the academia in the United States as well, where intellectuals have huge influences, but no political power. Many leaders in China have backgrounds in engineering or business, whereas scientists in modern China have a lot of power and influence. This is a characteristic of China. In the United States, we find more officials with law backgrounds who are more well-versed in debating on a problem, rather than knowing how to solve it.

***Life*:** Each can be regarded to have its own merits.

Elman: Right. There were unexpected outcomes from the revolution in China. Traditionally, the Chinese put knowledge and politics together, but after the revolution, the trend completely changed. It is interesting to note that before the Qing dynasty, intellectuals were very powerful and could influence significantly through the civil examinations; however, during the 20th century, even though people like Lu Xun 鲁迅 (1881–1936) were very influential in Literature, their influence was limited in other areas like social politics and law.

[Translated by the Editor and vetted by the Author.]

The Civil Examinations *keju* 科举, Textual Criticism *kaoju* 考据 and Science *kexue* 科学 in the History of China

Interview with Princeton University Professor Benjamin A. Elman by Reporter Chu Guofei of the *Chinese Social Sciences Today* (CSST)《中国社会科学报》, Originally in Chinese (2009)

Newspaper Editorial

The civil examination system, *kaojuxue* 考据学 (textual criticism), and the history of the development of modern science, are important areas in Sinology. Professor Elman advocates that these areas should be studied through the combined lenses of cultural, political, and social history, and this idea has caught the attention of academia. In his *A Cutural History of Civil Examinations in Late Imperial China* (University of California Press, 2000), Professor Elman narrated the development of the civil examination system from its establishment until its abolishment. In the West, it is considered a notable title in this field. In his research on the problems of science in the Ming and Qing China, Professor Elman avoided the "Europe-centered" approach usually adopted for the discourse about the rise of modern science, and studied instead the history of the development of modern science in China from 1600 to 1900 holistically, while at the same time presenting three stages in the transmission of Western knowledge into China — introduction, mediation, and fusion — noting the development and transformation of Chinese modern science in detail.

Professor Elman is currently at Fudan University serving as a Changjiang Visiting Chair Professor (*Tepin Changjiang jiangzuo jiaoshou* 特聘长江讲座教授), and our reporter interviewed him in Shanghai.

The Civil Examination System in China had its Value

The Talent-driving Civil Examination System of China

Reporter: For a very long period of time, the civil examinations had been regarded as a hindrance to the progress of the Chinese traditional cultural system. Do you agree with this interpretation?

Elman: Although the civil examination system in China was abolished in 1905, the tradition was passed down in a different form. Many countries nowadays have established their own examination systems which are actually adapted from the Chinese civil examination system. Even though the content has changed, the technique, methodology and format are all inherited from China. From this perspective, I do not agree that the Chinese civil examination system was backward. I think it was in fact advanced; it was only during late Qing era that it was always linked to the corrupt Manchu government, and thus was abolished together with all things related to the government. We can see now that the system of examination has carried on even after the abolishment of the civil examinations, such as the examination administration (*Kaoshi yuan* 考试院) established under the Republic of China, which in fact represented the modernization of the civil examination system. This civil system of examinations is actually very meaningful. We should understand more of its function and render a new judgment, rather than denying it totally, defining it as worthless.

China had creatively established the civil examination system in the Sui–Tang era, 581–907. Luoyang 洛阳 and Chang'an 长安 were the only two venues conducting the examinations, and all candidates regardless of their status had to go to the capitols to be tested; in the Song dynasty, more venues were allocated across the provinces; in the Ming dynasty, it expanded further to all 1,500 counties; in the Qing dynasty, civil examinations became very influential. That is why we say the civil examinations had all along been developing. China had been "precocious" in using examinations to attract talent, and scholars outside China were also aware of this fact.

Reporter: What do you think is the most positive significance of the civil examinations?

Elman: The objective of the civil examinations was to select talent, filtering the best candidates to serve the government. However, new problems arose in the Song–Ming era, when the number of candidates soared, and only a handful were selected, thus eventually reaching a 90% attrition rate. A system of such a high attrition is usually not very effective. A high failure rate like this suggests that the system itself is a failure.

However, we should not oversimplify this problem either. The system had after all nurtured a big group of talented individuals, even though the majority of them did not make it to the civil service through the examinations. These people, objectively speaking, still contributed to the advancement and progress of the society. Candidates of the examinations knew the language, both modern and ancient, and could read and write. Many who were unsuccessful at the examinations still had several options, such as Li Shizhen 李时珍 (1518–1593) who became a famous physician in late Ming era, while many became classical teachers. There were many others who denied that their initial intention to study was solely to sit for the civil examinations, and eventually ventured into Buddhist studies, and became Buddhists. What an interesting phenomenon! Even in the Yuan dynasty when the civil examinations were not widely available, the literati could still write dramas and novels, accomplishing remarkable achievements in Literature.

Reporter: In other words, the society sent a message to these people that though they were unsuccessful at the civil examinations, there were still many other options. This increased the level of freedom, space for maneuver and scope of choice significantly. What was the proportion of the population that sat for the civil examinations?

Elman: In the Ming dynasty, about two million, and it doubled in the Qing dynasty, to about four million. Though this is considered a small minority in the total population, it is still considerably large in aggregate relation to other countries. Among these candidates, 25,000 candidates qualified as *jinshi* 进士 in the Ming era, and the number was similar in the Qing era. These qualifiers made up less than 5% of the total number of candidates, while the remaining 95% had to find other options. This is very intriguing.

The Ever Evolving Civil Examinations in Times of Changes

Reporter: Do you think there is any gap between the civil examinations in the Ming-Qing era and the Sui-Tang era?

Elman: The civil examination system was not stagnant, the subjects were always changing. Before the Song dynasty, subjects during the Sui–Tang times were more inclined to poetry, and even more towards Literature. In the Song era, other subjects began to weigh more, as Classicism took precedence. After the occupation of the Middle Kingdom by the Mongols, the civil examinations were temporarily suspended, only resuming in the 14th to 15th century. In the Ming era, the *baguwen* 八股文 (8-legged essays) became a requirement for the examinations, taking the place of poetry. Examiners felt that the best knowledge for candidates to become officials should come from the Four Books and Five Classics, and should be related to the studies of morality. They felt that these should be the main subjects, and even though candidates still loved poetry, it was never included in the examinations until the Qianlong era, when poetry began to be reinstated while *baguwen* was still the requirement. During the mid-Qianlong period, the Four Books and Five Classics were still being tested while poetry again became part of the civil service syllabus.

Apart from poetry and the classics, *celun* 策论 (policy papers) were an important component of the civil examinations, where the questions were related to politics involving problems such as finance and economics. The format of *celun* examinations can be traced to the Han times, as recorded in the *Hanshu*《汉书》, when the emperor Han Wudi 汉武帝 asked Dong Zhongshu 董仲书 (179–104 B.C.) some questions, and although Dong's reply was verbal, it can be considered as the beginning of *celun*. This format existed in the Sui, Tang, Song, Ming and Qing dynasties, during the third segment of the examinations. We get to learn about some phenomena and problems through these *celun*, as discovered by Professor Zhang Qing and his Fudan students. The many debates in the late Qing *celun* regarding the attitude towards history centered on the question "What is history, and is Confucius's *Chunqiu*《春秋》a model"? Or should there be another model? *Celun* were overlooked by scholars mostly, while *baguwen* and poetry were usually the main concern when civil examinations were mentioned. *Celun* questions were in fact a very important component.

Reporter: You mentioned just now that *celun* appeared in the third segment of the examinations. However, many usually place emphasis on the first, rather than the second and third segments.

Elman: But the problem is, since when did the third segment start to be overlooked? In the Song–Ming period, the third segment was very important. It was usually the critical component when the top and second-place finishers were judged. In the Qing dynasty, especially during the 18th century, candidates felt that the examiners would not have the time to assess the *celun* due to the overwhelming number of candidates and so the focus was more on *baguwen* and poetry. Since these were shorter, using these two components as the evaluation criteria triumphed instead. Under this circumstance, the third segment eventually got overlooked. However, in its whole history, *celun* were very important in the oral examinations in Han and the examinations held in the capitols of the respective dynasties. Song, Ming and Qing dynasties all had *celun*, which were never dropped from the system and remained as a standard subject. People realize now that the content of the civil examinations was very rich and was not limited only to *baguwen*. Although *celun* were overlooked in the Qing era, they were considered important until the late Ming era, when candidates answered in about 4000 words, and examiners paid meticulous attention, commenting in great detail as well.

Reporter: You mentioned that the civil examinations in Sui–Tang times were more literature-inclined and eventually changed, but this old format began to be reinstated after mid-Qianlong period. What was the reason behind this series of change?

Elman: This is not easy to judge. On the one hand, the Manchus were very interested in poetry; on the other, from the technical point of view, *baguwen* and poems were shorter, which made them easier to be assessed by the examiners. Another factor is the criticism of the *Song–Ming Lixue* 宋明理学 (classical orthodoxy in the Song–Ming era). The Qing scholars felt that the poetry of Sui–Tang before the Song times was not influenced by Buddhism, and so they advocated resuming the writings prior to that, i.e., poetry.

The Civil Examinations were not the Only Ways to Enter Civil Service

Reporter: In the Qing era, how important was the civil examination system? Was it the only way to become officials?

Elman: Not so. The Qing dynasty also rejuvenated the civil examination system based on the experiences derived from the Yuan and Ming eras. The Yuan rulers were initially not keen to revive civil examinations, and they only started to do so slowly in a small scale after 1313. They did not trust the Han Chinese, and despite being agreeable to let them become local officials, they did not think it was necessary to have an overall dynastic selection system. The statistics are very telling: within the hundred plus years before the end of Yuan dynasty, there were only 1100 *jinshi*, but there were about 39,000 in Song, and Ming and Qing each had 25,000. It was in Ming era, when the Han Chinese revived the civil examinations completely, feeling that it was critical to select officials nationwide.

The Manchus believed the main factor causing the Yuan to be unsuccessful and having a relatively short reign was that they had not collaborated with the Han Chinese; the latter thus were not able to play an effective role in governance. After invading the Middle Kingdom in 1644–1645, the Manchus immediately inherited the civil examination system from Ming, and reinstated it in favor of collaborating with the Han Chinese. At the same time, they introduced translation examinations for Manchu, Chinese, and Mongol bannermen, though not on a large scale. The translation examinations included Manchu to Chinese and vice versa.

Contemporary Examination System is a New Form of Selection

Reporter: Having researched the civil examination system in early China, how do you view the contemporary examinations for the civil servants in China?

Elman: Many people feel that examination is a good strategy, a new form of selection. Personally, I am not very fond of it. I do not conduct examinations for my own classes, and just get students to write papers. I think the civil examinations served as a kind of system, a kind of skill. Why is there a need for examinations? It is believed that a selection mechanism is necessary to deal with too large a pool of talent. Whether it is effective and accurate, is another question. Candidates selected from the pre-modern Chinese civil examination system might not be the best, probably some even incompetent, whilst the genuine talents were eliminated. That is why a critical attitude toward it is very important.

The civil examinations in China had a long history, bearing a wealth of experience, and it is a very good reference to us. Some people in the United States are against the university entry examinations, advocating that an essay submission will do, but would this be a *baguwen*? If someone is successful, wouldn't others want to imitate his work? And so wouldn't everyone be writing the same thing? Wouldn't this become the new *baguwen*? All these problems are not new. Also, it is not easy for examiners to judge — we can use examinations to test ten students, but what about 1000 or 10,000 of them?

The civil examination system, technically, is highly commendable. Though it is not perfect, it is an essential system when it comes to talent selection. However, I do not think the form of examination is suitable for classes. It is not the best way to nurture talent. Those who perform well may not be the cleverest; they are only the ones who know the system better. Take the GRE for instance, some Chinese can outdo me at the examination, but when they come to America, they cannot handle the learning effectively. This is the problem of examinations systems, and Chinese in history had already discovered it. Another instance is that the Americans find many children from families with financial difficulties not performing well academically, where family income, cultural differences and social equality, are some of the factors. Chinese in the Ming dynasty already discovered this, noting that candidates from the North did not fare as well as their Southern peers, and so adopted the quota system — 40% for the Northern candidates, 60% for the Southern. The Chinese have abundant experience in this aspect. Examinations are an international issue. The current examination system is the form of selection for modern times, but not without limitations either.

Kaojuxue was the Impetus for the Development of Modern Science in China

Reporter: What effect did *kaojuxue* have on the development of modern science in China? What change has it brought to the modern science in China, and what were the underlying causes?

Elman: *Kaojuxue* is an interesting phenomenon, one which requires textual expertise to analyze problems. In the studies of the classics, the requirement for evidence is very strict in *kaojuxue*. *Kaojuxue* values the astronomy side of astrology, mathematics, etc., including them within its scope of studies, believing that these forms of knowledge will complement the studies of the classics. This point of view differs from that of Cheng–Zhu 程朱 (Cheng Yi 程颐, 1033–1107, and Zhu Xi 朱熹, 1130–1200) learning. In the late Ming and early Qing eras, the Jesuits brought Western astronomy to China, which enriched the materials of *kaojuxue*.

Kaojuxue had made major impact in many areas, such as, when Li Shizhen realized that the information on Chinese medicine was in a mess and full of mistakes. He then tidied it up and published his *Bencao gangmu*《本草纲目》(*Compendium of Materia Medica*). Many bore the same views as Li, having discovered problems in Chinese medical books such as *Huandi neijing*《黄帝内经》(*Inner Canon of the Yellow Emperor*), *Shanghan lun*《伤寒论》(*Treatise on Cold Damage Disorders*), etc. Take the *Huangdi neijing* for example, it was discovered to be written in the Han dynasty rather than by the Yellow Emperor. I feel that Buddhists and Daoists promoting the studies and tidying of their classics were also under the influence of *kaojuxue*. To date, nobody has understood thoroughly the relationship between *kaojuxue* and that of Buddism and Daoism. I believe they are related, as this was the trend at that time, and not unique to Confucianism.

Kaojuxue can be considered a revolution in methodology. People used this method to manage information, and trace the thoughts of the past. The outcome of *kaojuxue* differed from its initial intention. Scholars actually hoped to restore the truth of the classics: they traced the texts back to the Song dynasty, further back to the Han dynasty, and even back to the Zhou dynasty but where would this lead ultimately? The Sage? But the problem is — is the Sage truly a sage? Are Five Classics what they are? Everyone wanted to restore the truth, but would it materialize? The analysis of *kaojuxue* was to restore the ancient (*fugu* 复古), but the attitude changed in the 19th and 20th centuries, becoming skeptical of the ancient (*yigu* 疑古). Although the content had made tremendous change, the method still continued to be adopted. Chinese scholars like Gu Jiegang 顾颉刚 (1893–1980) were influenced extensively by *kaojuxue*, even though they were the modern intellectuals and historians. We should understand that the aim of *kaojuxue* in the 18th century was to restore the ancient and not to be skeptical, but the outcome was unexpected. With their methods, the classics were no longer classics, the sage no longer a sage. We can understand this in retrospect, but people back then were not able to.

Reporter: You mentioned the Jesuits just now. Many scholars think that the Jesuits brought modern science from the West to China, which had a significant impact on the development of modern science in China. However, judging from your views on *kaojuxue* above, you actually approached this topic from a different point of view, perceiving the development of modern science in China as a complete system, and advocating that the real impetus for this development was actually the rise of *kaojuxue* and revolutionary methods of research, rather than the influence from the West, which was only partially, a catalyst.

Elman: Yes, China had always had science since the beginning of time, and was only influenced later by the West and became synergized through advocating the Western sciences. The Chinese interpreted the natural science of the West as their *gewu zhizhi* 格物致知 (investigating into and extending knowledge), equating the *gezhi* 格致 of theirs to that of the West. The Jesuits then belonged to the Catholic Church, whose knowledge was not considered modern science yet. They were using the Greek philosophy from the 12th century to analyze matters. For example, they adopted the four elements, which were similar to the Chinese *wuxing* 五行 (five evolutive phases), and Chinese intellectuals thus thought that both systems, Chinese and Western, could be coordinated. The Chinese believed that Metal and Fire were reliable, but the Westerners placed more emphasis on Water and air, and both sides engaged in a heated academic debate. Of course, from our perspective today, they were both wrong. But we should notice that the communication between the two was to investigate contemporaneous science, and not modern science. The lessons learned from such debates enlightened Chinese on what science is. Our definition for science now can possibly be denied after 100 or 200 years of success. Every era has its limitations.

The Jesuits were motivated to promote new knowledge such as astronomy and medicine because they knew it would be popular in China. They realized it was not as popular in other places they had visited, such as Japan. The Jesuits hence brought early modern knowledge to China in their hope to win respect, which could help in their missions. This was a very clever strategy. The Jesuits did not agree with the Chinese view on *wuxing*, *yin yang* 阴阳, *wuji* 无极 (infinite) and *taiji* 太极 (Supreme Ultimate), but in order to interact with them and promote natural sciences from the West, they felt the need to debate on these issues. And therefore, academics of the West and China did not lack interaction. The West was not powerful enough in its political and military strength at that time, hence engaging through academic means was necessary. If the Westerners had not engaged the Chinese in discussions and debates, these Jesuits would have been driven away, which is completely different from the 19th century. We can therefore say that the Chinese could have actually said no to the earliest import of Western science; it had the freedom of choice and right of refusal.

Coming to the 19th century, the Jesuits came at the beginning of the industrial revolution, when Western learning was transforming to another level. And the content of discussion between the Jesuits and the Chinese was not contemporary science, but modern natural science instead. Science did not exist then, it was a juxtaposition of natural science and humanities, being then named "scientia". The Chinese translated it as *gezhi,* equating it to that of the Chinese tradition, which is an approach to understand "*li* 理" (principles). Therefore, the communication between the Jesuits and the Chinese was a very meaningful event in the history of cultural exchange globally.

The Characteristics of Different Eras being Reflected in the Transformation of the Theory of "Chinese Origin of Western Learning"

Reporter: Having touched on the clash of culture between the Chinese and the Westerners, what was the background of theorizing the "Chinese origins of Western learning" during the Ming–Qing transition?

Elman: The "Chinese origins of Western learning" meant that many things were spread from the Han and Tang dynasties to the West, such as gunpowder, the compass and papermaking, to the West via Arabia. It was when the Jesuits came to China that the Chinese realized what they had brought had existed in China since long ago, but the Han Chinese had not paid enough attention. It was under this circumstance that some Chinese proposed to restore Chinese learning again. Take Qian Daxin 钱大昕 (1728–1804) and Dai Zhen 戴震 (1724–1777) for instance. They engaged with the mathematics of the Jesuits, and discovered its relationship with the astronomy/astrological books in China. This advocacy of "restoring the past" is fairly reasonable to a certain extent.

On the other hand, when something new arrived in old China, it would definitely face a challenge: Would the Chinese accept it? But with the "Chinese origins of Western learning" theory, it made it relatively easier for them to accept it. The Chinese needed change, but not by denying themselves totally and embracing the West. By acknowledging the West as advanced but with its origins from China, this overlap reduced the Chinese resistance to reform. This thinking was therefore very popular from late Ming to the Qianlong eras. Until the late 19th century, it was gradually magnified and eventually taken as the claim that all things originated from China — "democracy" came from the *Rituals of Zhou* (*Zhouli*《周礼》), "science" and calculus too originated in China.

It was after the *Jiawu zhanzheng* 甲午战争 (Sino-Japanese War, 1894–1895), that the Chinese began to feel that the "Chinese origins of Western learning" theory did not make sense, and many Han Chinese now felt that China had nothing of its own, and that everything belonged to the West. We see a huge change in views. The Sino-Japanese War was a turning point — many believed in the "Chinese origins of Western learning" theory before that, but revolutionists and reformists began to feel that China had failed, and thus many began denying everything about the Chinese culture, thinking that the Chinese were inferior to the West in all aspects, from politics, economy to culture. The existence and transformation of the "Chinese origins of Western learning" theory happened under particular historical and social circumstances, reflecting the historical progression of the change in the self-confidence of the Chinese.

Science Did Not Play an Important Role in the Civil Examinations

Reporter: Holistically speaking, what role did science play in the civil examinations?

Elman: The syllabus of the civil examination system was always changing. In the Song dynasty, though, the main curriculum was not yet the teachings of Cheng-Zhu, but there was much mathematics-related content. But during the Ming era, there was a lot of content related to mathematics and astrology, etc., in the *celun*. Why? Because the late Ming people discovered inaccuracies in the calendar system, which now missed entries by one to two days. This affected the auspicious prediction of solar and lunar eclipses. This is why through the *celun*, we are able to see some natural science content retained in the civil examination syllabus. This was extended to the Qing examinations. During the times of the *Yangwu yundong* 洋务运动 (Foreign Affairs Movement) in the 80s and 90s of the 19th century, the civil examination system began to experience Western influences, where science could be included in the examinations. After the 1901 *Yihetuan yundong* 义和团运动 (Boxer Rebellion), the civil examination system underwent another major reform — the syllabus changed tremendously, with many newly added subjects, which included many aspects of *gezhixue* 格致学. Therefore, if the examinable subjects had undergone further reform, not only would science be included, it would be very substantial in the examinations.

The civil examinations from 1901 to 1904 involved political issues of both domestic and international matters, while the traditional questions from the Four Books and Five Classics came last. Examinations during this period were already improved, but it was too late, because most people already felt that the whole system should be abolished. There was a *Gezhi shuyuan* 格致书院 (Gezhi College) later in Shanghai, which only had science for its examinations. The Jiangnan Arsenal (*Jiangnan zhizao ju* 江南制造局) had many staff at that time who made use of the examinations of Gezhi College to get in touch with scientific information. Most of these people from late Qing era eventually influenced academia extensively. However, science was generally never regarded as the top priority in civil examinations, nor was due attention paid to it, and it became marginalized. Although dramatic changes took place in the late Qing era, it was too late. History can never turn back. I feel that this attitude of absolute denial of the roots of science in China was not right, but revolutionists wished to make use of this trend to reform everything and build a new China. The logic of revolution overtook the plan for improvement. This was a new direction in history.

[Translated by the Editor and vetted by the Author.]

The Story of China: Success in the 18th Century and Failure of the 19th Century

Interview with Princeton University Professor Benjamin Elman

Interview by Reporter Zhang Ying and Interns Zhao Lei and Zhang Yiwei of the *Southern Weekly* 《南方周末》, Originally in Chinese (2010)

"We have great respect for John Fairbank and historians of his generation, but as his students, we will definitely surpass them in the studies of Chinese history," said Princeton University Professor Benjamin Elman during his interview with the *Southern Weekly*. The Chinese version of *The Cambridge History of China: The Ch'ing Dynasty to 1800*, in which Elman too contributed, will be published in China soon.

Western studies of Chinese history earlier went through the "Impact-Response Model" of Fairbank, the "Tradition-Modernity Model" of Joseph Levenson, and now the latest trend in Chinese history studies in the United States is the "China-centered" approach. Fairbank is considered in the first generation of scholars and Elman is considered in the third.

Fairbank started his studies on China in the 30s in the last century. "There were many droughts and floods then, the standard of living was very low, and China was a third-world country. China in the Han, Song and Ming times was the center of the world, why did the great and prosperous Empire of China become so poor? Why did the *Yangwu yundong* 洋务运动 (Foreign Affair Movement) fail? Why did the *Xinhai geming* 辛亥革命 (Xinhai Revolution) fail? Why did the Communist Party succeed? The Communist Party became the leader after the failure of the Guomindang. The Communist Party then was the enemy of the United States, especially during the Korean War, and so the American government needed to analyze modern Chinese politics and economy in order to interact with the Chinese. The birth of *The Cambridge History of China* was related to this background as well," said Elman.

During the time of Levenson, China had not opened its doors to the world. Outsiders could only access information from Taiwan, Japan and the United States. "Levenson's research had a complete internal logic, but we realized that the Chinese did not think according to his conclusions, the Chinese in Levenson's research were the Chinese he invented, not real Chinese."

Elman migrated to the States from Germany at the age of one, and studied Chinese intensively in the East-West Center at the University of Hawai'i from 1966 to 1967. During the Sino-U.S. Cold War, Elman took part in the American program to study China, and went to Taiwan to study at National Taiwan University and Academia Sinica in 1973, where he met his future wife in 1984. From 1977, he went to study in Japan. He received his Ph.D. in Oriental Studies from the University of Pennsylvania in 1980 under the supervision of Professor Nathan Sivin, which was the beginning of his studies on the history of science and technology, civil examinations, and the legacy of the Qing dynasty in China.

In the early eighties of the last century, Elman was a visiting scholar at the Chinese Academy of Social Sciences, the Shanghai Academy of Social Sciences and Renmin University of China. This experience changed Elman's studies on the history of China, and likewise affected the research methodology and outlook of *The Cambridge History of China*, influencing the approach to Chinese history in American Sinology.

"We did not break the tradition, instead, we broke the traditional research methodology of that era," said Elman.

The Center of Globalization in the 18th Century was China and India

Southern Weekly: In *The Cambridge History of China*, why did you say that China was very successful in the 18th century?

Elman: This book discussed the success of China in politics, economy and in academic life in the 18th century. China learned and imitated a lot from the West at that time. Before the Opium War, the British had already cultivated the habit of drinking tea at two to three o'clock in the afternoon. They imitated the way Chinese appreciated tea, and bought much china, tea and silk. Chinese back then were already mixing politics and trade, sometimes in terms of tribute and bestowal. Although Japan refused to be a Qing tributary and Russia had special relations with the Manchu state, most other countries such as Great Britain had to pay tribute to China, which was the biggest economy in the international trade market, producing consumer products on a large scale.

In other words, China and India were the center of globalization then. Britain, Spain and Portugal were growing richer, but they did not yet produce anything to attract other countries' trading interests. In the *Cambridge History*, we even analyzed why China became the hub of world trade, the roles of *shidafu* 士大夫 (scholar-officials), what the society was like, what the popular culture was, the role of the Manchus, and the situation of women, in order to propose a new view on the Chinese history: China was not a failure in the 18th century, but successful and leading the way instead.

***Southern Weekly*:** Why did China go from success in the 18th century to failure in the 19th century?

Elman: To govern Xinjiang, Tibet, and Mongolia, the Manchus needed the support of the Han Chinese. They established military deployments at Dunhuang and Kashgar, and it was not easy to fight a war there. A stable economy was required, and therefore the Qing political system and agricultural production adopted creative measures, which stabilized the society, leading to a rise in population and increase in tax collection. The emperor could deploy soldiers to Tibet, Xinjiang and Mongol, accompanied by *shidafu* and the Jesuits who managed paintings and maps for them, whilst expanding the territory of the Qing dynasty.

In the 19th century, population became the biggest problem. It was already 300 million by 1800, and before the Taiping Heavenly Kingdom (*Taiping tianguo* 太平天国) era, it was almost 450 million. Catering to this massive population was still the backdated and traditional ways of agricultural production, which produced tremendous pressure on food production. This problem was unheard of in Tang, Song and Ming times. When the traditional political and economic methods failed, the government realized that new production methods were needed, and so the Chinese *shidafu* started to learn Western technology. China started its political reform as did the Japanese.

***Southern Weekly*: **Your conclusion will trigger a debate. Did population become a burden to China at that time?

Elman: True enough it did. Production in China had always been high, with lots of food, tea, silk, paper, and medicine, but it was lagging behind the rate of population growth. As per capita wealth among the people reduced, the general standard of living in Qing dropped. In the times of Qianlong, insightful individuals pointed out that resources were running out at the county levels with a population of 200,000 or more. The standard of living in China then was not too far below that of the West, but after the Opium War, it lowered relatively to Europe, which led to the rise in poverty. This was not obvious in the 18th century. From what we can see now, the gap between the West and China mainly occurred in the 19th century, and the population problem remained a very significant one.

***Southern Weekly*: **Why do you attribute the turning point of China's decline to the Taiping Heavenly Kingdom instead of the Opium War?

Elman: When was the beginning of China's problems? Many scholars attributed it to the Opium War, but I disagree as do many of my colleagues. I think it should be the Taiping Heavenly Kingdom. The Opium War was of course important, but its impact on China internally was not big and was restricted to Canton mainly. The root of the problem was the Taiping Heavenly Kingdom, when millions of Chinese died, and the rebels captured the Jiangnan area, which was then the economic center of China. Shanghai, Suzhou, Hangzhou and Yangzhou were occupied a few times by both sides, having severely drained the vitality of the Qing government, and resulting in the strength of the dynasty weakening. After the Taiping times, both the *shidafu* and Qing government realized the need to reform the political system, which gave birth to the Foreign Affairs Movement.

The industrial revolution in China had already begun at that time, between 1865 and 1890, when Shanghai established the Jiangnan Machine Manufacturing General Bureau (*Jiangnan jiqi zhizao zongju* 江南机器制造总局), Nanjing established the Jinling Machinery Manufacturing

Bureau (*Jinling jiqi zhizao zongju* 金陵机器制造局), and Western schools were set up. Thereafter, the Lanzhou Textile Bureau and the Shanghai Machinery Textile Bureau were established, the Imperial University of Peking was founded, and Western schools were set up in the provinces. Then the navy *yamen* 衙门 (also known as the Board of Admiralty) was established and procured warships, and the South–North Telegraph Office, the Kaiping Mining Administration, and the Hubei Hanyang Arsenal were also established. The whole country began a process of mechanization and industrialization, the outlook of the dynasty's development was promising; the Foreign Affairs Movement was very successful. It was after 1895 that it hit a snag — because of the *Jiawu zhanzheng* 甲午战争 (Sino-Japanese War, 1894–1895).

Japanese Fabricated the Story of "Backward China"

Westerners thought China would win the war over Korea, since the Chinese navy was twice as powerful as the Japanese, but the Chinese navy was divided into four parts — Northern (Port Arthur to Tianjin), Southern (Guangzhou), Jiangnan (Yangzi delta), and Fujian (Fuzhou) fleets, which did not collaborate in action. During the Sino-French War of 1884–1885, the Northern Navy did not render support to the Fuzhou one, and during the Sino-Japanese War, the Fuzhou navy did not extend any help to Li Hongzhang 李鸿章 (1823–1901) and the Beiyang Fleet. If the Chinese navy had worked together, the Japanese would not have dared to invade China. During the Sino-Japanese War, the cannons of the Chinese were poorly equipped. The Sino-Japanese War was not fought between two backward nations: the Chinese had 65 warships, and its navy ranked eighth in the world; Japan only had 32 warships, and held a world ranking of 11. Both had very modern warships, mostly steam-driven.

This war had the whole world watching, as it was the first modernized sea battle. Never before had steam powered navies faced off. The Japanese warships were very light, faster in speed, with powerful engines, and fast-firing guns and cannon; Chinese ships were very heavy, slower in movement and shooting. The Europeans and Americans then realized that speed was more advantageous, and that ships should be lighter. This had direct influence on their plans for establishing naval fleets.

After the Sino-Japanese War, historians and revolutionists attributed the backwardness of China as dating all the way back to the Han, Tang and Ming dynasties, with the problem lying with the orthodox teachings of Cheng–Zhu 程朱 (Cheng Yi 程颐, 1033–1107, and Zhu Xi 朱熹, 1130–1200) learning, and for many other reasons. Japan felt they were now the "second China", the most powerful nation in East Asia. Japan now took center stage and China retreated from Asian dominance. The critique then was very interesting, while the Japanese began to say that being a traditional nation was the cause of China's corruption and backwardness. Many Chinese at that time started to study in Japan, such as Lu Xun 鲁迅 (1881–1936), who learned of this new perspective — taking the Meiji Reformation of Japan as the model and thinking China was outdated. After the Sino-Japanese War, the Han revolutionists needed a backward China for their revolution. However, if China had won the war, history would have been different.

***Southern Weekly*:** Fairbank's views and conclusions in *The Cambridge History of China* are very different from yours.

Elman: Fairbank studied China in the 1950s and 1960s, from the perspective of China lagging behind and its bleak outlook for its economy, politics, culture and society. This in a way reflected how China was perceived then.

The 18th and 19th centuries seem like two different stories. The one in the former century China was perceived as a success; the one in the latter century China as a failure. This contrast is not a mistake, just an issue of perspective. We should understand why Fairbank saw China in the 19th century as backward and traditional. This story is not created by the Westerners but the Japanese. The first Sino-Japanese War was in late Ming, when Japan occupied Korea. The Ming government supported Korea and defeated the Japanese, which led to the Japanese returning to their homeland. After the 1894–1895 Sino-Japanese War, Japan felt that it could surpass China. Their students studied in the West, becoming the most Westernized nation in East Asia, and Chinese then went to Japan to learn.

That was Fairbank's point of view in his historical studies as well: China as very backward, and Japan as very advanced. Japan had a huge influence in this aspect, as they were well-versed in the language, history, arts, and culture of China, understanding a lot more and deeper than the Westerners. I am currently researching the Sino-Japanese relationship in the 18th century, when China was powerful and Japan weak; not until the 19th century, did it become the other way round. Especially after the Sino-Japanese War, Japan began to construct a superior perspective of itself and provided historical evidence. Before China opened its doors, Westerners went to Japan for experts on China. Japanese have studied China for hundreds of years, and possess abundant historical information. This is why before the opening up of China, the Japanese were a major influence on the development of Western Sinology. The Kyoto School in 1950s completed a new history of China, with the narrative focusing on the Song dynasty as a magnificent one, which declined from then on. China was beyond redemption when it came to the Qing dynasty. We do not see it this way now. We follow historical references, research objectively, and amend the previous historical studies with a new attitude.

The problem in the 21st century is the reinforcement of nationalism and ethnic consciousness when countries like Korea, Japan, Vietnam, which had complex ties with China, are discovering their own characteristics, and claiming to be the true China. The Koreans say they have retained a lot of Chinese traditional culture, and that their traditional ways outnumber those of Chinese. Now that China is growing stronger, these countries that were influenced by China in history are getting cold feet, thus starting to emphasize their own uniqueness and difference.

The Expansion by Land and Sea

***Southern Weekly*:** China in the 18th century was expanding by land, while many nations in the West were taking the marine approach. How do you view the differences?

Elman: The Western region of China at that time, including Xinjiang, was not developed economically. Britain was expanding in the Far East, mainly in India, having acquired all of India's productivity. Its influences infiltrated China, although they were not able to occupy China. Great Britain still managed to utilize the productivity of Shanghai, Guangzhou, etc., to open up the Chinese market. Britain expanded through the sea since it was near the waters, but though China did have coastlines, it did not show interest in expanding its territories overseas to the same extent.

Countries in the West were relatively small when compared to China. China's main income came from its internal economy, and from taxes like the *zu-yong-diao* 租庸调 (grain-labor-cloth tax). By comparison, the total economy of the Western countries was small, which meant that changes in the external economy had a huge impact on domestic affairs. For example, the size of Fujian was almost the same as Britain, and though the benefits from trade were important to the province, when put within the context of the whole of China, it was less obvious. There were more than 300 million people nationwide, 95% in agriculture — farmers were the foundation of the country. And although Britain placed importance on agriculture as well, the influence of the foreign trade was considered more important.

***Southern Weekly*:** Chinese scholars used to think that the benchmark of China lagging behind was expansion through the sea. Continental civilization was defeated by maritime civilization. How do you view this?

Elman: What failed was the Chinese navy not civilization. After the Sino-Japanese War, the Qing navy was unable to pick themselves up. But now, the Chinese navy has revived. If China had won the war, without the burden of those war debts, the navy would have been able to develop continuously. After defeating China, the Japanese developed the ambition to control the Pacific. If not for this, China would have influenced the Western Pacific on a large scale. We can say that once the Sino-Japanese War and Japanese–Russian War of 1904–1905 were over, Japan began to emerge as a threat to other Asian countries, even the United States, with the scale of its navy rising to third or fourth after being 11th in the world in 1894. These developments prompted Britain to consider Japan as an ally to face other nations in the Far East. War made China lose its most important strategic port at Port Arthur. If Port Arthur and Weihaiwei had continued to belong to China, all would have been different.

Although the Foreign Affairs Movement failed for the time being, the influence to the nation was far-reaching. During the Republican times, the Jiangnan Arsenal was manufacturing airplanes. So in conclusion, if not for the external threats, internal problems could have been tackled. When both operated at the same time, the nut was especially hard to break.

The Jesuits Amended the Chinese Calendar

***Southern Weekly*:** What, in your opinion, was the influence the Jesuits had on Chinese culture?

Elman: First and foremost is the calendar. Matteo Ricci (1552–1610), who studied mathematics under his teacher Christopher Clavius (1538–1612), a famous European mathematician at the University of Rome, and others were assigned to China in late Ming era in the 16th century. They noticed that the Chinese calendar was inaccurate and behind that of Rome. They felt that it was an opportunity, as Jesuits would need the consent of the Chinese to preach Christianity.

They believed that helping the Chinese would bring accessibility to their mission. This is the intriguing part about the Jesuits, as they did not make accommodations in other countries. Apart from the reform of the late Ming Calendrical Studies, they also helped the Ming government enhance its standards of technological knowledge, having influenced how the Chinese manufactured porcelain, built observatories, and brought books on astronomy and mathematics to China. They knew to read and write the Chinese language, and like the Chinese, they translated some Latin into Chinese. For instance, for their highest order of knowledge "*scientia*", Ricci consulted with Li Zhizao 李之藻 (1565–1630) and Xu Guangqi 徐光启 (1562–1633), who thought "*scientia*" was similar to *gewu zhizhi* 格物致知 (investigating into and extending knowledge). Because of Ricci, Chinese converts also stressed "*gezhixue*" 格致学 in Ming China. *Gezhi* at that time was a very important concept and was closely related to the astronomy that Jesuits brought with them.

***Southern Weekly*:** The science that the Jesuits brought was dated compared to the latest advancements in Europe then. What was the cause for this gap?

Elman: There was a vast difference between the science in China and the West in the 16th and 17th centuries. Both were influenced by the Arabs several centuries earlier. The Chinese realized that the new Jesuit mathematics "algebra" (*jiegenfang* 借根方) was not much different from the "heavenly element notation" (*tianyuan shu* 天元术) and "four elements notation" (*siyuan shu* 四元术) in Song–Yuan times. The Chinese in the early Ming era already had calculation methods, but they were not of much use, as they were too complex for practical artisanal solutions. They realized in late Ming era, that the Westerners used square roots to solve many astronomical and mathematical problems, and the Chinese could now revive it from the pre-Jesuit *tianyuanshu* and *siyuanshu*.

Scholasticism was the core of the Jesuits learning, that is, the "Four Elements" of Aristotle (384–322 B.C.) from ancient Greece; and *wuxing* 五行 (five evolutive phases) was the basis of Chinese thinking. These were not the elements of modern science. During the 18th century, mathematics and industry began to be inseparable, and Newton (1643–1727) and Leibniz (1646–1716) invented the differential (*weifen* 微分) and integral (*jifen* 积分) for the calculus. The Jesuits in the 17th century were more advanced, but in the 18th century they fell behind the Protestants in Europe who were more advanced in science and technology. Jesuits never used the differential and integral calculus to solve problems, and it was not brought to China until the 19th century by the Protestants.

The missions that came to China in the 18th century were mainly Jesuits, and like Rome in southern Europe, they were weak in the knowledge of the Differential and Integral Calculus. Later, when they were criticized in Europe for their religious policies, their order was banned by the Pope in Europe. In late Qianlong times, China lost its window to the West, and so new science and physics during the Newtonian times did not manage to reach China until the 19th century.

***Southern Weekly*:** The Christian Protestant missions then came to China.

Elman: It was not until the Opium War was over, that the Christian missions came to China. But the Qing dynasty was not stable then, and in addition, the warfare resulting from the Taiping Heavenly Kingdom affected China's acceptance of influence from the West. China was still at the age of pre-modern weapons, whereas weapons in the West were already industrialized.

Following the Opium War, John Fryer (1839–1928) translated the Differential and Integral Calculus into Chinese with the help of Li Shanlan

李善兰 (1810–1882), when new materials and techniques for modern engineering were beginning to be transmitted to China after 1865. China began to shift from handicrafts to industrialization. The Christian missions set up modern hospitals and schools, brought knowledge such as modern architecture and urban planning with them, thereby increasing the influence of modern science in late Qing China.

[Translated by the Editor and vetted by the Author.]

Bibliography

Adas, Michael, *Machines as the Measure of Men: Science, Technology, and Ideologies of Western Dominance* (Ithaca: Cornell University Press, 1989).

Alitto, Guy, *The Last Confucian: Liang Shu-ming and the Chinese Dilemma of Modernity* (Berkeley: University of California Press, 1979).

Andrews, Bridie, "Medical Lives and the Odyssey of Western Medicine in Early Twentieth-Century China", paper presented at the History of Science Society Annual Meeting, San Diego, CA, November 8, 1997.

Andrews, Bridie, "Tailoring Tradition: The Impact of Modern Medicine on Traditional Chinese Medicine", in Viviane Alleton and Alexeï Volkov, eds., *Notions et Perceptions du Changement en Chine* (Paris: Collège de France, Institut des Hautes Études Chinoises, 1994), 152.

Andrews, Bridie, "Traditional Chinese Medicine as Invented Tradition", *Bulletin of the British Association for Chinese Studies* 6 (1995): 6–15.

Andrews, Bridie, "Tuberculosis and the Assimilation of Germ Theory in China, 1895–1937", *Journal of the History of Medicine and Allied Sciences* 52, 1 (1997): 142–143.

Andrews, Bridie, *The Making of Modern Chinese Medicine, 1850–1960* (Vancouver: University of British Columbia Press, 2014).

Anonymous, "The New Novel Before the New Novel", in Judith Zeitlin and Lydia Liu, eds., *Writing and Materiality in China: Essays in Honor of Patrick Hanan* (Cambridge, MA: Harvard University Asia Center, 2003), 317–340.

Ayers, John and Kerr, Rose, *Blanc De Chine: Porcelain from Dehua* (Chicago: Art Media Resources, 2002).

Bai Limin, "Mathematical Study and Intellectual Transition in the Early and Mid-Qing", *Late Imperial China* 16.2 (1995): 23–61.

Ban Gu 班固, comp., *Hanshu* 《汉书》 (History of the Former Han Dynasty) (7 Vols. Beijing: Zhonghua Bookstore, 1962, and Taipei: Shixue chubanshe, 1974).

Barber, W. T. A., "A Chinese examination paper", *North-China Herald*, July 7, 1888: 15.

Barrow, John, *Travels in China* (London: T. Cadell and W. Davies, 1804).

Bastid, Marianne, *Educational Reform in Early 20th-Century China*, translated by Paul J. Bailey (Ann Arbor: University of Michigan China Center, 1988).

Benedict, Carol, "Policing the Sick: Plague and the Origins of State Medicine in Late Imperial China", *Late Imperial China* 14, 2 (December 1993): 60–77.

Bennett, Adrian, *John Fryer: The Introduction of Western Science and Technology into Nineteenth-Century China* (Cambridge: Harvard University Research Center, 1967).

Berinstein, Dorothy, "Hunts, Processions, and Telescopes: A Painting of an Imperial Hunt by Lang Shining (Giuseppe Castiglione)", *RES: Anthropology and Aesthetics* 35, Intercultural China (Spring 1999): 170–184.

Beurdeley, Michel and Raindre, Guy, *Qing Porcelain* (New York: Rizzoli, 1986).

Biggerstaff, Knight, "Shanghai Polytechnic Institution and Reading Room: An Attempt to Introduce Western Science and Technology to the Chinese", *Pacific Historical Review* 25 (May 1956): 127–134.

Biggerstaff, Knight, *The Earliest Modern Government Schools in China* (Ithaca, New York: Cornell University Press, 1961), 165–166.

Bodde, Derk, "Prison Life in Eighteenth Century Peking", *Journal of the American Oriental Society* 89 (April–June 1969): 311–333.

Bodde, Derk, "The Chinese Cosmic Magic known as Watching for the Ethers", in Soren Egerod and Elise Glahn, eds., *Studia Serica Bernhard Karlgren Dedicata, Sinological Studies Dedicated to Bernhard Karlgren on His Seventieth Birthday October Fifth 1959* (Copenhagen: Ejnar Munksgaard, 1959), 14–35.

Boorman, Howard and Howard, Richard, eds., *Biographical Dictionary of Republican China* (New York: Columbia University Press, 1967).

Bréard, Andrea, *Re-Kreation eines mathematischen Konzeptes im chinesischen Diskurs: "Reihen" vom 1. bis zum 19. Jahrhundert* (Stuttgart: Franz. Steiner, 1999).

Brokaw, Cynthia, *The Ledgers of Merit and Demerit: Social Change and Moral Order in Late Imperial China* (Princeton, NJ: Princeton University Press, 1991).

Brook, Tim, *The Confusions of Pleasure: Commerce and Culture in Ming China* (Stanford: Stanford University Press, 1999).

Buck, Peter, *American Science and Modern China, 1879–1936* (Cambridge; New York: Cambridge University Press, 1980).

Campany, Robert F., *Strange Writing: Anomaly Accounts in Early Medieval China* (Albany: SUNY Press, 1996).

Carlson, Ellsworth, *The Kaiping Mines, 1877–1912* (Cambridge: Harvard East Asian Monographs, 1971).

Chan Sin-wai, trans., *An Exposition of Benevolence: The Jen-hsueh of T'an Ssu-t'ung* (Hong Kong: The Chinese University Press, 1984).

Chan Wing-tsit, *Instructions for Practical Living and Other Neo-Confucian Writings by Wang Yang-ming* (New York: Columbia University Press, 1963).

Chan, Wellington, "Government, Merchants, and Industry to 1911", in John Fairbank and Kwang-Ching Liu, eds., *The Cambridge History of China*, Volume 11, Part 2 (Cambridge: Cambridge University Press, 1980), 422–429.

Chen Li 陈澧, *Dongshu dushu ji* 《东塾读书记》(Chen Li's Reading Notes) (Taipei: Commercial Press, 1970).

Chen Minsun, "Ferdinand Verbiest and the Geographical Works by Jesuits in Chinese 1584–1674", in John Witek, S.J., ed., *Ferdinand Verbiest, S.J. (1623–1688): Jesuit Missionary, Scientist, Engineer, and Diplomat* (Nettetal: Steyler Verlag, 1994), 123–164.

Chen Yuanlong 陈元龙, "Fan-li" 凡例 (Outline), in *Gezhi jingyuan* 《格致镜源》(Mirror Origins of Investigating Things and Extending Knowledge). In *Siku quanshu* 《四库全书》 (Complete Collection in the Imperial Four Treasuries), (Reprint, Taipei: Commercial Press, 1983–1986), Vol. 1031, 1–3.

Ch'en, Kenneth, "Matteo Ricci's Contribution to, and influence on, Geographical Knowledge in China", *Journal of the American Oriental Society* 59, 3 (September 1939), 325–359.

Chia, Lucille, "Mashaben: Commercial Publishing in Jianyang, Song–Ming", in Paul Jakov Smith and Richard von Glahn, eds., *The Song–Yuan–Ming Transition in Chinese history* (Cambridge: Harvard University Asia Center, 2003).

Chia, Lucille, "Printing for Profit: The Commercial Printers of Jianyang, Fujian (Song–Ming)" (Ph.D. Dissertation, Columbia University, 1996).

Chou, Min-chih Maynard, "Science and Value in May Fourth China: The Case of Hu Shih" (University of Michigan Ph.D. dissertation in History, 1974), 23–35.

Chow Tse-tsung, *The May 4th Movement: Intellectual Revolution in Modern China* (Cambridge: Harvard University Press, 1960).

Chu Pingyi, "Ch'eng-Chu Orthodoxy, Evidential Studies and Correlative Cosmology: Chiang Yung and Western Astronomy", *Philosophy and the History of Science: A Taiwanese Journal* 4.2 (1995): 71–108.

Chu Pingyi, "Remembering Our Grand Tradition: The Historical Memory of the Scientific Exchanges between China and Europe", *Historia Scientiarum* 41 (2003): 193–215.

Chu Pingyi, "Scientific Dispute in the Imperial Court: The 1664 Calendar Case", *Chinese Science* 14 (1997): 7–34.

Chu Pingyi, "Tongguan Tianxue, Yixue yu Ruxue: Wang Honghan yu Ming–Qing Zhi Ji Zhongxi Yixue de Jiaohui" 〈通贯天学、医学与儒学：王宏翰与明清之际中西医学的交会〉, *Lishi yuyan yanjiu suo jikan* 《历史语言研究所集刊》 70, 1 (1999): 165–201.

Chu Pingyi, "Western Astronomy and Evidential Study: Tai Chen on Astronomy", in Yung Sik Kim and Francesca Bray, eds., *Current Perspectives in the History of Science in East Asia* (Seoul: Seoul National University Press, 1999), 144.

Clulow, Adam, *The Company and the Shogun: The Dutch Encounter with Tokugawa Japan* (New York: Columbia University Press, 2014).

Clunas, Craig, "Ming and Qing Ivories: Useful and Ornamental Pieces", in William Watson, ed., *Chinese Ivories from the Shang to the Qing* (London: Oriental Ceramic Society, British Museum Publications, 1984), 122.

Clunas, Craig, *Superfluous Things: Material Culture and Social Status in Early Modern China* (Cambridge, UK: Polity Press, 1991).

Cohen, Paul, "Christian Missions and Their Impact to 1900", in Denis Twitchett and John Fairbank, eds., *The Cambridge History of China*, Volume 10, *Late Ch'ing, 1800–1911*, Part 1 (Cambridge: Cambridge University Press, 1978), 574–575.

Cohen, Paul, *Between Tradition and Modernity: Wang T'ao and Reform in Late Ch'ing China* (Cambridge: Council on East Asian Studies, Harvard University, 1987).

Cohn, Bernard, *Colonialism and Its Form of Knowledge: The British in India* (Chicago: University of Chicago Press, 1996).

Cong Xiaoping, *Teachers' Schools and the Making of the Modern Chinese Nation-state, 1897–1937* (Vancouver: University of British Columbia Press, 2007).

Craig, Albert, "Science and Confucianism in Tokugawa Japan", in Marius Jansen, ed., *Changing Japanese Attitudes Toward Modernization* (Princeton, NJ: Princeton University Press, 1965), 139–142.

Crossley, Pamela Kyle, *A Translucent Mirror: History and Identity in Qing Imperial Ideology* (Berkeley: University of California Press, 1999).

Croizier, Ralph, *Traditional Medicine in Modern China: Science, Nationalism, and the Tensions of Cultural Change* (Cambridge, MA: Harvard University Press, 1968).

Curtis, Emily, "Glass for the Qing Court: The Jesuit Workshop", paper presented at the colloquium on "Art Brokering for China: The Missionary Connection", sponsored by the UCLA Center for Chinese Studies in conjunction with the Southern California China Colloquium, May 4, 2002.

Curtis, Emily, "Plan of the Emperor's Glassworks", *Ars Asiatiques* (Paris) 56 (2001): 81–90.

Dagenais, Ferdinand, "Organizing Science in Republican China (1914–1950)", presented at the "Ideology and Science Symposium", Center for Chinese Studies, University of California, Berkeley, October 19, 2001.

Dagenais, Ferdinand, *John Fryer's Calendar: Correspondence, Publications, and Miscellaneous Papers with Excerpts and Commentary* (Berkeley: Center for Chinese Studies, University of California, Berkeley, 1999).

Dai Yi 戴逸 *et al.*, *Jiawu zhanzheng yu Dong-Ya zhe*ngzhi 《甲午战争与东亚政治》(The 1895 War and East Asian Politics) (Beijing: Zhongguo shehui kexue chubanshe, 1994).

Dai Zhen 戴震, "Dai xiansheng xingzhuang"〈戴先生行狀〉(Activities and Character of Mister Dai), in *Dai Zhen wenji*《戴震文集》(Dai Zhen's Collected Essays), (Hong Kong: Zhonghua shuju, 1974), 251–260.

Daston, Lorraine, "Marvelous Facts and Miraculous Evidence in Early Modern Europe", *Critical Inquiry* 18 (1991): 93–124.

Daston, Lorraine, "The Nature of Nature in Early Modern Europe", *Configurations* 6 (1998): 149–172.

Day, John, "The Search for the Origins of the Chinese Manuscript Copies of Matteo Ricci's Maps", *Imago Mundi* 47 (1995).

de Heer, Philip, *The Care-Taker Emperor: Aspects of the Imperial Institution in Fifteenth-Century China as Reflected in the Political History of the Reign of Chu Ch'i-yü* (Leiden: E.J. Brill, 1986).

Deane, Thatcher E., "The Chinese Imperial Astronomical Bureau: Form and Function of the Ming Dynasty 'Qintianjian' from 1365 to 1627" (Seattle: University of Washington Ph.D. dissertation in History, 1989).

Dear, Peter, "Jesuit Mathematical Science and the Reconstitution of Experience in the Early Seventeenth Century", *Studies in History and Philosophy of Science* 18 (1987): 135–141.

d'Elia, Pasquale M., "Recent Discoveries and New Studies (1938–60) on the World Map in Chinese of Father Matteo Ricci S.J.", *Monumenta Serica* 20 (1961): 82–164.

Ding Wenjiang 丁文江, *Liang Rengong xiansheng nianpu changbian chugao*《梁任公先生年谱长编初稿》(First Draft of Mr. Liang Qichao's chronological biography) (2 vols. Taipei, Shijie shuju, 1972), 551–574.

Dixue qianshi 《地学浅释》 (Shanghai: Jiangnan Arsenal, 1873).

Dunne, George H.S.J., *Generation of Giants: The Story of the Jesuits in China in the Last Decades of the Ming Dynasty* (Notre Dame: University of Notre Dame Press, 1962).

Duyvendak, Jan Julius Lodewijk, *Ma Huan Re-examined* (Amsterdam: Noord-Hollandsche uitgeversmaatschappij, 1933).

Elisonas, Jurgis (George Elison), "Christianity and the Daimyo", in John Hall, ed., *The Cambridge History of Japan.* Volume 4. *Early Modern Japan* (Cambridge: Cambridge University Press, 1991).

Elliot, Mark, "The Limits of Tartary: Manchuria in Imperial and National Geographies", *Journal of Asian Studies* 59, 3 (August 2000): 603–646.

Elliot, Mark, *The Manchu Way: The Eight Banners and Ethnic Identity in Late Imperial China* (Stanford: Stanford University Press, 2001).

Elman, Benjamin A., "From Pre-modern Chinese Natural Studies to Modern Science in China", in Michael Lackner and Natascha Vittinghoff, eds., *Mapping Meanings: The Field of New Learning in Late Qing China* (Leiden: E. J. Brill, 2004), 25–73.

Elman, Benjamin A., "Geographical Research in the Ming–Ch'ing Period", *Monumenta Serica* 35 (1981–1983): 1–18.

Elman, Benjamin A., "Jesuit Scientia and Natural Studies in Late Imperial China", *Early Modern History: Contacts, Comparisons, Contrasts* 6, 3 (Fall 2002): 1–24.

Elman, Benjamin A., "Naval Warfare and the Refraction of Qing Nineteenth Century Industrial Reforms into Failure", *Modern Asian Studies*, 38, 2 (Fall 2003): 283–326.

Elman, Benjamin A., "Rethinking 'Confucianism' and 'Neo-Confucianism' In Modern Chinese History", in Benjamin A. Elman, John Duncan, and Herman Ooms, eds., *Rethinking Confucianism: Past and Present in China, Japan, Korea, and Vietnam* (Los Angeles: UCLA Asia Pacific Monograph Series, 2002), 518–554.

Elman, Benjamin A., "The Formation of 'Dao Learning' as Imperial Ideology During the Early Ming Dynasty", in Theodore Huters, R. Bin Wong, and Pauline Yu, eds., *Culture and State in Chinese History: Conventions, Accommodations, and Critiques* (Stanford: Stanford University Press, 1997), 58–82.

Elman, Benjamin A., "The Hsueh-hai T'ang and the Rise of New Text Scholarship in Canton", *Ch'ing-shih wen-t'i* (now *Late Imperial China*), 4, 2 (December 1979): 51–82.

Elman, Benjamin A., "Wang Kuo-wei and Lu Hsun: The Early Years", *Monumenta Serica*, 34 (1979–1980): 389–401.

Elman, Benjamin A., *A Cultural History of Civil Examinations in Late Imperial China* (Berkeley: University of California Press, 2000).

Elman, Benjamin A., *Civil Examinations and Meritocracy in Late Imperial China 1400–1900* (Cambridge: Harvard University Press, 2013).

Elman, Benjamin A., "Towards a History of Modern Science in Republican China", in Benjamin A. Elman and Jing Tsu, eds., *Science and Technology in Modern China: 1880s to 1940s* (Leiden: E. J. Brill, 2014), 15–38.

Elman, Benjamin A. and Woodside, Alex, eds., *Education and Society in Late Imperial China, 1600–1900* (Berkeley: University of California Press, 1994).

Elman, Benjamin A., *From Philosophy to Philology: Social and Intellectual Aspects of Change in Late Imperial China* (Cambridge: Harvard University, Council on East Asian Studies, 1984).

Elvin, Mark, *The Structure of the Chinese Past* (Stanford: Stanford University Press, 1975).

Epstein, Maram, "Engendering Order: Structure, Gender, and Meaning in the Qing Novel Jinghua yuan", *Chinese Literature: Essays, Articles, and Reviews* 18 (December 1996): 105–131.

Evans, Nancy, "The Canton T'ung-wen Kuan: A Study of the Role of Bannermen in One Area of Self-Strengthening", *Papers on China* (Harvard), 22A (1969): 89–103.

Feng Youlan (Fung Yu-lan), "Why China Has No Science — An Interpretation of the History and Consequences of Chinese Philosophy", *International Journal of Ethics* 32, 3 (April 1922): 237–263.

Feuerwerker, Albert, "Economic Trends in the Late Ch'ing Empire, 1870–1911", in John Fairbank and Kwang-Ching Liu, eds., *The Cambridge History of China*, Vol. 11, part 2, 59–68.

Findlin, Paula, *Possessing Nature: Museums, Collecting, and Scientific Culture in Early Modern Italy* (Berkeley: University of California Press, 1994).

Foret, Phillipe, *Mapping Chengde: The Qing Landscape Enterprise* (Honolulu: University of Hawai'i Press, 2000).

Foss, Theodore, "A Western Interpretation of China Jesuit Cartography", in Charles Ronan, S.J., and Bonnie Oh, eds., *East Meets West: the Jesuits in China, 1582–1773* (Chicago: Loyola University Press, 1988), 209–251.

Frederick W. Mote, "The T'u-mu Incident of 1449", in Frank Kierman, Jr., and John Fairbank, eds., *Chinese Ways in Warfare* (Cambridge: Harvard University Press, 1974), 243–272.

Frumer, Yulia, "Clocks and Time in Edo Japan" (Princeton University Ph.D. dissertation in the History of Science, 2012).

Fryer, John, "An Account of the Department for the Translation of Foreign Books at the Kiangnan Arsenal, Shanghai", *North-China Herald*, January 29, 1880.

Fryer, John, "Chinese Prize Essays: Report of the Chinese Prize Essay Scheme in connection with the Chinese Polytechnic Institution and Reading Rooms, Shanghai, for 1886 and 1887", *North-China Herald*, January 25, 1888: 100.

Fryer, John, "Report of the Chinese Scientific Book Depot, Shanghai, 1887", *North-China Herald*, December 28, 1887: 702–703.

Fryer, John, "Second Report of the Chinese Prize Essay Scheme in connection with the Chinese Polytechnic Institution and Reading Rooms, Shanghai, From July 1887 to July 1889", *North-China Herald*, July 20, 1889: 85–86.

Fuchs, Walter, "Der Jesuiten-Atlas der Kanghsi-Zeit", *Monumenta Serica*, Monograph Series 4 (1943): 60–75.

Fuchs, Walter, "Materialen zur Kartographie der Mandju-Zeit, part 1", *Monumenta Serica*, 1 (1935–1936): 395–396.

Fuchs, Walter, "Materialen zur Kartographie der Mandju-Zeit, part 2", *Monumenta Serica*, 3 (1938): 189–231.

Fujitsuka Chikashi (Rin) 藤塚鄰, *Nichi Sen Shin no bunka kōryū*《日鮮清の文化交流》(Cultural interactions among Japan, Chosŏn Korea, and Qing China) (Tokyo: Chūbunkan shoten, 1947).

Fung, Allen, "Testing the Self-Strengthening: The Chinese Army in the Sino-Japanese War of 1894–95", *Modern Asian Studies* 30, 4 (1996): 1007–1031.

Furth, Charlotte, *Ting Wen-chiang: Science and China's New Culture* (Cambridge: Harvard University Press, 1970).

Gardner, Daniel, *Chu Hsi and the Ta-hsueh: Neo-Confucian Reflection on the Confucian Canon* (Cambridge: Harvard University Council on East Asian Studies, 1986).

Geertz, Hildred, "An Anthropology of Religion and Magic, I", *Journal of Interdisciplinary History* 6, 1 (Summer 1975): 71–89.

Gegu yaolun 《格古要论》 (Essential Criteria of Antiquities), in Hu Wenhuan 胡文焕 (fl. c. 1596), comp., *Gezhi congshu* 《格致丛书》 (Collectanea of Works Investigating into and Extending Knowledge), Ming Wanli edition, 1573–1619; microfilm (Taipei: National Library of Taiwan, Rare Books Collection, Vol. 25).

Gernet, Jacques, *China and the Christian Impact* (Cambridge: Cambridge University Press, 1982).

Gezhi huibian 《格致汇编》 (Reprint in 6 vols. Nanjing: Guji shudian, 1992).

Gezhi shuyuan keyi 《格致书院课艺》, Fryer Private Library, University of California, Berkeley (2 vols. Shanghai: Shanghai Polytechnic, 1886–1893).

Giles, Lionel, "Translations from the World Map of Father Ricci", *Geographical Journal* 52, 6 (1918), 367–385, and 53, 1 (1919), 19–30.

Giles, Lionel, *An Alphabetical Index to the Chinese Encyclopedia, Ch'in Ting Ku Chin T'u Shu Chi Ch'eng* (Reprint, Taipei: Chengwen Publishing Co., 1969).

Glacken, Clarence J., *Traces on the Rhodian Shore: Nature and Culture in Western Thought from Ancient Times to the End of the Eighteenth Century* (Berkeley: University of California Press, 1967).

Goldschmidt, Asaf, *The Evolution of Chinese Medicine: Song Dynasty, 960–1200* (London; New York: Routledge, 2009).

Goodman, Grant, *Japan: the Dutch Experience* (London; Dover, N.H.: Athlone Press, 1985).

Goodrich, L.C. et al., eds., *Dictionary of Ming Biography* (2 vols. New York: Columbia University Press, 1976).

Grieder, Jerome, *Hu Shih and the Chinese Renaissance: Liberalism in the Chinese Revolution, 1917–1937* (Cambridge: Harvard University Press, 1970).

Guangxu chao donghua lu 《光绪朝东华录》 (Records from within the Eastern Gate during the Guangxu reign) (Shanghai: Zhonghua shuju, 1909).

Gujin tushu jicheng 《古今图书集成》 (Synthesis of Books and Illustrations Past and Present), 1728 edition (Shanghai: Zhonghua shuju, 1934).

Guy, R. Kent, *The Emperor's Four Treasuries: Scholars and the State in the Late Ch'ien-lung Era* (Cambridge: Harvard Council on East Asian Studies, 1987).

Hahn, Roger, "Laplace and the Mechanistic Universe", in David C. Lindberg and Ronald Numbers, eds., *God & Nature: Historical Essays on the Encounter between Christianity and Science* (Berkeley: University of California Press, 1986), 266.

Hahn, Roger, *The Anatomy of a Scientific Institution: The Paris Academy of Sciences, 1666–1803* (Berkeley: University of California Press, 1971).

Hamaguchi Fujiō 濱口富士雄, *Shindai kōkyogaku no shisōshiteki kenkyū* 《清代考據學の思想史的研究》 (Research on the Intellectual History of Qing Dynasty Evidential Studies) (Tokyo: Kokusho kankōkai, 1994).

Han Qi 韩琦, "Bai Jin de Yijing yanjiu he Kangxi shidai de xixue zhongyuanshuo" 〈白晋的易经研究和康熙时代的西学中源说〉 (Joachim Bouvet's Study of the Changes Classic and the Theory of the Chinese Origins of Western Learning in the Kangxi Period), *Hanxue yanjiu* 《汉学研究》 (Taiwan) 16, 1 (1998): 185–200.

Han Qi 韩琦, "Gewu qiongli yuan yu Mengyangzhai: shiqi shiba shiji zhi Zhong-Fa kexue jiaoliu" 〈格物穷理院与蒙养斋：十七十八世纪之中法科学交流〉 (The Academy of Science and the Mengyangzhai: Sino-French Scientific Exchange in the Seventeenth and Eighteenth Centuries), *Faguo hanxue* 《法国汉学》 (French Sinology) (Beijing) 4 (2000): 302–324.

Han Qi 韩琦, *Zhongguo kexue jishu de xichuan ji qi yingxiang*《中国科学技术的西传及其影响》(The Western transmission of Chinese technologies and their influence) (Beijing: Hebei chubanshe, 1999).

Hanan, Patrick, "The Missionary Novels of Nineteenth-Century China", *Harvard Journal of Asiatic Studies* 60, 2 (December 2000): 440–441.

Harrell, Paula, *Sowing the Seeds of Change: Chinese Students, Japanese Teachers, 1895–1905* (Stanford: Stanford University Press, 1992).

Harris, Steven, "Transposing the Merton Thesis: Apostolic Spirituality and the Establishment of the Jesuit Scientific Tradition", *Science in Context* 3.1 (1989): 29–65.

Hart, Roger, "Beyond Science and Civilization: A Post-Needham Critique", *East Asian Science, Technology, and Medicine* 16 (1999): 88–114.

Hart, Roger, "Local Knowledges, Local Contexts: Mathematics in Yuan and Ming China", paper presented at the Song–Yuan–Ming Transitions Conference, Lake Arrowhead, CA (June 5–11, 1997).

Hart, Roger, *Imagined Civilizations: China, the West, and Their First Encounter* (Baltimore: The Johns Hopkins University Press, 2013).

Hart, Roger, *The Chinese Roots of Linear Algebra* (Baltimore: Johns Hopkins University Press, 2011).

Hartwell, Robert, "Financial Expertise, Examinations, and the Formulation of Economic Policy in Northern Song China", Journal of Asian Studies, 30.2 (1971): 281–314.

Hartwell, Robert, "Historical-Analogism, Public Policy, and Social Science in Eleventh- and Twelfth-Century China", *American Historical Review* 76, no. 3 (June 1971): 690–727.

Hashimoto Keizō 橋本敬造, "Rekisho Kōsei no seiritsu"〈歷象考成の成立〉(The formation of the compendium of observational and computational astronomy), in Yabuuchi Kiyoshi 藪内清 and Yoshida Mitsukuni 吉田光邦, eds., *Min Shin jidai no kagaku gijutsu shi*《明清時代の科學技術史》(History of science and technology in the Ming and Qing periods) (Kyoto: Research Institute for Humanistic Studies, 1970), 49–92.

Hatano Yoshihiro, "The Response of the Chinese Bureaucracy to Modern Machinery", *Acta Asiatica* 12 (1968): 13–28.

Hatano Yoshihiro 波多野善大, *Chūgoku kindai kōgyō shi no kenkyū*《中國近代工業史の研究》(Research on the history of modern Chinese industry) (Kyoto, 1960).

Heilbron, J. L., *Electricity in the 17th and 18th Centuries: A Study of Early Modern Physics* (Berkeley: University of California Press, 1979).

Henderson, John, "The Assimilation of the Exact Sciences into the Ch'ing Confucian Tradition", *Journal of Asian Affairs* 5.1 (Spring 1980): 15–31.

Hevia, James Louis, "Looting Beijing: 1860, 1900", in Lydia Liu, ed., *Tokens of Exchange: The Problem of Translation in Global Circulations* (Durham: Duke University Press, 1999), 192–199.

Hevia, James Louis, *Cherishing Men From Afar: Qing Guest Ritual and the Macartney Embassy of 1793* (Durham and London: Duke University Press, 1995).

Ho Peng Yoke, *Li, Qi, and Shu: An Introduction to Science and Civilization in China* (Hong Kong: Hong Kong University Press, 1985).

Hoe, Jock, "Zhu Shijie and his *Jade Mirror of the Four Unknowns*", in *First Australian Conference on the History of Mathematics: Proceedings of a Conference at Monash University* (Clayton, Australia) 6 & 7 (November 1980): 105.

Horng Wann-sheng 洪万生, "Tongwen'guan suanxue jiaoxi Li Shanlan" 〈同文馆算学教习李善兰〉 (Li Shanlan's mathematical teaching in the Translation School), in Yang Tsui-hua 杨翠华, and Huang Yi-long 黄一农, eds., *Jindai Zhongguo keji shi lunji* 《近代中国科技史论集》 (Taipei: Academia Sinica, Taiwan, Institute of Modern History, and National Qinghua University, Institute of History, 1991), 215–259.

Horng Wann-sheng, "Chinese Mathematics at the Turn of the 19th Century", in Cheng-hung Lin and Daiwie Fu, eds., *Philosophy and Conceptual History of Science in Taiwan* (Netherlands: Kluwer Academic Publishers, 1993), 183–190, especially 186.

Horng Wann-sheng, "Li Shan-lan: The Impact of Western Mathematics in China During the Late 19th Century" (City University of New York Ph.D. dissertation in History, 1991), 16–17.

Hoskin, Keith, "Examinations and the Schooling of Science", in Roy MacLeod, ed., *Days of Judgement: Science, Examinations, and the Organization of Knowledge in Late Victorian England* (Driffield, N. Humberside: Studies in Education, 1982).

Hostetler, Laura, "Qing Connections to the Early Modern World: Ethnography and Cartography in Eighteenth-Century China", *Modern Asian Studies* 34, 3 2000): 623–662.

Hostetler, Laura, *Qing Colonial Enterprise: Ethnography and Cartography in Early Modern China* (Chicago: University of Chicago Press, 2001).

Hsia, Florence C., *Sojourners in a Strange Land: Jesuits and their Scientific Missions in Late Imperial China* (Chicago: University of Chicago, 2009).

Hsia, Florence C., "Some Observations on the *Observations*: The Decline of the French Jesuit Scientific Mission in China", in Éric Brian, ed., *Revue de synthèse: Les Jésuites dans le monde moderne Nouvelles approches* (Paris: CNRS, 1999), 305–333.

Hsia, R. Po-chia, "The Catholic Mission and Translations in China, 1583–1700", in Peter Burke and R. Po-chia Hsia, eds., *The Cultural History of Translation in the Early Modern World* (Cambridge: Cambridge University Press, 2007), 39–51.

Hsiao Kung-chuan, *A Modern China and a New World: Kang Yu-wei, Reformer and Utopian, 1858–1927* (Seattle: University of Washington Press, 1975).

Hu Minghui, "Cosmopolitan Confucianism: China's Different Road to Modern Science (1664–1830)" (Los Angeles: UCLA Ph.D. dissertation in History, 2003).

Hu Minghui, "Provenance in Contest: Searching for the Origins of Jesuit Astronomy in Early Qing China, 1664–1705", *The International History Review* 24, 1 (March 2002): 1–36.

Hu Mingjie, "Merging Chinese and Western Mathematics: The Introduction of Algebra and the Calculus in China, 1859–1903" (Princeton: Princeton University Ph.D. dissertation in History, 1998), 252–253.

Hu Shi, "The Scientific Spirit and Method in Chinese Philosophy", in Charles A. Moore, ed., *The Chinese Mind* (Honolulu: University of Hawai'i Press, 1967), 199–222.

Hu Shi 胡适, "Qingdai xuezhe de zhixue fangfa" 〈清代学者的治学方法〉 (Qing dynasty scholarly methods), in *Hushi wencun*《胡适文存》(4 vols. Taipei: Yuandong dashu gongsi, 1968), Vol. 1, 383–412.

Huangchao wenxian tongkao 《皇朝文献通考》, in *Shitong* 《十通》 (The ten comprehensive encyclopedias of civil and military governance) (Shanghai: Commercial Press, 1935–1937).

Huang Yi-long and Chang Chih-ch'eng, "The Evolution and Decline of the Ancient Practice of Watching for the Ethers", *Chinese Science* 13 (1996): 82–85, 90–96, 104.

Huang Yi-long, "Court Divination and Christianity in the K'ang-hsi Era", *Chinese Science* 10 (1991): 1–20.

Huang Yi-long 黄一农, "Qingchu tianzhujiao yu huijiao tianwenjia de chengdou" 〈清初天主教与回族天文家的争斗〉(The Struggle Between Catholic and Muslim Astronomers in the Early Qing). *Jiuzhou xuekan*《九州学刊》, 5.3 (1993): 47–69.

Huff, Toby E., *Intellectual Curiosity and the Scientific Revolution: A Global Perspective* (Cambridge: Cambridge University Press, 2011).

Huff, Toby E., *The Rise of Early Modern Science: Islam, China, and the West* (Second edition. Cambridge University Press, 1993/2003). See also the author's review of the first edition in *American Journal of Sociology*, (November 1994): 817–819.

Huffman, James, "Commercialization and Changing World of the Mid-Meiji Press", in Helen Hardacre and Adam Kern, eds., *New Directions in the Study of Meiji Japan* (Leiden: E. J. Brill, 1992), 574–579.

Hummel, Arthur, ed., *Eminent Chinese of the Ch'ing Period* (Washington: U.S. Government Printing Office, 1943).

Hutchison, Keith, "Supernaturalism and the Mechanical Philosophy", *History of Science* 21 (1983): 297–299.

Hymes, Robert, "Not Quite Gentlemen? Doctors in Song and Yuan", *Chinese Science* 7 (1986): 11–85.

Itō Shūichi 伊藤秀一, "Kindai Chūgoku ni okeru kagaku gijutsu no chii" 〈近代中国における科学技術の地位〉 (The status of modern Chinese science and industry), *Tōyō gakujutsu kenkyū* 《東洋学術研究》 5, 5–6 (1967): 65–77.

Jacob, Margaret, *Scientific Culture and the Making of the Industrial West* (New York: Oxford University Press, 1997).

Jami, Catherine, "Western Devices for Measuring Time and Space: Clocks and Euclidian Geometry in Late Ming and Ch'ing China", in Chun-chieh Huang and Erik Zürcher, eds., *Time and Space in Chinese Culture* (Leiden: E. J. Brill, 1995), 169–200.

Jami, Catherine, "Western Influence and Chinese Tradition in an Eighteenth-Century Chinese Mathematical Work", *Historia Mathematica* 15 (1988): 311–331.

Jami, Catherine, "From Louis XIV's Court to K'ang-hsi's Court: An Institutional Analysis of the French Jesuit Mission to China (1688–1722)", in Hashimoto Keizō et al., eds., *East Asian Science: Tradition and Beyond* (Osaka: Kansai University Press, 1995), 493–499.

Jami, Catherine, "Learning Mathematical Sciences during the Early and Mid-Ch'ing", in Benjamin A. Elman and Alexander Woodside, eds., *Education and Society in Late Imperial China, 1600–1900* (Berkeley: University of California Press, 1994).

Jensen, Lionel, *Manufacturing Confucianism: Chinese Traditions and Universal Civilization* (Durham: Duke University Press, 1997).

Ji Yun 纪昀 et al., comps., *Siku quanshu* 《四库全书》 (Complete Collection in the Imperial Four Treasuries), (Reprint, Taipei: Commercial Press, 1983–1986).

Ji Yun 纪昀 et al., comps., *Siku quanshu zongmu* 《四库全书总目》 (General Catalog of the Complete Collection in the Imperial Four Treasuries), (Reprint, Taipei: Yiwen Press, 1974).

Ji Yun 纪昀 et al., comps., *Siku quanshu zongmu* 《四库全书总目》 (General Catalog of the Complete Collection in the Imperial Four Treasuries) (Reprint, Taipei: Commercial Press, 1983–1986).

Ji Wende 计文德, *Cong Siku quanshu tanjiu Ming Qing jian shuru zhi Xixue*《从四库全书探究明清间输入之西学》(Inquiring into the importation of Western learning during the Ming–Ch'ing transition from the point of view of the Complete Collection in the Imperial Four Treasuries) (Taipei: Han Mei tushu youxian gongsi, 1991).

Jia Sheng, "The Origins of the Science Society of China, 1914–1937" (Ithaca: Cornell University Ph.D. dissertation in History, 1995).

Jiao Hong 焦竑 and Wu Daonan 吴道南, and others, comps., *Zhuangyuan ce*《状元策》(Policy Essays by *Optimi*), (1733 *Huaidetang*《怀德堂》edition. "Zongkao" 总考 (General Overview), 15a).

Johnston, Alastair, *Cultural Realism: Strategic Culture and Grand Strategy in Chinese History* (Princeton: Princeton University Press, 1995).

Kamachi, Noriko, "The Chinese in Meiji Japan: Their Interaction with the Japanese before the Sino-Japanese War", in Akira Iriye, ed., *The Chinese and the Japanese: Essays in Political and Cultural Interactions* (Princeton: Princeton University Press, 1980), 69–72.

Kangxi jixia gewu bian《康熙几暇格物编》(Compilation on Investigating Things for Kangxi in Leisure), (Tongxuezhai 通学斋 edition).

Keenan, Barry, "Beyond the Rising Sun: The Shift in the Chinese Movement to Study Abroad", in Laurence Thompson, ed., *Studia Asiatica* (San Francisco: Chinese Materials Center, 1975), 157–169.

Keene, Donald, "The Sino-Japanese War of 1894–1895 and Its Cultural Effects in Japan", in Donald Shively, ed., *Tradition and Modernization in Japanese Culture* (Princeton: Princeton University Press, 1971), 122–123.

Keizō Hashimoto, *Hsu Kuang-ch'i and Astronomical Reform* (Osaka: Kansai University Press, 1988).

Kennedy, Thomas, "Chang Chih-tung and the Struggle for Strategic Industrialization: The Establishment of the Hanyang Arsenal, 1884–1895", *Harvard Journal of Asiatic Studies* 33 (1973): 154–182.

Kennedy, Thomas, "The Establishment and Development of the Kiangnan Arsenal, 1860–95" (Columbia University Ph.D. dissertation in History, 1968).

Kennedy, Thomas, *The Arms of Kiangnan: Modernization in the Chinese Ordnance Industry, 1860–1895* (Boulder: Westview Press, 1978).

Kexue yu rensheng guan《科学与人生观》(2 vols. Shanghai: Dongya tushuguan, 1923).

Kim, K. H., *Japanese Perspectives on China's Early Modernization: A Bibliographical Survey* (Ann Arbor, MI: University of Michigan, Center for Chinese Studies, 1974).

Kiyosi Yabuuti, "Chinese Astronomy: Development and Limiting Factors", in Shigeru Nakayama and Nathan Sivin, eds., *Chinese Science: Explorations of an Ancient Tradition* (Cambridge: The MIT Press, 1973), 98–99.

Kuo Ting-yee and Liu Kwang-Ching, "Self-Strengthening: the pursuit of Western technology", in Denis Twitchett and John Fairbank, eds., *The Cambridge History of China*, Volume 10, *Late Ch'ing, 1800–1911*, Part 1 (Cambridge: Cambridge University Press, 1978), 519–537.

Kwok, D.W.Y., *Scientism in Chinese Thought, 1900–1950* (New Haven: Yale University Press, 1965).

Lach, Donald F., *Asia in the Making of Europe. Volume II. A Century of Wonder, Book 3: The Scholarly Disciplines* (Chicago: University of Chicago Press, 1977).

Lach, Donald F., *China in the Eyes of Europe: The Sixteenth Century* (Chicago: Phoenix Books, 1968).

Lai Chi-kong, "Li Hung-chang and Modern Enterprise: The China Merchants' Company, 1872–1885", in Samuel Chu and Kwang-Ching Liu, eds., *Li Hung-chang and China's Early Modernization* (Armonk, New York: M.E. Sharpe, Inc., 1994), 216–247.

Lam, Lay-Yong, "Chu Shih-chieh's *Suan-hsueh ch'i-meng* (Introduction to Mathematical Studies)", *Archive for History of Exact Sciences*, 21, 1 (1979): 1–31.

Lam, Lay-Yong, *A Critical Study of the Yang Hui Suan Fa, a Thirteenth-Century Mathematical Treatise* (Singapore University Press, 1977).

Lam, Peter, "The Glasswork of the Qing Imperial Household Department", in *Elegance and Radiance: Grandeur in Qing Glass, The Arthur K. F. Lee Collection* (Hong Kong: Art Museum, Chinese University of Hong Kong, 2000).

Lam, Peter, *Elegant Vessels for the Lofty Pavilion: The Zande Lou Gift of Porcelain with Studio Marks* (Hong Kong: Art Gallery, Chinese University of Hong Kong, 1993).

Latour, Bruno, *Science in Action: How to Follow Scientists and Engineers Through Society* (Cambridge: Harvard University Press, 1987).

Lee, Victoria, "The Arts of the Microbial World: Biosynthetic Technologies in Twentieth-Century Japan" (Princeton: Princeton University Ph.D. dissertation in the History of Science, 2014).

Legge, James, trans, *The Four Books* (Reprint, New York: Paragon, 1966).

Lei Hsiang-lin, *Neither Donkey nor Horse: Medicine in the Struggle over China's Modernity* (Chicago: University of Chicago Press, 2014).

Leibo, Steven, *Transferring Technology to China: Prosper Giquel and the Self-Strengthening Movement* (Berkeley: University of California Press, 1985).

Leung, Angela, "Medical Learning from the Song to the Ming", in Paul Smith and Richard von Glahn, eds., *The Song–Yuan–Ming Transition in Chinese History* (Cambridge, MA: Harvard Asian Center Monograph, 2003), 374–398.

Levenson, Joseph, *Confucian China and Its Modern Fate: A Trilogy* (Berkeley: University of California Press, 1968).

Li Bin 李斌, "Xishi wuqi dui Qingchu zuozhan fangfa de yingxiang"〈西式武器对清初作战方法的影响〉(The impact of Western style weapons on combat techniques in the early Qing), *Ziran bianzheng fa tongxun* 《自然辩证法通讯》24.4 (2002): 45–53.

Li Chi, *The Travel Diaries of Hsu Hsia-k'o* (Hong Kong: Chinese University, 1974).

Li Ju-chen, *Flowers in the Mirror,* abridged translation by Lin Tai-yi (Berkeley, University of California Press, 1965).

Li Ruzhen, *Jinghua yuan* 《镜花缘》 (Flowers in the Mirror) (Taipei: Xuehai chubanshe, 1985).

Li, San-pao, "Letters to the Editor in John Fryer's *Chinese Scientific Magazine*, 1876–1892: An Analysis", *Jindaishi yanjiu suo jikan*《近代史研究所集刊》(Academia Sinica, Taiwan) 4 (1974).

Li Shuangbi 李双璧, "Cong 'gezhi' dao 'kexue': Zhongguo jindai keji guan de yanbian guiji"〈从格致到科学：中国近代科技观的演变轨迹〉(From Gezhi to Kexue: The steps in the change in modern Chinese views of science and technology), *Guizhou shehui kexue*《贵州社会科学》137 (1995.5): 105–107.

Li Yan and Du Shiran, *Chinese Mathematics: A Concise History*, translated by John Crossley and Anthony Lun (Oxford: Clarendon Press, 1987).

Liang Jun 梁峻, *Zhongguo gudai yizheng shilue*《中国古代医政史略》(Brief history of ancient Chinese state medicine) (Huhehot: Inner Mongolia People's Press, 1995).

Liang Qichao 梁启超, "Gezhixue yange kaolue"〈格致学沿革考略〉(Brief account of the evolution of Gezhi studies), in *Yinbingshi wenjii*《饮冰室文集》(8 vols. Taipei: Zhonghua shuju, 1970).

Liang Qichao, *Ouyou xinyinglu jielu*《欧遊心影录节录》, in Liang Qichao, *Yinbingshi zhuanji*《饮冰室专集》(10 vols. Taipei: Zhonghua shuju, 1972).

Libbrecht, Ulrich, *Chinese Mathematics in the Thirteenth Century: The Shu-shu Chiu-chang of Ch'in Chiu-shao* (Cambridge: MIT Press, 1973).

Lin Mousheng, *A Guide to Chinese Learned Societies and Research Institutes* (New York: China Institute in America, 1936).

Lin Qingzhang 林庆彰, *Qingchu de chunjing bianweixue*《清初的群经辨伪学》(Study of Forged Classics in the Early Qing) (Taipei: Wenjin Press, 1990).

Lin Tongyang, "Ferdinand Verbiest's Contribution to Chinese Geography and Cartography", in John Witek, S.J., ed., *Ferdinand Verbiest, S.J. (1623–1688): Jesuit Missionary, Scientist, Engineer, and Diplomat* (Nettetal: Steyler Verlag, 1994), 135–164.

Lin Yü-sheng, "The Origins and Implications of Modern Chinese Scientism in Early Republican China: A Case Study — The Debate on 'Science vs. Metaphysics' in 1923", in *Zhonghua min'guo chuqi lishi yantaohui lunwenji*《中华民国初期历史研讨会论文集》*1912–1927* (2 vols. Taipei: Academia Sinica, Taiwan, Institute of Modern History, 1985).

Lindberg, David C., "The Transmission of Greek and Arabic Learning to the West", in David C. Lindberg, ed., *Science in the Middle Ages* (Chicago: University of Chicago Press, 1978), 52–90.

Liu, Cary, "The Qing Dynasty Wen-yuan-ko Imperial Library: Architecture and the Ordering of Knowledge" (Ph.D. Dissertation, University of Princeton, 1997).

Liu Kwang-Ching and Smith, Richard, "The Military Challenge: The North-West and the Coast", in John Fairbank and Kwang-Ching Liu, eds., *The Cambridge History of China*, Volume 11, *Late Qing, 1800–1911*, Part 2 (Cambridge: Cambridge University Press, 1980), 202–273.

Liu Kwang-Ching, "Nineteenth-Century China", in Ping-ti Ho and Tang Tsou, eds., *China in Crisis* (2 vols. Chicago: University of Chicago Press, 1968), Vol. 1, 93–178.

Liu, Lydia, "Robinson Crusoe's Earthenware Pot", *Critical Inquiry* 25 (Summer 1999): 728–757.

Liu, Lydia, *Translingual Practice: Literature, National Culture, and Translated Modernity — China 1900–1937* (Stanford: Stanford University Press, 1995).

Liu Shipei 刘师培, "Gewu jie"〈格物解〉(Explication of investigating things), in *Zuo'an waiji*《左庵外集》, in *Liu Shenshu xiansheng yishu*《刘申叔先生遗书》(4 vols. Taipei: Daxin shuju, 1965).

Liu Xi 刘熙, "Preface" 序 to the *Shiming*《释名》(Explication of Names), in the *Gezhi congshu*《格致丛书》(Collectanea of Works Investigating Into and Extending Knowledge) (Late Ming Wanli edition preserved in the National Library of Taiwan).

Lloyd, Geoffrey and Sivin, Nathan, *The Way and the Word: Science and Medicine in Early China and Greece* (New Haven: Yale University Press, 2003).

Lo Jung-pang. "The Decline of the Early Ming Navy", *Oriens Extremus* 5, 2 (1958), 147–168.

Loehr, George, "The Sinicization of Missionary Artists and Their Works at the Manchu Court during the Eighteenth Century", *Cahiers D'histoire Mondiale* 8 (1963): 795–803.

Loewe, Michael, *Early Chinese Texts: A Bibliographical Guide* (Berkeley: University of California, Society for the Study of Early China, 1993).

Lu Jian 陆坚 and Wang Yong 王勇, eds., *Zhongguo dianji liuchuan yu yingxiang* 《中国典籍流传与影响》(Transmission and influence of Chinese books) (Hangzhou: Hangzhou daxue chubanshe 杭州大学出版社, 1990).

Lu Shiyi 陆世仪 (1611–1672), "Sibianlu lunxue"〈思辨录论学〉(Discussions of Learning from the Record of Thought and Debate), in Wei Yuan 魏源, ed. *Huangchao jingshi wenbian* 《皇朝经世文编》 (Collected Writings on Statecraft from the Qing Dynasty), (Reprint of 1827 and 1873 edition, Taipei: World Bookstore, 1964), 3.7–3.9.

Lu Xun, *Selected Works of Lu Hsun*, translated by Yang Hsien-yi and Gladys Yang (Beijing: Foreign Languages Press, 1956).

Luk, Bernard Hung-kay, "A Study of Giulio Aleni's *Chih-fang wai-chi*", *Bulletin of Oriental and African Studies* 40 (1977), 58–84.

Luo Shilin 罗士林, *Siyuan yujian xicao* 《四元玉鉴细草》(Jade Mirror of the Four Unknowns with Detailed Calculations) (Reprint of 1836 edition. Shanghai: Shangwu yinshuguan, 1937).

Macartney, George, *An Embassy to China; being the journal kept by Lord Macartney during his embassy to the Emperor Ch'ien-lung, 1793–1794*, edited with an introduction and notes by J. L. Cranmer-Byng (London: Longmans, 1962).

MacKinnon, Stephan, *Power and Politics in Late Imperial China* (Berkeley: University of California Press, 1980).

Mancall, Mark, *Russia and China: Their Diplomatic Relations to 1728* (Cambridge: Harvard University Press, 1971).

Martin, W. A., *The Awakening of China* (London: Hodder and Stoughton, 1907).

Martzloff, Jean-Claude, *A History of Chinese Mathematics*, translated by Stephen Wilson (New York: Springer-Verlag, 1997).

Masao Watanabe, *The Japanese and Western Science* (Philadelphia: University of Pennsylvania Press, 1990).

Masini, Federico, "Using the works of the Jesuit missionaries in China to study the Chinese language: a research project", paper presented at the International Conference "Translating Western Knowledge into Late Imperial China", Göttingen University, December 6–9 (1999).

McCord, Edward, *The Power of the Gun: The Emergence of Modern Chinese Warlordism* (Berkeley: University of California Press, 1980).

Menegon, Eugenio, *Un solo Cielo: Giulio Aleni S.J.* (*1582–1649*) (Rome: Grafo, 1994).

Meng Yue, "Hybrid Science *versus* Modernity: The Practice of the Jiangnan Arsenal", *East Asian Science, Technology, and Medicine* 16 (1999): 13–52.

Menzies, Gavin, *1421: The Year China Discovered America* (New York: HarperCollins, 2002).

Métailé, Georges, "The *Bencao gangmu* of Li Shizhen — An Innovation for Natural History?", in Elisabeth Hsu, ed., *Innovation in Chinese Medicine* (Cambridge: Cambridge University Press), 221–261.

Mills, J. V. G., trans., and Ptak, Roderich, rev., annot. and ed., *Hsing-ch'a-sheng-lan: The Overall Survey of the Star Raft by Fei Hsin* (Amsterdam, Noord-Hollandsche uitgeversmaatschappij, 1933).

Millward, James A., *Beyond the Pass: Economy, Ethnicity, and Empire in Qing Central Asia, 1759–1864* (Stanford: Stanford University Press, 1998).

Morohashi Tetsuji 諸橋轍次, *Dai kanwa jiten*《大漢和辭典》(Great Han-Japanese Dictionary), 13 vols. (Tokyo: Taishūkan shoten, 1957–1960).

Mosca, Matthew W., *From Frontier Policy to Foreign Policy: The Question of India and the Transformation of Geopolitics in Qing China* (Stanford: Stanford University Press, 2013).

Moulder, Francis, *Japan, China, and the Modern World Economy: Toward a Reinterpretation of East Asian Development, ca. 1600 to ca. 1918* (Cambridge: Harvard University Press, 1977).

Mungello, Donald, *Curious Land: Jesuit Accommodation and the Origins of Sinology* (Honolulu: University of Hawai'i Press, 1985).

Needham, Joseph and others, *Science and Civilisation in China* (Multi-volumes. Cambridge: Cambridge University Press, 1954).

Needham, Joseph, *Science and Civilisation in China*, Vol. 3 (Cambridge: Cambridge University Press, 1959).

Needham, Joseph, *Science and Civilisation in China,* Vol. 4, Part 3 (Cambridge: Cambridge University Press, 1954).

Ng Wai-ming, "The *I Ching* in the Adaptation of Western Science in Tokugawa Japan", *Chinese Science* 15 (1999): 94–115.

Numata Jirô, *Western Learning: A Short History of the Study of Western Science in Early Modern Japan*, translated by R.C.J. Bachofner (Tokyo: The Japan-Netherlands Institute, 1992), 60–95.

Onogawa Hidemi 小野川秀美, *Shimmatsu seiji shisō kenkyū*《清末政治思想研究》(Research on late Qing political thought) (Tokyo: Misuzu shobō, 1969).

Pagani, Catherine, *"Eastern Magnificence & European Ingenuity": Clocks of Late Imperial China* (Ann Arbor: University of Michigan Press, 2001).

Pan Jixing, "The Spread of Georgius Agricola's *De Re Metallica* in Late Ming China", *T'oung Pao* 57 (1991): 108–118.

Paret, Peter, *Understanding War* (Princeton: Princeton University Press, 1992).

Park, Katharine and Daston, Lorraine, "Unnatural Conceptions: The Study of Monsters in Sixteenth-Century France and England", *Past and Present* 92 (August 1981): 20–54.

Percival, Sir David, trans., *Chinese Connoisseurship, the Ko Ku Yao Lun: The Essential Criteria of Antiquity* (London: Faber, 1971).

Perdue, Peter, "Boundaries, Maps, and Movement: Chinese, Russian, and Mongolian Empires in Early Modern Central Eurasia", *The International History Review* 20, 2 (June 1998): 263–286.

Perdue, Peter, "Military Mobilization in Seventeenth and Eighteenth-Century China, Russia and Mongolia," *Modern Asian Studies*, 30, 4 (1996): 757–793.

Perdue, Peter, *China Marches West: The Qing Conquest of Central Eurasia* (Cambridge: Belknap Press of Harvard University Press, 2005).

Peterson, Willard, "'Chinese Scientific Philosophy' and Some Chinese Attitudes Towards Knowledge about the Realm of Heaven-and-Earth", *Past and Present* 87 (May 1980): 20–30.

Peterson, Willard, "Fang I-chih: Western Learning and the 'Investigation of Things'", in Wm. Theodore de Bary et al., eds., *The Unfolding of Neo-Confucianism* (New York: Columbia University Press, 1975), 369–411.

Peterson, Willard, "Western Natural Philosophy Published in Late Ming China", *Proceedings of the American Philosophical Society* 117.4 (August 1973): 295–322.

Peyrefitte, Alain, *The Immobile Empire*, translated by Jon Rothschild (New York: Knopf, 1992), which is based on his *L'Empire immobile ou Le Choc des mondes* (Paris: Fayard, 1989).

Pomeranz, Kenneth, *The Great Divergence: Europe, China, and the Making of the Modern World Economy* (Princeton: Princeton University Press, 2001).

Pong, David, "Keeping the Foochow Navy Yard Afloat: Government Finance and China's Early Modern Defense Industry, 1866–75", *Modern Asian Studies* 21, 1 (February 1987): 121–152.

Pong, David, *Shen Pao-chen and China's Modernization in the Nineteenth Century* (Cambridge: Cambridge University Press, 1994).

Porter, David, "Sinicizing Early Modernity: The Imperatives of Historical Cosmopolitanism", *Eighteenth-Century Studies* 43, 3 (Spring 2010): 299–306.

Powell, Ralph, *The Rise of Chinese Military Power* (Princeton: Princeton University Press, 1955).

Prakash, Gyan, *Another Reason: Science and the Imagination of Modern India* (Princeton: Princeton University Press, 1999).

Pusey, James R., *China and Charles Darwin* (Cambridge: Harvard University Press, 1983).

Qi Qizhang 戚其章, ed., *Jiawu zhanzheng jiushi zhounian ji'nian lunwen ji* 《甲午战争九十周年纪念论文集》(Essays in rememberance of the 99th anniversary of the 1894 War) (Ji'nan: Ji-Lu shushe, 1986).

Qian Daxin 钱大昕, *Qianyantang wenji* 《潜研堂文集》(Collected Writings from the Hall of Subtle Research) (Taipei: Commercial Press, 1968).

Qian Gang 钱钢, *Haizang: Jiawu zhanzheng 100 nian* 《海葬：甲午战争 100 年》(Sea catasorophe: The 100 year anniversary of the 1894 War) (Taipei: Fengyun shidai chuban gongsi, 1994).

Qian Zhongshu 钱钟书 and Zhu Weizheng 朱维铮, eds., *Wanguo gongbao wenxuan* 《万国公报文选》, (Beijing: Sanlian shudian, 1998).

Qingchao tongdian 《清朝通典》(Complete Institutions of the Qing Dynasty), in *Shitong*《十通》(Ten Comprehensive Encyclopedias), (Shanghai: Commercial Press, 1936).

Quan Hansheng 全汉昇, "Qingmo de 'xixue yuan chu Zhongguo' shuo"〈清末的西学源出中国说〉(The Late Qing Theory that the Origin of Western Studies Came from China), *Lingnan xuebao*《岭南学报》, 4.2 (June 1935): 57–102.

Quan Hansheng 全汉昇, "Jiawu zhanzheng yiqian de Zhongguo gongyehua yundong"〈甲午战争以前的中国工业化运动〉, *Lishi yuyan yanjiusuo jikan*《历史语言研究所集刊》25, 1 (1954): 77–78.

Rawlinson, John, *China's Struggle for Naval Development, 1839–1895* (Cambridge: Harvard University Press, 1967).

Reardon-Anderson, James, *The Study of Change: Chemistry in China, 1840–1949* (Cambridge: Cambridge University Press, 1991).

"Review Symposia" on Needham's work in *Isis* 75, 1 (1984): 171–189.

Reynolds, David, "Redrawing China's Intellectual Map: Images of Science in Nineteenth-Century China", *Late Imperial China* 12, 1 (June 1991): 47–48.

Reynolds, David, "The Advancement of Knowledge and the Enrichment of Life: The Science Society of China and the Understanding of Science in the Early Republic" (Madison: University of Wisconsin Ph.D. dissertation in History, 1986).

Reynolds, Douglas, *China, 1898–1912: The Xinzheng Revolution and Japan* (Cambridge: Council on East Asian Studies, Harvard University Press, 1993).

Richter, Giles, "Entrepreneurship and Culture The Hakubunkai Publishing Empire in Meiji Japan", in Helen Hardacre and Adam Kern, eds., *New Directions in the Study of Meiji Japan* (Leiden: E. J. Brill, 1992), 590–602.

Robinson, David, *Martial Spectacles of the Ming Court* (Cambridge, MA: Harvard University Asia Center, 2013).

Rogaski, Ruth, *Hygienic Modernity: Meanings of Health and Disease in Treaty-Port China* (Berkeley: University of California Press, 2004).

Ross, Sydney, "Scientist: The Story of a Word," *Annals of Science* 18, 2 (1962): 65–71.

Ruan Yuan 阮元, *Yanjingshi ji* 《揅经室集》 (Collection from the Studio for the Investigation of Classics), 3 vols. (Taipei: Shijie shuju, 1964).

Ruan Yuan, "Chouren zhuan fanli" 〈畴人传凡例〉 (Conventions in Biographies of mathematical astronomers), in Ruan Yuan, *Chouren zhuan* 《畴人传》 (Biographies of mathematical astronomers) (Taipei: Shijie shuju, 1962), 1–5.

Ruan Yuan, ed., *Huang Qing jingjie* 《皇清经解》 (Qing exegesis of the Classics) (20 vols. Taipei, Fuxing Reprint of the 1892 edition, 1961).

Rudolph, Richard, "Early China and the West: Fertilization and Fetalization", in Rudolph and Schuyler Cammann, *China and the West: Culture and Commerce* (Los Angeles: Clark Library, 1977), 4–5.

Saitō Tetsurō 齊藤哲郎, "Chi no ryōan — Chūgoku no 'kagaku to jinseikan' ronsō o chūshin ni" 〈知の兩岸—中國の'科學と人生観'論爭を中心に〉, (Two sides of knowledge: Centering on the debate between Chinese science and the view of human life) *Chūgoku — shakai to bunka* 《中國—社會と文化》 8 (1993): 133–155.

Sakade Yoshinobu 坂出祥伸, "Kaisetsu — Mindai nichiyō ruisho ni tsuite" 〈解說 — 明代日用類書について〉 (Explanations — Concerning Ming Dynasty Daily Use Compendia) in Sakade Yoshinobu et al., eds., *Chūgoku nichiyō ruisho shūsei* 《中國日用類書集成》 (Collection of Chinese Daily Use Compendia), (14 vols. Tokyo: Kyūko shoin, 1999ff), Vol. 1, 7–30.

Sato Ken'ichi, "Re-evaluation of *Tengenjutsu* or *Tianyuanshu*: in the context of comparison between China and Japan", *Historia Scientiarum* 5, 1 (1995): 57–67.

Schneider, Laurence, "Genetics in Republican China", in John Z. Bowers, J. William Hess, and Nathan Sivin, eds., *Science and Medicine in Twentieth-Century China: Research and Education*, 3–29.

Schorr, Adam, "The Trap of Words: Political Power, Cultural Authority, and Language Debates in Ming Dynasty China" (Ph.D. Dissertation, UCLA, 1994).

Scott, Rosemary, "Eighteenth Century Overglaze Enamels: The Influence of Technological Development on Painting Style", in *Colloquies on Art & Archaeology in Asia* (London: Percival David Foundation of Chinese Art, 1987).

Screech, Timon, "Rethinking the Visual Revolution in Edo: Inoue Masashige and International Visual Culture in the Years after the Prohibition of the Seas", in *The Scientific Eye and the Visual Wonders in Edo* (Tokyo: Suntory Museum of Art, 2014), 14–20.

Serruys, Henry, *Sino-Mongol Relations during the Ming*, in *Melanges Chinois et Bouddhiques, Vol. 1: The Mongols in China during the Hung-wu Period, 1368–1398*, and *Vol. 2: The Tribute System and the Diplomatic Missions* (1967) (Brussels: Institut Belge des Hautes etudes Chinoises, 1959).

Shang Zhicong 尚智丛, "1886–1894 nianjian jindai kexue zai wan Qing zhishi fenzi zhong de yingxiang"《1886–1894 年间近代科学在晚清知识份子中的影响》, (The impact of the period from 1886 to 1894 of modern science on late Qing intellectuals) *Qingshi yanjiu* 《清史研究》(August 2001): 72–82.

Shapin, Stephen and Schaffer, Simon, *Leviathan and the Air-Pump: Hobbes, Boyle, and the Experimental Life* (Princeton: Princeton University Press, 1989).

Shapin, Steven, *A Social History of Truth: Civility and Science in Seventeenth-Century England* (Chicago: University of Chicago Press, 1994).

Shek, Richard, "Some Western Influences on T'an Ssu-t'ung's Thought", in Paul Cohen and John Schrecker, eds., *Reform in Nineteenth-Century China* (Cambridge, MA: Harvard University Research Center, 1976).

Shenbao 《申报》 (Tongzhi 12th year, intercalary 6th month, 29th day; August 21, 1873).

Shen, Grace, *Unearthing the Nation: Modern Geology and National Identity in Republican China, 1911–1949* (Chicago: University of Chicago Press, 2013).

Shi Quan 石泉, *Jiawu zhanzheng qianhou zhi wan-Qing zhengju* 《甲午战争前后之晚清政局》(The late Qing political situation before and after the 1894 War) (Beijing: Sanlian shudian, 1997).

Shumpei Okamoto, "Background of the Sino-Japanese War, 1894–1895", in *Impressions of the Front: Woodcuts of the Sino-Japanese War, 1894–1895* (Philadelphia: Museum of Art, 1983), 13.

Siu, Victoria, "Castiglione and the Yuanming Yuan Collections", *Orientations* (Hong Kong) 19 (November 1988): 72–79.

Sivin review of Buck, *American Science & Modern China*, in *Journal of Asian Studies*, 1981, 231.

Sivin, Nathan, "Wang Hsi-shan (1628–1682)", in Charles Gillispie, ed., *Dictionary of Scientific Biography* (New York: Scribner's Sons, 1970–1978), Vol. 14, 159–168.

Sivin, Nathan, "Copernicus in China", in *Colloquia Copernica II: Études sur l'audience de la théorie héliocentrique* (Warsaw: Union Internationale d'Historie et Philosophie des Sciences, 1973), 63–114.

Sivin, Nathan, "Introduction", in Nathan Sivin, ed., *Science & Technology in East Asia*, (New York: Science History Publications, 1977), xi–xv.

Sivin, Nathan, "Science and Medicine in Chinese History", in Paul Ropp, ed., *Heritage of China: Contemporary Perspectives on Chinese Civilization* (Berkeley: University of California Press, 1990), 164–196.

Sivin, Nathan, "Why the Scientific Revolution did not take place in China — or didn't it?", reprinted in Nathan Sivin, *Science in Ancient China: Researches and Reflections* (Brookfield: Variorium, 1995), VII, 45–66.

Sivin, Nathan, "Li Shih-chen", in Charles Gillispie, ed., *Dictionary of Scientific Biography* (New York: Charles Scribner's and Sons, 1973), Vol. 8, 390–398.

Sivin, Nathan, "The Myth of the Naturalists", in Nathan Sivin ed., *Medicine, Philosophy, and Religion in Ancient China: Researches and Reflections* (Brookfield, VT: Variorum, 1995), 1–29.

Smith, Richard, "Foreign Training and China's Self-Strengthening: The Case of Feng-huang-shan", *Modern Asian Studies* 10, 2 (1976): 195–223.

Smith, Richard, "Mapping China's World: Cultural Cartography in Late Imperial Times", in Yeh Wen-hsin, ed., *Landscape, Culture and Power in Chinese Society* (Berkeley: University of California, Berkeley, Center for East Asian Studies, 1998), 62–68.

Smith, Richard, "Reflections on the Comparative Study of Modernization in China and Japan: Military Aspects", *Journal of the Hong Kong Branch of the Royal Asiatic Society* 16 (1976): 11–23.

Smith, Richard, *Chinese Maps: Images of All Under Heaven* (Oxford: Oxford University Press, 1996).

Spence, Jonathan, *Emperor of China: Self-portrait of Kangxi* (New York: Vintage Books, 1974).

Spence, Jonathan, *To Change China: Western Advisers in China, 1620–1960* (Middlesex: Penguin Books, 1980).

Ssu-yü Teng and Biggerstaff, Knight, comps., *An Annotated Bibliography of Selected Chinese Reference Works*, 2nd edition (Cambridge: Harvard University Press, 1971).

Ssu-yü Teng and Fairbank, John, eds., *China's Response to the West: A Documentary Survey, 1839–1923* (Cambridge: Harvard University Press, 1979).

Standaert, Nicolas S.J., "The Investigation of Things and the Fathoming of Principles (*Gewu qiongli*) in the Seventeenth-Century Contact Between Jesuits and Chinese Scholars", in John W. Witek, S.J., ed., *Ferdinand Verbiest (1622–1688): Jesuit Missionary, Scientist, Engineer and Diplomat* (Nettetal: Steyler Verlag, 1994), 395–420.

Standaert, Nicolas S.J., *Handbook of Christianity in China, Volume One: 635–1800* (Leiden; Boston: Brill, 2001–2010).

Stolberg, Eva-Maria, "Interracial Outposts in Siberia: Nerchinsk, Kiakhta, and the Russo-Chinese Trade in the Seventeenth/Eighteenth Centuries", *Journal of Early Modern History* 4, 3–4 (December 1999): 322–336.

Su Jing 苏精, *Qingji Tongwen guan ji qi shisheng*《清季同文馆及其师生》(Qing foreign language schools and their teachers and students) (Taipei: Shanghai yinshua chang, 1985).

Symbols of Power: Masterpieces from the Nanjing Museum (Anaheim: Bowers Museum, 2002).

Szczesniak, Boleslaw, "The Seventeenth Century Maps of China: An Inquiry into the Compilations of European Cartographers", *Imago Mundi* 13 (1956): 116–136.

Taiping yulan《太平御览》(Encyclopedia of the Taiping Era, 976–983), reprint of *Sibu congkan*《四部丛刊》edition (Taipei: Zhonghua Bookstore, 1960).

Takehiko Hashimoto, "Introducing a French Technological System: The Origin and Early History of the Yokosuka Dockyard", *East Asian Science, Technology, and Medicine* 16 (1999): 53–65.

Tambiah, Stanley J., *Magic, Science, Religion, and the Scope of Rationality* (Cambridge: Cambridge University Press, 1990).

Tan Sitong, "Xing suanxue yi"〈兴算学议〉(Proposal to promote mathematics), in *Tan Sitong quanji*《谭嗣同全集》(2 vols. Beijing: Zhonghua shuju, 1981).

Tan Tai, "Chouren jie"〈畴人解〉(Explanation of mathematical astronomers), in Ruan Yuan, *Chouren zhuan*《畴人传》(Biographies of mathematical astronomers) (Taipei: Shijie shuju, 1962), 1–4.

Tang Caichang 唐才常, "Zhuzi yulei yiyou Xiren gezhi zhi li tiaozheng"〈朱子语类已有西人格致之理条〉(Master Zhu's records of conversations already contained Westerner's notions of science as *gezhi*), in *Tang Caichang ji*《唐才常集》(Beijing: Zhonghua shuju, 1980).

The Library of Philip Robinson, Part II (London: Sotheby's Auction Catalogue, 1988).

"The Progress of Foreign Studies" in the *North-China Herald*, April 14, 1893: 513–514.

Thiriez, Régine, "The Qianlong Emperor's European Palaces", in *The Delights of Harmony: The European Palaces of the Yuanmingyuan & the Jesuits at the 18th Century Court of Beijing* (Worcester: Cantor Art Gallery, College of Holy Cross, 1994).

Thomas, Keith, *Man and the Natural World: Changing Attitudes in England, 1500–1800* (London: Penguin Books, 1983).

Thomas, Stephen, *Foreign Intervention and China's Industrial Development, 1870–1911* (Boulder: Westview Press, 1984).

Ting Wen-chiang, "On Hsu Hsia-k'o (1586–1641); Explorer and Geographer", *New China Review* 3, 5 (October 1921): 325–337.

Trigault Nichola, and Gallagher, Louis J. S.J., trans., *China in the Sixteenth Century: The Journals of Matteo Ricci: 1583–1610.* (New York: Random House, 1953).

Tsien Tsuen-hsuin, "A History of Bibliographical Classification in China", *The Library Quarterly* 22.4 (October 1952): 309–310.

Tsien Tsuen-hsuin, "Western Impact on China Through Translation", *Far Eastern Quarterly* 13 (1954): 305–327.

Turnbull, David, "Cartography and Science in Early Modern Europe: Mapping and the Construction of Knowledge Spaces", *Imago Mundi*, 48 (1996): 5–24.

Vagnoni, Alphonso, *Kongji gezhi* 《空际格致》 (Investigation of the Atmosphere), (Reproduction of the 1633 version in *Tianzhujiao dongchuan wenxian sanpian* 《天主教东传文献三篇》, Taipei: Wenjin chubanshe 文津出版社, 1979).

van de Ven, Hans J., *Breaking with the Past: The Maritime Customs Service and the Global Origins of Modernity in China* (New York: Columbia University Press, 2014).

van de Ven, Hans J., "War in the Making of Modern China", *Modern Asian Studies* 30, 4 (1996): 737–756.

Vertente, Christine et al., *The Authentic Story of Taiwan: An Illustrated History, based on Ancient Maps, Manuscripts, and Prints* (Knokke, Belgium and Taipei: Mappamundi Publishers and SMC Publishing Inc., 1991).

Virgin, Louise, "Japan at the Dawn of the Modern Age", in *Japan at the Dawn of the Modern Age: Woodblock Prints from the Meiji Era, 1868–1912* (Boston: Museum of Fine Arts, 2001).

Vittinghoff, Natascha, "Unity Vs. Uniformity: Liang Qichao and the Invention of a 'New Journalism' for China", *Late Imperial China* 23, 1 (June 2002): 91–143.

Waldron, Arthur, *The Great Wall of China: From History To Myth* (Cambridge: Cambridge University Press, 1990).

Waley-Cohen, Joanna, "China and Western Technology in the Late Eighteenth Century", *American Historical Review,* 98.5 (1993): 1525–1544.

Wallis, Helen, "The Influence of Father Ricci on Far Eastern Cartography", *Imago Mundi* 19 (1965), 35–45.

Wang Baoping 王宝平, "Mindai no kakushoka Ko Bunkan ni kan suru kosatsu〈明代の刻書家胡文煥に關する考察〉(An Inquiry into the Ming Dynasty Printer Hu Wenhuan)", *Kyūko*《汲古》36 (1999): 47–57.

Wang Baoping 王宝平, "Riben Hu Wenhuan congshu jingyan lu"〈日本胡文焕丛书经眼录〉(Record of Seeing Hu Wenhuan's Collectanea in Japan), in Lu Jian 陆坚 and Wang Yong 王勇, eds., *Zhongguo dianji liuchuan yu yingxiang*《中国典籍流传与影响》(Hangzhou: Hangzhou daxue chubanshe 杭州大学出版社, 1990), 322–347.

Wang Baoping 王宝平, "Zhongguo Hu Wenhuan congshu jingyan lu"〈中国胡文焕丛书经眼录〉(Record of Viewing Hu Wenhuan's Collectanea in China), *Zhong Ri wenhua luncong*《中日文化论丛》(1991): 6–25.

Wang Bing 王冰, "Jindai zaoqi Zhongguo he Riben zhi jian de wulixue jiaoliu"〈近代早期中国和日本之间的物理学交流〉(Early modern exchanges in physics between Japan and China) *Ziran kexueshi yanjiu*《自然科学史研究》15, 3 (1996): 227–233.

Wang Chang 王昶, *Chunrongtang ji*《春融堂集》(Precis of the Shanghai Polytechnic Academy) (Collection from the Hall of Cheerful Spring), (1807 edition).

Wang Ermin 王尔敏, *Shanghai Gezhi shuyuan zhilue*《格致书院志略》(Hong Kong: Zhongwen daxue chuban she, 1980).

Wang Ermin 王尔敏, *Qingji bing gongye de xingqi*《清季兵工业的兴起》(The rise of Qing industrial armaments) (Taipei: Institute of Modern History, Academia Sinica, 1972).

Wang Fan-shen, "The 'Daring Fool' Feng Fang (1500–1570) and His Ink Rubbing of the Stone-inscribed *Great Learning*", *Ming Studies* 35 (August 1995): 74–91.

Wang Gungwu, *Community and Nation: China, Southeast Asia and Australia* (St. Leonards, Australia: Allen and Unwin, 1992).

Wang Hui, "From Debates on Culture to Debates on Knowledge: Zhang Junmai and the Differentiations of Cultural Modernity in 1920's China", paper presented at the Workshop "Reinventions of Confucianism in the 20th Century", sponsored by the UCLA Center for Chinese Studies under the auspices of the University of California Pacific Rim Research Program, Los Angeles, January 31, 1998.

Wang Hui, "The Fate of 'Mr. Science' in China: The Concept of Science and Its Application in Modern Chinese Thought", *positions: east asia cultures critique* 3, 1 (Spring 1995): 21–23.

Wang Ping 王萍, "Ruan Yuan yu *Chouren zhuan*" 〈阮元与畴人传〉 (Ruan Yuan and the Biographies of mathematical astronomers), *Jindai shi yanjiusuo jikan* 《近代史研究所集刊》 (1973): 601–611.

Wang Ping 王萍, "Qingchu lisuanjia Mei Wending" 〈清初历算家梅文鼎〉 (Mei Wending — Early Qing Calendrical Specialist), *Jindaishi yanjiusuo jikan* 《近代史研究所集刊》 (Taipei, Academia Sinica) 2 (1971): 313–324.

Wang Renjun 王仁俊, "Lueli" 〈略例〉 (Brief precedents), in *Gezhi guwei* 《格致古微》 (Wuchang: Zhixuehui 质学会, 1897 edition).

Wang Tao 王韬, ed., *Gezhi keyi huibian* 《格致课艺汇编》 (Shanghai: Polytechnic, 1897).

"Wang Yangming nianpu" 〈王阳明年谱〉 (Chronological Biography of Wang Yangming), in Wang *Yangming quanji* 《王阳明全集》 (Taipei: Kaozheng Press, 1973).

Wang Yangzong 王扬宗, "Gezhi huibian zhi Zhongguo bianjizhe kao" 〈格致汇编之中国编辑者考〉, (An examination of the editorship of the Scientific Journal) *Wenxian* 《文献》 63 (January 1995): 237–243.

Wang Yangzong 王扬宗, "1850 niandai zhi 1910 nian Zhongguo yu Riben zhi jian kexue shuji de jiaoliu shulue" 〈1850 年代至 1910 年中国与日本之间科学书籍的交流述略〉 (Summary of the exchange of science texts between China and Japan from 1850 to 1910), *Tōzai gakujutsu kenkyūjo kiyō* 《东西学术研究所纪要》 (Kansai University) 33 (March 2000): 139–152.

Wang Y. C., *Chinese Intellectuals and the West, 1872–1949* (Chapel Hill: University of North Carolina Press, 1966).

Wei Yun'gong 魏允恭, ed., *Jiangnan zhizao ju ji* 《江南制造局记》 (2 vols. Taibei: Wenhai chuban she, 1969).

Weisheipl, James, "Classification of the Sciences in Medieval Thought", *Medieval Studies* 27 (1965): 54–90.

Widmer, Eric, *The Russian Ecclesiastical Mission in Peking during the Eighteenth Century* (Cambridge: Harvard University East Asian Research Center, 1976).

Witek, John, ed., *Ferdinand Verbiest (1623–1688): Jesuit Missionary, Scientist, Engineer, and Diplomat* (Nettetal: Steyler Verlag, 2001).

Wong, Sam and Wong, Valerie, "Kuang Qizhao and the Significance of the 1870s and 1880s for China's Modernization", forthcoming.

Wong Young-tsu, *A Paradise Lost: The Imperial Garden Yuanming Yuan* (Honolulu: University of Hawai'i Press, 2001).

Wright, David, "Careers in Western Science in Nineteenth-Century China: Xu Shou and Xu Jianyin", *Journal of the Royal Asiatic Society*, third series, 5 (1995): 49–90.

Wright, David, "John Fryer and the Shanghai Polytechnic: making space for science in nineteenth-century China", *British Journal of History of Science* 29 (1996): 15.

Wright, David, "The Great Desideratum: Chinese Chemical Nomenclature and the Transmission of Western Chemical Concepts", *Chinese Science* 14 (1997): 35–70.

Wright, David, "The Translation of Modern Western Science in Nineteenth-Century China, 1840–1895", *Isis* 89 (1998): 671.

Wright, David, "Yan Fu and the Tasks of the Translator", in Michael Lackner, Iwo Amelung, and Joachim Kurtz, eds., *New Terms for New Ideas: Western Knowledge & Lexical Change in Late Imperial China* (Leiden: E. J. Brill, 2001), 235–255.

Wright, David, *Translating Science: The Transmission of Western Chemistry into Late Imperial China, 1840–1900* (Leiden: E. J. Brill, 2000).

Wright, Mary, *The Last Stand of Chinese Conservatism: The T'ung-chih Restoration, 1862–1874* (Stanford: Stanford University Press, 1957).

Wu, Shellen, *Underground Empires: Coal and China's Entry into the Modern World Order* (Stanford: Stanford University Press, 2015).

Wu Yili, *Reproducing Women: Medicine, Metaphor, and Childbirth in Late Imperial China* (Berkeley: University of California Press, 2010).

Wylie, Alexander, *Notes on Chinese Literature* (Reprint, Taipei: Book Shop Limited, 1970; original. Shanghai: American Presbyterian Press, 1867).

Xi Yufu 席裕福, comp., *Huangchao zhengdian leizuan*《皇朝政典类纂》(Classified Materials on Qing Dynasty Government Regulations) (Reprint, Taipei: Shenwu Press 1969).

Xiong Mingyu 熊明遇, *Gezhi cao*《格致草》(Draft for Investigating Things and Extending Knowledge) (1648 edition in the U.S. Library of Congress Asian Library).

Xiong Yuezhi 熊月之, "Gezhi huibian yu xixue chuanbo" 《格致汇编》与西学传播〉 (The Chinese Science Magazine and the transmission of Western learning), *Shanghai yanjiu luncong* 《上海研究论丛》 1 (1989): 65–66.

Xiong Yuezhi 熊月之, *Xixue dongjian yu wan Qing shehui*《西学东渐与晚清社会》 (Shanghai: Renmin chuban she, 1994).

Xu Guangtai (Hsu Kuang-tai) 徐光台, "In the Name of 'Gewu Qiongli' (The Investigation of Things and the Fathoming of Principles: The Transmission of Western Learning in Late Ming and Early Qing)". Unpublished 1999 mss.

Xu Guangtai, "Mingmo Qingchu xifang gezhixue de chongji yu fanying" 〈明末清初西方格致学的冲击与反应〉 (The Impact and Reaction to Western Natural Studies in the Late Ming and Early Qing), in *Shibian chunti yu geren*《世变群体与个人》(Taipei: Taiwan National University, 1996), 236–258.

Xu Guangtai, "Ruxue yu kexue: yige kexueshi guandian de tantao" 〈儒学与科学：一个科学史观的探讨〉, *Qinghua xuebao* 《清华学报》, New Series, 26, 4 (December 1996): 369–392.

Yamada Keiji 山田慶兒, *Shushi no shizengagu*《朱子の自然學》(Zhu Xi's natural studies) (Tokyo: Iwanami, 1978).

Yan Fu 严复, "Jiuwang juelun"〈救亡决论〉(On what determines rescue or perishing), in *Wuxu bianfa ziliao*《戊戌变法》(Beijing: Shenzhou guoguang she, 1953), 360–371.

Yang Boda, "An Account of Qing Dynasty Glassmaking", in *Scientific Research in Early Chinese Glass* (Corning, New York: The Corning Museum of Glass, 1991).

Yang Boda 杨伯达, "Qingdai boli gaishu"〈清代玻璃概述〉(Overview of Qing dynasty glassware), *Gugong bowuyuan yuankan*《故宫博物院院刊》(1983): 13–16.

Yang Tsui-hua, 杨翠华, "Ren Hongjun yu Zhongguo jindai de kexue sixiang yu shiye" 〈任鸿隽与中国近代的科学思想与事业〉(Ren Hongjun and Chinese modern science and thought), *Jindaishi yanjiusuo jikan* 《近代史研究所季刊》24 (June 1995): 297–326.

Yang Tsui-hua, "The Development of Geology in Republican China, 1912–1937", in Lin Cheng-hung and Fu Daiwie, eds., *Philosophy & Conceptual History of Science in Taiwan* (Dordrecht: Kluwer Academic Publishers, 1993), 221–244.

Yee, Cordell, "Traditional Chinese Cartography and the Myth of Westernization", in J. B. Harley and David Woodward, eds., *The History of Cartography. Volume 2. Book 2: Cartography in the Traditional East and Southeast Asian Societies*, 170–202.

Yu Wang Luen, "Knowledge of Mathematics and Science in Ching-Hua-Yuan", *Oriens Extremus* 21, 2 (1974): 217–236.

Yu Weigang 于为刚, "Hu Wenhuan yu *Gezhi congshu*" 〈胡文焕与《格致丛书》〉(Hu Wenhuan and the Collectanea of Works Investigating into and Extending Knowledge), *Tushuguan zazhi* 《图书馆杂志》 4 (November 1982): 63–65.

Yü Ying-shih, "Some Preliminary Observations on the Rise of Ch'ing Confucian Intellectualism", *Tsing Hua Journal of Chinese Studies*, New Series 11.1 and 2 (December 1975): 105–146.

Yuanming yuan《圆明园》(Lofty Pavilion), 1 (November 1981).

Zeng Jianli 曾建立, "Gezhi guwei yu wan Qing 'Xixue Zhongyuan' lun"〈《格致古微》与晚清西学中源论〉(The *Gezhi guwei* and the late Qing Xixue Zhongyuan lun), *Zhongzhou xuekan* 《中州学刊》(November 2000): 146–150.

Zhang Binglin 章炳麟, "Zhizhi gewu zhengyi" 〈致知格物正义〉(Correct meaning of extending knowledge and investigating things), in *Zhang Taiyan quanji*《章太炎全集》 (Shanghai: Renmin chuban she, 1982), Vol. 5, 60–62.

Zhang Hongsheng 张鸿声, "Qingdai yiguan kaoshi ji tili"〈清代医官考试及体例〉(Qing dynasty examinations for medical officials with examples), *Zhonghua yishizazhi*《中华医史杂志》(*Chinese Journal of Medical History)* 25.2 (1995): 95–96.

Zhang Huang 章潢, *Tushu bian* 《图书编》 (Discerning charts and books) (Reprint, Taipei: Chengwen chuban she, 1971).

Zhang Junmai, "Rensheng guan"〈人生观〉(View of human life), in *Kexue yu rensheng guan*《科学与人生观》 (2 vols. Shanghai: Dongya tushuguan, 1923), I, 4–10.

Zhang Qiong, "About God, Demons, and Miracles: The Jesuit Discourse on the Supernatural in Late Ming China", *Early Science and Medicine* 4, 1 (February 1999): 1–36.

Zhang Qiong, "Demystifying *Qi*: The Politics of Cultural Translation and Interpretation in the Early Jesuit Mission to China", in Lydia Liu, ed., *Tokens of Exchange: The Problem of Translation and Interpretation in the Early Jesuit Mission to China* (Durham: Duke University Press, 1999, 74–106).

Zhang Qiong, "Nature, Supernature, and Natural Studies in Sixteenth- and Seventeenth-Century China" (Unpublished paper presented at the Colloquium sponsored by the Center for the Cultural Studies of Science, Medicine, and Technology, UCLA History Department, Los Angeles, November 16, 1998).

Zhang Rong, "Imperial Glass of the Yongzheng Reign", in *Elegance and Radiance: Grandeur in Qing Glass, The Arthur K. F. Lee Collection* (Hong Kong: Art Museum, Chinese University of Hong Kong, 2000), 64.

Zhang Yufa 张玉法, *Jindai Zhongguo gongye fazhan shi, 1860–1916*《近代中國工業發展史, 1860–1916》 (Taipei: Guiguan tushu gongsi, 1992).

Zheng Yi, "Xu Xiake's Travel Notes: Motion, Records and Genre Change", in Yi Zheng and Ofer Gal, eds., *Motion and Knowledge in the Changing Early Modern World: Orbits, Routes and Vessels* (Dordrecht: Springer-Verlag, 2014), 31–45.

Zhongwai shiwu cewen leibian dacheng 《中外时务策问类编》(1903 edition), 16.1a–b (essay prepared by Zhong Tianwei), 18.2b–3a.

Zhongxi wenjian lu 《中西闻见录》, Vol. 1, 403–404 (1872.2: 10a–b), 411–418 (1872.2: 14a–17b), 481–487 (1873.3: 16a–19b).

Zhu Yizun 朱彝尊, *Jingyi kao* 《经义考》(Critique of Classical Studies) (Shanghai: Zhonghua shuju, 1927–1935).

Zhu Zhenheng 朱震亨, *Gezhi yulun* 《格致馀论》 (Views on Extending Medical Knowledge), in the *Siku quanshu* 《四库全书》 (Complete Collection in the Imperial Four Treasuries) (Reprint, Taipei: Commercial Press, 1983–1986, Vol. 746–638).

Zhu Zhifan 朱之蕃, "Xu" 序 (Preface), to the *Baijia mingshu* 《百家名书》 (Famous Works of the Hundred Schools), (Late Ming Wanli edition, circa 1603, 5b–6a).

Zhuzi yulei 《朱子语类》 (Conversations with Master Zhu [Xi] Classified Topically), (1473 edition. Reprint, Taipei: Zhongzheng Bookstore).

Zürcher, Erik, "Renaissance Rhetoric in Late Ming China: Alfonso Vagnoni's Introduction to his *Science of Comparison*", in Federico Masini, ed., *Western Humanistic Culture Presented to China by Jesuit Missionaries (XVII–XVIII centuries)* (Rome: Institutum Historicum S.I., 1996), 331–359.

Index